1

FERROELECTRICITY
IN CRYSTALS -
With Preliminaries on Crystallography,

Atomic Bonding, Lattice Vibrations,

Specific Heats of solids,

Thermal Expansion of Crystals

& Defects in Crystals

By

Prof. S. Devanarayanan, Ph.D., D.Sc.

ISBN 978-1537326450

September 2016

CreateSpace Independent Publishing

(Amazon.com)

Dedication

In memory of my **Alma Maters**,

viz., University College, Thiruvananthapuram,

and Indian Institute of Science, Bangalore

| Ko nveesa te Paada-saroja-bhaajaam
Sudurlabhor∫theshu chaturshva peeha
Tathaapi naaham pravrunomi bhuman
Bhavat-padaambhoja nishevanotsuka: ||
(Srimad Bhaagavatam, Sk III, Chap 4, Sl 15)
Thus spoke Sri Bhagavaan to Uddhava

Preface

During 1950s experimental crystal physics has emerged from a few research laboratories in Universities into a wide world of practice. There developed rapidly quantum electronics, quantum and non-linear optics, production of semiconductor instruments, piezo-technology, acoustics, *etc.* , all involve single crystals and their singular peculiarities as well as the use of new crystal-physical phenomena and other allied discoveries rapidly. My research career in crystal physics started in 1963, and got trained, in the field of ferroelectric crystals, under Prof. I.S. Zheludev (Moscow University) who was a Visiting Professor in the Department of Physics of the Indian Institute of Science, Bangalore.

The specific features of crystals arise from the symmetry and anisotropy of the crystalline medium. Solid State Physics is largely devoted to the study of crystals and of electrons in crystals. J.F. Nye in "Symmetry and Physical Properties of Crystals" was concerned only with their external form and symmetry relationships among the various coefficients that describe the physical properties. After 1913 following Bragg's discovery of X-ray diffraction, the word "crystal" was first used to refer to ice and then to quartz, until the late mid- 20th Century.

I hope that this review will inspire further theoretical and experimental work to understand the nature of the FE phenomenon to compare the experimental results more satisfactorily with theory.

The encouraging support provided by Mrs. Chitra Devanarayanan, Mr. Ajith S. Devan, Mrs. D. Aparna Gayathri, Mr. S. Sumesh and Mrs. Sowmya Shankar are acknowledged herewith.

Valuable comments and suggestions are welcome, and the same may be communicated through chsd1976@gmail.com.

&&*&*&*&*&*&*&*&*&*

CONTENTS

Chapter 1

PRELIMINARIES - 1
CRYSTALLLOGRAPHY

Chapter 1

PRELIMINARIES - 1

CRYSTALLLOGRAPHY

"The true purpose of education is to train the min *d to think,*
for that reason it is priceless" - Albert Einstein

1.1 Crystals and Lattices:

1.1.1 Crystals

The shortest definition of a crystal is that 'a **crystalline state** is a solid composed of atoms arranged in an orderly repetitive array' or 'atomic arrays that are periodic in space are called crystals' The key feature of crystals is order. .

A class of solids showing neither reticular nor granular structure is termed as non-crystalline or <u>amorphous.</u>

Different arrangements also include varying degrees of rotational and parity inversion symmetry.

A mineral is a **'naturally occurring homogeneous solid'** with a **definite (but not generally fixed) chemical composition** and a **'highly ordered atomic arrangement',** usually formed by an **inorganic process.** One of the consequences of this ordered internal arrangement of atoms is that all crystals of the same mineral look similar.

Geometrical crystallography is the study of the different possibilities and laws that govern crystals.

Nicolas Steno (1669) *Law of constancy of interfacial angles* - angles between corresponding crystal faces of the same mineral have the same angle, even if the crystals are 'distorted' as illustrated by the cross-sections through 3 quartz crystals.

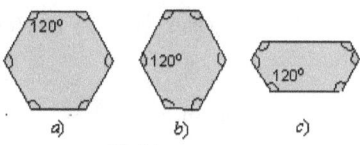

Fig 1.1

Wollaston (1989) designed a single circle optical goniometer for measurements accurate to 1' of arc, which was followed by design of a two-circle goniometer. Soon came the polarizing microscope. Von Laue and others laid the beginning of the field of solid state physics. W.L. Bragg (1913) determined for the first time the crystal structure of Halite (NaCl, *etc.*). The structural unit of inorganic crystals contains more than 100 atoms or molecules. In intermetallic compound, for example $NaCd_2$ has its structural unit contains 1192 atoms. In a protein crystal its structural unit has $\sim 10^4$ atoms.

1.1.2 Crystalline Lattice

Lattices demonstrate *Discrete Translation Symmetry*

1.1.3 Motif (Basis)

The **motif** is a list of the atoms associated with each lattice point, along with their fractional coordinates relative to the lattice point. Since each lattice point is, by definition, identical, if the motif is added to each lattice point, one will generate the entire structure.

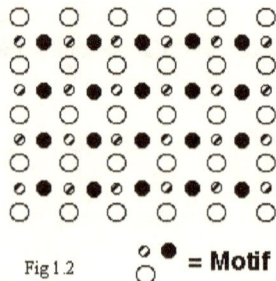

Fig 1.2 = Motif

1.2. Periodicity in Crystals

1.2.1 Space Lattice

A lattice is defined by three fundamental translation vectors, $\hat{a}_1$, $\hat{a}_2$ and $\hat{a}_3$ such that the atomic arrangement look the same in every respect when viewed from a point r, as when viewed from another point r'

$$r' = r + u_1\hat{a}_1 + u_2\hat{a}_2 + u_3\hat{a}_3$$

(1.1)

where u_1, u_2, u_3 are integers. The set of points defined by equation above for all values of u_1, u_2, u_3 define a lattice.

$$\text{Lattice} + \text{Motif (basis)} = \text{Crystal structure}$$

(1.2)

1.2.2 Primitive Lattice Cell

A parallelepiped defined by primitive lattice vectors, $\hat{a}_1$, $\hat{a}_2$ and $\hat{a}_3$, is called a primitive cell.

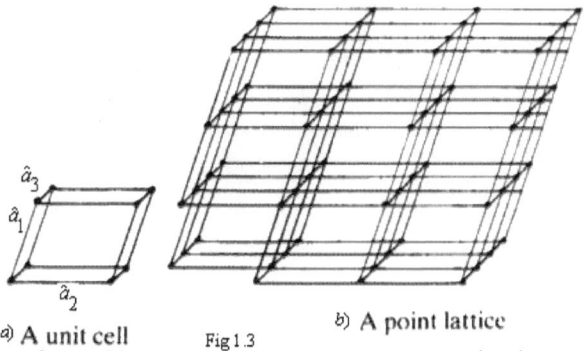

a) A unit cell   Fig 1.3   b) A point lattice

This cell has the shape of a parallelepiped, with its corner point having adjacent edges as the fundamental vectors $\hat{a}_1$, $\hat{a}_2$ and $\hat{a}_3$. It is clear that there is no cell of smaller volume that could serve as a building block of the structure.

> Ideally, the most stable arrangement of polyhedra in a crystal will be that which will MINIMIZE the ENERGY per unit volume

i.e., i) preserves electrical neutrality,

ii) satisfies the directionality and discreteness of all covalent bond,

iii) minimizes strong ion-ion repulsion, and

iv) packs the atoms as CLOSELY as possible, consistent with i), ii) and iii) above.

1.2.3 Lattice Translation Vector, $\vec{T}$

$$\vec{T} = u\,\hat{a}_1 + v\,\hat{a}_2 + w\,\hat{a}_3$$

(1.3)

In shorthand, lattice vectors are written in the form:

$$T = [uvw]$$

(1.4)

Negative values are not prefixed with a minus sign. Instead a bar is placed above the number to denote that the value is negative:

Thus $\vec{T} = -u\,\hat{a} + v\,\hat{b} - w\,\hat{c}$

(1.5)

This lattice vector would be written in the form:

$$T = [\bar{u}v\bar{w}]$$

(1.6)

Lattice directions are written the same way as lattice vectors, in the form [UVW].

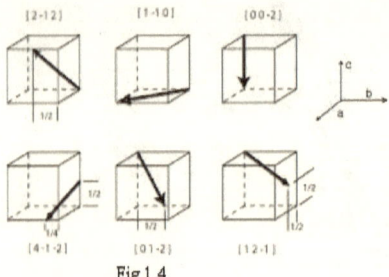

Fig 1.4

Many crystal systems have elements of symmetry. In these systems, certain sets of directions are symmetrically equivalent to each other. The set of directions that are symmetrically related to the direction [uvw] are written <uvw>.

1.2.4 Fundamental Lattice Types, Bravais lattices

A distinct type of lattice is called a Bravais lattice.

a) In 2-D space there are five types of Bravais lattices.

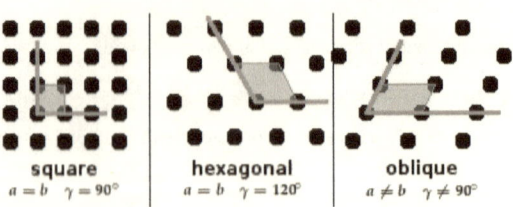

Fig 1.5

In 3-D, there are fourteen types of Bravais lattices.

1.2.5 The Primitive Unit Cell

The most common types of unit cell is the **primitive (P) unit cell** with one lattice point per unit cell;

1.2.5.1 The **face centred** (F) unit cell

An additional lattice points **at the centre of each face** and four lattice points per unit cell; and

1.2.5.3 The **body centred** (I) unit cell

It has a lattice point in the middle of the unit cell and two lattice points per unit cell.

1.2.5.4 Other cell types are the C face centred unit cell and the rhombohedral unit cell.

1.3.   CRYSTAL SYSTEM

13.1 Crystal Class

Table 1.1

Sl. No	Bravais Lattice type	Lattice cell Parameters	Crystal System / Characteristic Symmetry
1)	Primitive Cubic (P)	$a = b = c$ $\alpha = \beta = \gamma = 90^{0}$	Cubic
2)	Face Centered Cubic (F)	$a = b = c$ $\alpha = \beta = \gamma = 90^{0}$	Four 3-fold axes along $a+b+c,\ -a+b+c$
3)	Body Centered Cubic (I)	$a = b = c$ $\alpha = \beta = \gamma = 90^{0}$	Four 3-fold axes along $a+b+c, -a+b+c$
4)	Primitive Orthorhombic(P)	$a \neq b \neq c$ $\alpha = \beta = \gamma = 90^{0}$	Orthorhombic 3 mutually perpendicular
5)	Face Centered Orthorhombic(C)	$a \neq b \neq c$ $\alpha = \beta = \gamma = 90^{0}$	2-fold rotation or
6)	Face Centered Orthorhombic(F)	$a \neq b \neq c$ $\alpha = \beta = \gamma = 90^{0}$	perpendicular roto-inversion axes
7)	Body Centered Orthorhombic(I)	$a \neq b \neq c$ $\alpha = \beta = \gamma = 90^{0}$	along $a,\ b$ and c
8)	Primitive Tetragonal(C)	$a = b \neq c$ $\alpha = \beta = \gamma = 90^{0}$	Tetragonal A single 4-fold
9)	Body Centered Tetragonal (I)	$a = b \neq c$ $\alpha = \beta = \gamma\ 90^{0}$	rotation or $roto-inversion$
10)	Simple Monoclinic (P)	$a \neq b \neq c$ $\alpha = \gamma = 90^{0},$ $\beta \neq 90^{0}$	axis along c Monoclinic A single 2-fold
11)	B-Face Centred Monoclinic (C)	$a \neq b \neq c$ $\alpha = \gamma = 90^{0}, \beta \neq 90^{0}$	rotation or roto-inv along b
12)	Hexagoal (P)	$a \neq b \neq c$ $\alpha = \beta = 90^{0}, \gamma = 120^{0}$	A Hexagonal , a 6-fold rotation or a roto-inversion, c
13)	Triclinic (P)	$a \neq b \neq c$ $\alpha \neq \beta \neq \gamma \neq 90^{0}$	Triclinic axis in any direction
14)	Primitive Rhombohedral (P)	$a = b = c$ $\alpha = \beta = \gamma \neq 90^{0}$	Trigonal 3 – fold axis along c

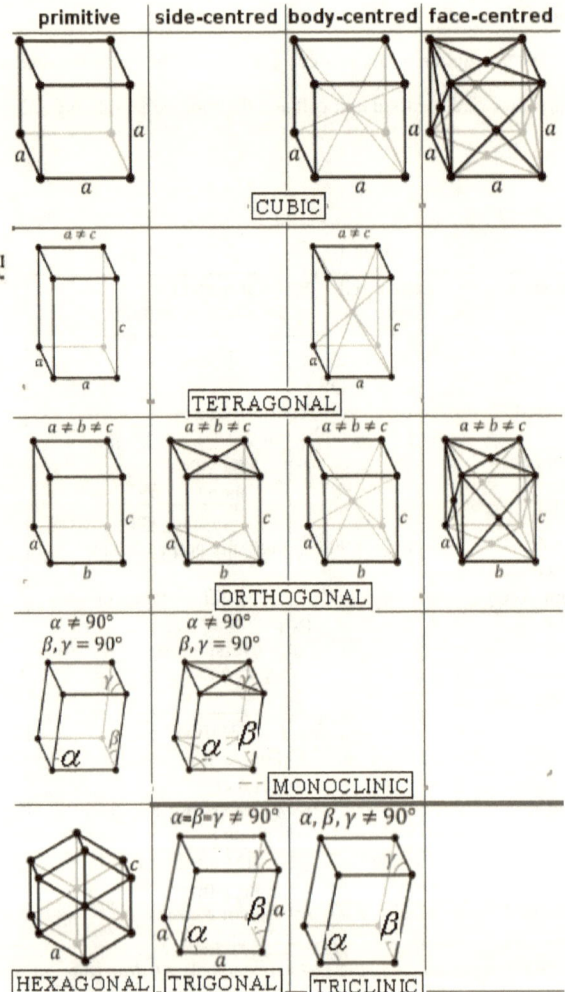

Fig 1.6 14 Bravais Lattices and 7 Crystal Classes.

1.3.2 Lattice Geometry

To define the geometry of the unit cell in 3-D dimensions, choose a right-handed set of crystallographic axes, x, y, and z, which point along the edges of the unit cell. The origin of our coordinate system is at one of the lattice points

1.3.2.1. Unit Cell

If you know the motif, an easy way to find the number of atoms per unit cell is to multiply the number of atoms in the motif by the number of lattice points in the unit cell.

The unit cell which has one lattice point at each vertex is called a <u>primitive cell</u>. The three edges of a primitive cell define a 'lattice', but from any given lattice an infinite number of primitive cells can be selected. <u>A primitive cell is a minimum volume cell</u>. There is a density of one lattice point per primitive cell.

The number of lattice points in unit cell is denoted by symbol Z. So

Z = <u>Number of formula units in a unit cell</u>.

For a general plane (hkl) the number of lattice planes intersection in the unit cell is $(h+k+l)$. Then <u>the number of atoms</u> in the (111) plane

atoms in (111) plane

$$= \frac{(\text{\# of atoms in the crystal}) \cdot (\text{\# of molecules / unit cell, Z})}{(h+k+l)=3} \qquad (1.7)$$

1.3.2.2 Lattice parameters (Unit cell parameters)

The length of the unit cell along the x, y, and z direction are defined as a, b, and c. Alternatively, the sides of the unit cell in terms of vectors a, b, and c.

Volume,

$$\boxed{V = |\vec{a} \cdot \vec{b} \wedge \vec{c}|} \qquad (1.8)$$

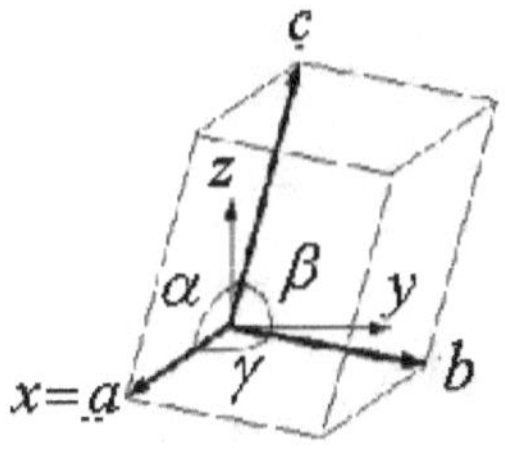

Fig 1.7

$a, b, c, \alpha, \beta, \gamma$ are collectively known as the **lattice parameters** (often also called 'unit cell parameters', or just 'cell parameters').

1.3.2.3 <u>Wigner-Seitz primitive</u> cell

A primitive unit cell may also be constructed as follows

(i)   Start with an array of points in the (direct) lattice,

(ii)   Connect any one lattice point to all the neighbouring lattice points with lines.

(iii) At the mid-point of these lines draw normals (if one started out with a two dimensional lattice) or normal-planes (if started out with a 3-D lattice). The smallest area (or volume) enclosed in this way is called the Wigner-Seitz primitive cell of the direct lattice. All space may be filled up without leaving any gap by joining these Wigner Seitz primitive cells.

Fig 1.8 Wigner-Seiz primitive cell

1.4   Some common Simple Crystal Structures

1.4.1   Simple Cubic (SC)

This structure is relatively rare amongst the metallic elements and only ^{209}Po appears to crystallize in the SC structure at room temperature and pressure. This is likely because the packing efficiency and coordination number for this structure are low at 0.52 and 6, respectively. Crystallographic data shows the length of a side of the unit cell to be 3.34 $\mathring{A}$.

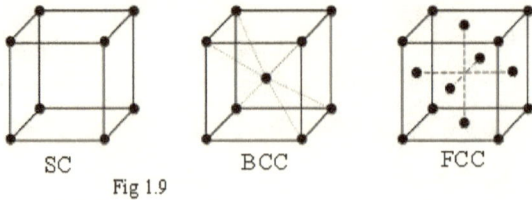

SC   BCC   FCC

Fig 1.9

1.4.2   The Body-Centred Cubic (BCC) lattice

BCC is another common crystal structure adopted by metallic solids such as Fe, Cr, Mo and W.

1.4.3   The Face-Centred Cubic (FCC) lattice

This structure is very common amongst metallic elements because it maximizes nearest neighbor interactions (coordination number of 12). The unit cell has a packing efficiency of 0.74. The FCC structure is also known as the cubic close-packed (CCP) structure

1.4.4   The Sodium Chloride (NaCl) Structure

This cell can be described as a simple FCC lattice with a two atom (Na, Cl) basis or two interpenetrating FCC lattices, one of Na and one of Cl, displaced from each other by 0.5

of the body diagonal. This is a common structure for ionic compounds including LiH, KCl, PbS, AgBr, MgO and MnO.

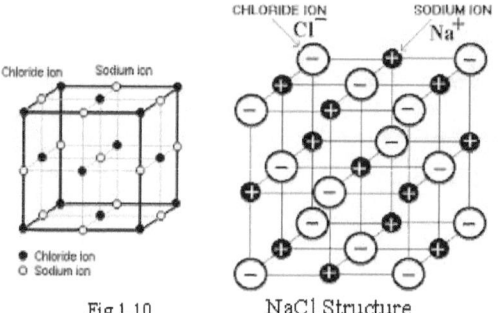

Fig 1.10   NaCl Structure

1.4.5   The Diamond Structure

Adopted by C (diamond), Si, Ge and grey Sn, which have a strong covalent bonding tendency. The structure is composed of C atoms with tetrahedral bonds. The tetrahedron has the geometric shape of a pyramid with four triangular faces forming isosceles triangles (has the same side length). The tetrahedral (T_d) molecule has four atoms. All the bond angles from the center atom are $109.5°$ The structure is a FCC lattice with two atoms associated with each lattice point, one atom at $(0,0,0)$ and another at $(\frac{1}{4},\frac{1}{4},\frac{1}{4})$.

These two atoms form a basis of diamond structure, and two atoms are the same. Many semiconductors, such as silicon (Si), Germanium (Ge), and Gallium Arsenide (GaAs), diamond structure.

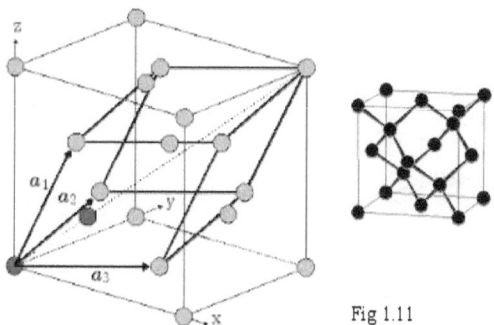

Fig 1.11

1.4.6   The conventional CaF_2 (fluorite) unit cell.

The fluoride ions form a FCC lattice while the Ca ions are placed in a simple cubic arrangement in the tetrahedral holes. The anti-fluorite structure has the atomic positions reversed.

1.4.7   The cubic ZnS unit cell (zinc blende)

This is closely related to the diamond unit cell, and has two interpenetrating FCC lattices, one lattice is composed of Zn atoms and the other of S atoms.

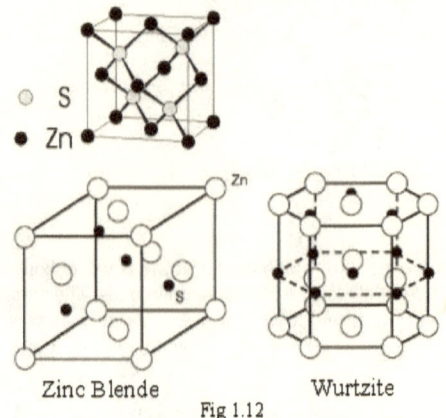

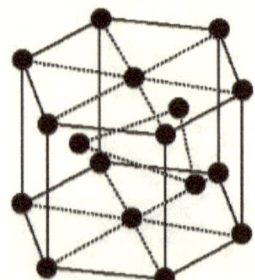

Zinc Blende   Wurtzite
Fig 1.12

1.4.8   Hexagonal Close Packed (HCP) Structure

Fig 1.13 HCP

Number of atoms per unit cell of a cubic lattice,

$$n = \frac{a^3 \rho N_A}{M}$$

(1.9)

Table 1.2

Type	SC	BCC	FCC	HCP	NaCl	Diamond
		Simple Crystal structure Data				
Coordination # N	6	8	12	12		
NN distance 2r	a	$\frac{a\sqrt{3}}{2}$	$\frac{a\sqrt{2}}{2}$		$\frac{a\sqrt{3}}{8}$	
Lattice constant, a	2r	$\frac{4r}{\sqrt{3}}$	$\frac{4r}{\sqrt{2}}$	2r	a	
# of atoms / cell, n	$(\frac{1}{8}x8)$	$\{(\frac{1}{8}x8)+1\}$	4	6	8	
# of lattice points, Z	1	2	4			
V of all atoms in cell,v	$(\frac{4}{3}\pi r^3)x1$	$(\frac{4}{3}\pi r^3)x\,2$	$(\frac{4}{3}\pi r^3)x4$	$\left[\pi a^3\right]$	$(\frac{8}{3}\pi r^3)$	
Volume of unit cell, V		$a^3=(2r)^3$ $\frac{64r^3}{3\sqrt{3}}$	$\frac{64r^3}{2\sqrt{2}}$	$\frac{3\sqrt{3}a^2c}{2}$	$(\frac{4r}{\sqrt{23}})^3$	
APF $\frac{v}{V}$	$\left[\frac{\pi}{6}=52\%\right]$	$\left[\frac{\sqrt{3}}{8\pi}=68\%\right]$	$\left\{\frac{\pi}{3\sqrt{2}}=74\%\right\}$	$\left\{\frac{\pi}{3\sqrt{2}}=74\%\right\}$	$\left\{\frac{\sqrt{3}}{8\pi}=68\%\right\}$	$\left(\frac{\pi\sqrt{3}}{16}=34\%\right)$
Example	Po	{Cu,Al,Pb,Ag}	Mg	-	(LiH, KCl, PbS, MgO,MnO)	[Ge, Si, GaAs]

1.5 SYMMETRY OPERATIONS

A crystal structure is described in terms of four (4) important information:

i)   the type of lattice,
ii)   the crystal axes $\hat{a}, \hat{b}$ & $\hat{c}$
iii)   the basis, and
iv)   the symmetry operations.

Each operation is performed relative to a point, line, or plane - called a symmetry element.

1.5.1 What is a symmetry element?

A *symmetry element* is a geometrical entity (a line, plane or point) with respect to which one or more symmetry operations may be carried out

1.5.2. The set of 5 kinds of symmetry operations associated with the symmetry of a

molecule.

They are: rotation, reflection, and inversion

1.5.3 Proper rotation

Rotation of a motif about a central point through an angle α in an anticlockwise

direction, and repeated an integer n times to bring the motif to the starting position, such that

$$\alpha = 2\pi / n \qquad\qquad\qquad\qquad (1.10)$$

The geometrical locus about which the operations of repetition take place is called the symmetry element, called the $n-$ fold axis (C_n). Here $n = 2\pi / \alpha$.

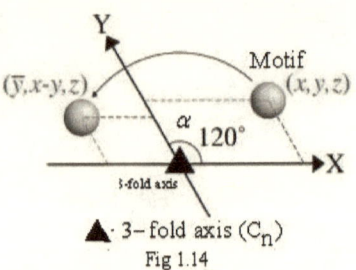

3$-$fold axis (C$_n$)

Fig 1.14

Table 1.3

n-fold axis(C_n)	$\alpha° = 2\pi^C / n$
2 – fold axis	$\alpha = 180°$
3-fold-axis	$\alpha = 120°$
4-fold-axis	$\alpha = 90°$
6-fold-axis	$\alpha = 30°$

It can be seen that the symmetry element has to obey the lattice translational symmetry, so that

distance between two neighbouring lattice points $= a \; Cos \; (\alpha = 2\pi^c / n)$ (1.11)

is violated if $n = 5$.

Table 1.4

Notation	Symmetry Element	Symmetry Operation	Description
Symmetry Elements and Operations:			
1) E	Identity	Zero	Nothing changes.
2) i	Centre of symmetry or inverted centre	Inversion	Projects the object through the centre (inverts about the centre)
3) C_n	n-fold proper axis of rotation	Rotation	Rotates $(360/n)°$ in the clock-wise or anticlockwise direction about the axis
4) $\sigma_h, \sigma_v, \sigma_d$	Mirror plane	Reflection	Reflects across a plane$\perp$, $\parallel$ and Diagonal to principal axis
5) S_n	n-fold improper axis of rotation with a plane of reflection	Rotation followed by a reflection	Rotates $(360/n)°$ in the clockwise or anticlockwise direction about the axis followed by a reflection across a plane perpendicular to the rotation axis

$\hat{\sigma}_h$ (*horizontal* plane); in a plane $\perp$ principal axis,

$\hat{\sigma}_d$ (*dihedral* plane); in a plane containing and $\parallel$ principal axis and bisecting lower order axes, *viz.* dihedral plane of symmetry

e.g. $\hat{\sigma}_{xy}(x, y, z) \rightarrow (x, y, -z)$

Note: $\hat{\sigma}^{2n} = \hat{E}$, n = integer

$$\hat{i}\ (x, y, z) \rightarrow (-x, -y, -z) \tag{1.12}$$

A tetrahedral Structure has the following 24 symmetry operations: 1 E, $3C_2$, $8C_3$ (= $4C_3$ + $4C_2^{-1}$), 6σ, and 6 S_4 (= $3S_4 + 3S_4^{-1}$).

1.5.4 Enantiomorphism:

Whatever discussed above is for a motif which itself has no symmetry. If the motif, such as a foot wear like shoe, which has two forms – left handed and right handed, *i.e.*, asymmetric then each in the pair is called enantiomorph of the other..

1.5.6 Habit: External form of a crystal is called habit.

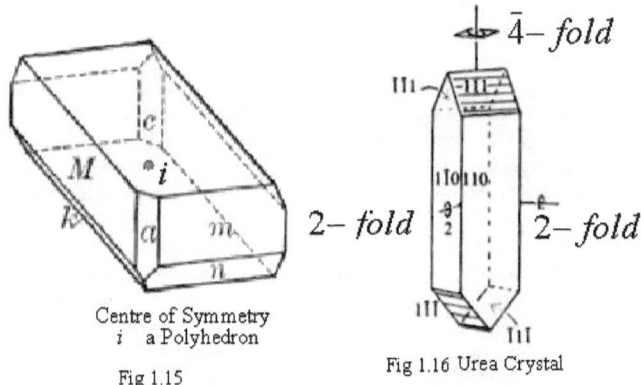

Centre of Symmetry
i a Polyhedron

Fig 1.15

Fig 1.16 Urea Crystal

1.5.7 Improper Rotations: $\hat{S}_n^{\,k}$

Rotation about an n-fold axis followed by reflection through a plane perpendicular to

Improper n-fold Rotation, $\hat{S}_n^{\,k}$, k = 1,....., n

When k = 1, n = 1 $\hat{S}_n^{\,k} = \hat{\sigma}$, Reflection Operation

When k = 1, n = 2 $\hat{S}_n^{\,k} = \hat{i}$, Inversion Operation

1.5.8 What is a symmetry operation?

Each of these **Symmetry Operations** is associated with a Symmetry Element which is a point, a line, or a plane about which the operation is performed such that the crystal orientation and position before and after the operation are indistinguishable

It defines the movement which results in a lattice indistinguishable from the original.

1.5.9 Combinations of Symmetry Operations

According to Group theory (Devanarayanan, 2013, in Quantum Chemistry, Chap 12) when one symmetry element A_α is properly combined with another B_β, then a third element $C_{-\gamma}$ is automatically generated, such that $A_\alpha.B_\beta = C_{-\gamma}$.

i.e., if any two rotations are given then the third is implied.

Stereograms of some of the 32 point groups is reproduce in Fig.1.17

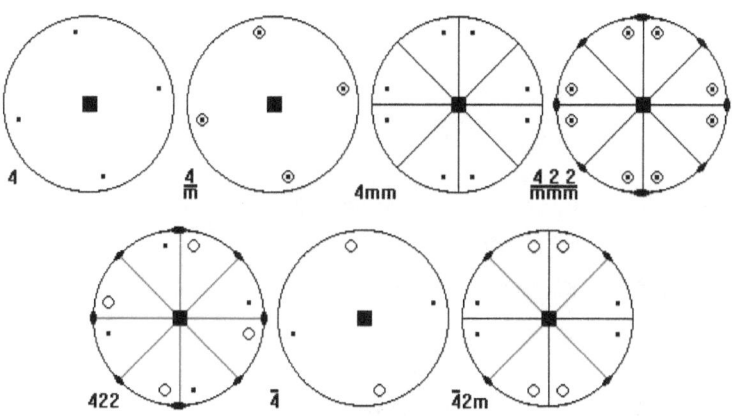

Fig 1.17 Stereograms

. 1.5.10. What is a Point Group?

All symmetry elements of a finite object (rigid body), passing by a point, define the total symmetry of the object, which is known as the *point group symmetry* of the object.

A Point Group consists of a set of all the elements of symmetry possessed by a lattice and which intersect at a common point. *i.e.,* no translations, everything operates about Centre of Mass of the lattice

There are many *symmetry point groups*, but in crystals they must be consistent with the crystalline periodicity (repetition by translation). On the other hand, for instance, the symmetry axes of order **5** (**5**-fold axes) are not possible in crystals and therefore only *32 point groups* are allowed in the crystalline state of matter. These **32 crystallographic point groups** are also known in Crystallography as the *32* **crystal classes**.

> Point Group • Crystal Translational Periodicity = 32 Crystal Classes

> 32 Crystal Classes •CentreofSymmetry = 11 Laue Groups

> Crystal Translational Periodicity • 32 Crystal Classes = 14 BRAVAIS LATTICES

Symbols of point groups are as per Hermann-Maguin for Crystallographers and Schoeflies by spectroscopists.

Table 1.5

| 32 CRYSTAL CLASSES | | | | SYSTEM |
G	$G \cup G$	$H \cup (G \cup H)$		System
1	$\bar{1}$	-		Triclinic
2	$\frac{2}{m}$	m		Monoclinic
3	$\bar{3}$	-		Hexagonal
4	$\frac{4}{m}$	$\bar{4}$		Tetragonal
6	$\frac{6}{m}$	$\bar{6}$		Hexagonal
222	$\frac{2}{m}\frac{2}{m}\frac{2}{m}$	2mm		Orthorhombic
322	$3\frac{2}{m}\frac{2}{m}$	3mm		Hexagonal
422	$\frac{4}{m}\frac{2}{m}\frac{2}{m}$	4mm and $\bar{4}$2m		Tetragonal
622	$\frac{6}{m}\frac{2}{m}\frac{2}{m}$	6mm and $\bar{6}$2m		Hexagonal
332	$\bar{3}$	$\bar{3}\frac{2}{m}$	-	Cubic
432	$\frac{4}{m}\bar{3}\frac{2}{m}$	$\bar{4}$3m		Cubic

For point group C_{3V} the matrices corresponding to its symmetry operations are

$$\mathbf{C_{3v}} \quad \hat{E} \quad \hat{C}_3^{\,1} \quad \hat{C}_3^{\,2} \quad \sigma_v \quad \sigma_v' \quad \sigma_v''$$

$$\begin{pmatrix} 1 & 0 & 0 \\ 0 & 1 & 0 \\ 0 & 0 & 1 \end{pmatrix} \begin{pmatrix} -\frac{1}{2} & \frac{\sqrt{3}}{2} & 0 \\ -\frac{\sqrt{3}}{2} & -\frac{1}{2} & 0 \\ 0 & 0 & 1 \end{pmatrix} \begin{pmatrix} -\frac{1}{2} & -\frac{\sqrt{3}}{2} & 0 \\ \frac{\sqrt{3}}{2} & -\frac{1}{2} & 0 \\ 0 & 0 & 1 \end{pmatrix} \begin{pmatrix} -1 & 0 & 0 \\ 0 & 1 & 0 \\ 0 & 0 & 1 \end{pmatrix} \begin{pmatrix} \frac{1}{2} & -\frac{\sqrt{3}}{2} & 0 \\ -\frac{\sqrt{3}}{2} & -\frac{1}{2} & 0 \\ 0 & 0 & 1 \end{pmatrix} \begin{pmatrix} \frac{1}{2} & \frac{\sqrt{3}}{2} & 0 \\ \frac{\sqrt{3}}{2} & -\frac{1}{2} & 0 \\ 0 & 0 & 1 \end{pmatrix}$$

1.6. Crystal Systems and Bravais Lattices, Point Groups

and Number of Symmetry elements

The rotational symmetry of a crystal places constraints on the shape of the conventional unit cell we choose to describe the structure. On this basis we divide all structures into one of 7 crystal systems. For example, for crystals with 4 fold symmetry it will always be possible to choose a unit cell that has a square base with $a = b$ and $\gamma = 90°$:

Table 1.6 THE SEVEN CRYSTAL SYSTEMS AND
FOURTEEN BRAVAIS LATTICES

Crystal system	Point group	Restrictions on axes or angles of unit cell	Bravais lattices with lattice type
1 Triclinic	1, $\bar{1}$	None	
2 Monoclinic	2, m, 2/m	$\alpha = \gamma = 90°$	
3 Orthorhombic	222, mm2, mmm	$\alpha = \beta = \gamma = 90°$	
4 Tetragonal	4, $\bar{4}$, 4/m, 422, 4mm, $\bar{4}$2m, 4/mmm	$a = b$ $\alpha = \beta = \gamma = 90°$	
5 Trigonal	3, $\bar{3}$, 32, 3m, $\bar{3}$m	$a = b = c$ $\alpha = \beta = \gamma$ or $a = b$ $\alpha = \beta = 90°$ $\gamma = 120°$	
Hexagonal	6, $\bar{6}$, 6/m, 622, 6mm, $\bar{6}$m2, 6/mmm	$a = b$ $\alpha = \beta = 90°$ $\gamma = 120°$	
6 Cubic	23, m3, 432, $\bar{4}$3m, m3m	$a = b = c$ $\alpha = \beta = \gamma = 90°$	

P - primitive ; I - body-centred ; A - A-centred ; B-Centred ; C-C-centred ,
F: Face-centred ; R: Rhombohedral

Table 1.7 THE THIRTY-TWO POINT GROUPS AND THEIR CLASS STRUCTURES

Groups		No. of elements	No. of classes	Representative elements and the number of elements in each class† of the *Point group*
International Symbol	Schoenflies Symbol			
1	C_1	1	1	$E(1)$
2	C_2	2	2	$E(1)$; $C_2(1)$
3	C_3	3	3	$E(1)$; $C_3^1(1)$, $C_3^2(1)$
4	C_4	4	4	$E(1)$; $C_4^1(1)$, $C_4^2(1)$, $C_4^3(1)$
6	C_6	6	6	$E(1)$; $C_6^1(1)$, $C_6^2(1)$, $C_6^3(1)$, $C_6^4(1)$, $C_6^5(1)$
$\frac{1}{m}$	C_{1h} (or C_s)	2	2	$E(1)$; $\sigma_h(1)$
$\frac{2}{m}$	C_{2h}	4	4	$E(1)$; $C_2(1)$, $C_2\sigma_h(1)$, $\sigma_h(1)$
$\frac{3}{m}$ (or $\bar{6}$)	C_{3h}	6	6	$E(1)$; $C_3^1(1)$, $C_3^2(1)$, $\sigma_h(1)$, $C_3^1\sigma_h(1)$, $C_3^2\sigma_h(1)$
$\frac{4}{m}$	C_{4h}	8	8	$E(1)$; $C_4^1(1)$, $C_4^2(1)$, $C_4^3(1)$, σ_h, $\sigma_h C_4^1(1)$, $\sigma_h C_4^2(1)$, $\sigma_h C_4^3(1)$
$\frac{6}{m}$	C_{6h}	12	12	$E(1)$; $C_6^1(1)$, $C_6^2(1)$, $C_6^3(1)$, $C_6^4(1)$, $C_6^5(1)$, $\sigma_h(1)$, $C_6^1\sigma_h(1)$, $C_6^2\sigma_h(1)$, $C_6^3\sigma_h(1)$, $C_6^4\sigma_h(1)$, $C_6^5\sigma_h(1)$
2 mm	C_{2v}	4	4	$E(1)$; $C_2(1)$, $\sigma_v(1)$, $\sigma_v(1)$
3 mm	C_{3v}	6	3	$E(1)$; $C_3(2)$, $\sigma_v(3)$
4 mm	C_{4v}	8	5	$E(1)$; $C_4^2(1)$, $C_4(2)$, $\sigma_v(2)$, $\sigma_v(2)$
6 mm	C_{6v}	12	6	$E(1)$; $C_6^3(1)$, $C_6(2)$, $C_6^2(2)$, $\sigma_v(3)$, $\sigma_e(3)$
$\bar{2}$	S_2 (or C_i)	2	2	$E(1)$; $S_2(1)$
$\bar{4}$	S_4	4	4	$E(1)$; $S_4(1)$, $S_4^2(1)$, $S_4^3(1)$
$\bar{6}$	S_6 (or C_{3i})	6	6	$E(1)$; $S_6(1)$, $S_6^2(1)$, $S_6^3(1)$, $S_6^4(1)$, $S_6^5(1)$
222	D_2 (or V)	4	4	$E(1)$; $C_2(1)$, $C_2'(1)$, $C_2''(1)$
322	D_3	6	3	$E(1)$; $C_3(2)$, $C_2(3)$
422	D_4	8	5	$E(1)$; $C_4^2(1)$, $C_4(2)$, $C_2'(2)$, $C_2(2)$
622	D_6	12	6	$E(1)$; $C_6^3(1)$, $C_6(2)$, $C_6^2(2)$, $C_2(3)$, $C_2(3)$
$\frac{2}{m}\frac{2}{m}\frac{2}{m}$ (or mmm)	D_{2h}	8	8	$E(1)$; $C_2(1)$, $C_4(1)$, $C_2(1)$, $C_2(1)$, $C_2\sigma_h(1)$, $\sigma_v(1)$, $\sigma_v(1)$
$\bar{6}m2$	D_{3h}	12	6	$E(1)$; $C_3(2)$, $C_2(3)$, $C_3\sigma_h(2)$, $\sigma_h(1)$, $\sigma_v(3)$
$\frac{4}{m}\frac{2}{m}\frac{2}{m}$ (or $\frac{4}{mmm}$)	D_{4h}	16	10	$E(1)$; $C_4^2(1)$, $C_4(2)$, $C_2(2)$, $C_2(2)$, $\sigma_h(1)$, $C_4\sigma_h(2)$, $C_4^2\sigma_h(1)$, $\sigma_v(2)$, $\sigma_e(2)$
$\frac{6}{m}\frac{2}{m}\frac{2}{m}$ (or $\frac{6}{mmm}$)	D_{6h}	24	12	$E(1)$; $C_6^3(1)$, $C_6(2)$, $C_6^2(2)$, $C_2(3)$, $C_2(3)$, $C_6\sigma_h(2)$, $\sigma_h(1)$, $C_6^3\sigma_h(1)$, $C_6^2\sigma_h(2)$, $\sigma_v(3)$, $\sigma_e(3)$
$\bar{4}2m$	D_{2d}	8	5	$E(1)$; $C_2(1)$, $\sigma_d(2)$, $C_2(2)$, $C_2\sigma_d(2)$
$3\frac{2}{m}$	D_{3d}	12	6	$E(1)$; $I(1)$, $C_3(2)$, $C_3I(2)$, $C_2(3)$, $C_2I(3)$
23 (or 322)	T	12	4	$E(1)$; $C_2(3)$, $C_3(4)$, $C_3^2(4)$
$\frac{2}{m}3$ (or $m3$)	T_h	24	8	$E(1)$; $C_2(3)$, $C_3(4)$, $C_3^2(4)$, $I(1)$, $C_2I(3)$, $C_3I(4)$, $C_3^2I(4)$
$\bar{4}3m$	T_d	24	5	$E(1)$; $C_2(3)$, $C_3(8)$, $\sigma_d(6)$, $S_4(6)$
432	O	24	5	$E(1)$; $C_2(8)$, $C_4(6)$, $C_2(6)$, $C_2^2(3)$
$\frac{4}{m}\bar{3}\frac{2}{m}$	O_h	48	10	$E(1)$; $I(1)$, $C_3(8)$, $C_3I(8)$, $C_4(6)$, $C_4I(6)$, $C_2(6)$, $C_2I(6)$, $C_2^2(3)$, $C_2^2I(3)$

† The number of elements in each class is denoted by the numeral in the bracket.

Table 1.8 **CRYSTALLOGRAPHIC POINT GROUPS**

Crystal system	Schoenflies symbol	Hermann–Mauguin symbol	# of elements or Order of group	Laue group
1. Triclinic	C_1	1	1	$\bar{1}$
	C_i (S_2)	$\bar{1}$	2	
2. Monoclinic	C_2	2	2	$2/m$
	C_s (C_{1h})	m $(1/m)$	2	
	C_{2h}	$2/m$	4	
3. Orthorhombic	D_2 (V)	222	4	mmm
	C_{2v}	$mm2$	4	
	D_{2h}	mmm	8	
4. Tetragonal	C_4	4	4	$4/m$
	S_4	$\bar{4}$	4	
	C_{4h}	$4/m$	8	
	D_4	422	8	$4/mmm$
	C_{4v}	$4mm$	8	
	D_{2d}	$\bar{4}2m$	8	
	D_{4h}	$4/mmm$	16	
5. Trigonal	C_3	3	3	$\bar{3}$
	C_{3i}	$\bar{3}$	6	
	D_3	32	6	$3m$
	C_{3v}	$3m$	6	
	D_{3d}	$\bar{3}m$	12	
Hexagonal	C_6	6	6	$6/m$
	C_{3h} (S_6)	$\bar{6}$ $(3/m)$	6	
	C_{6h}	$6/m$	12	
	D_6	622	12	$6/mmm$
	C_{6v}	$6mm$	12	
	D_{3h}	$\bar{6}m2$	12	
	D_{6h}	$6/mmm$	24	
6. Cubic	T	23	12	$m3$
	T_h	$m3$	24	
	O	432	24	$m3m$
	T_d	$43m$	24	
	O_h	$m3m$	48	

1.7 Crystallographic Point Groups and Representative Crystals

Table 1.9 CRYSTALLOGRAPHIC POINT GROUPS

Basic Definition and Samples

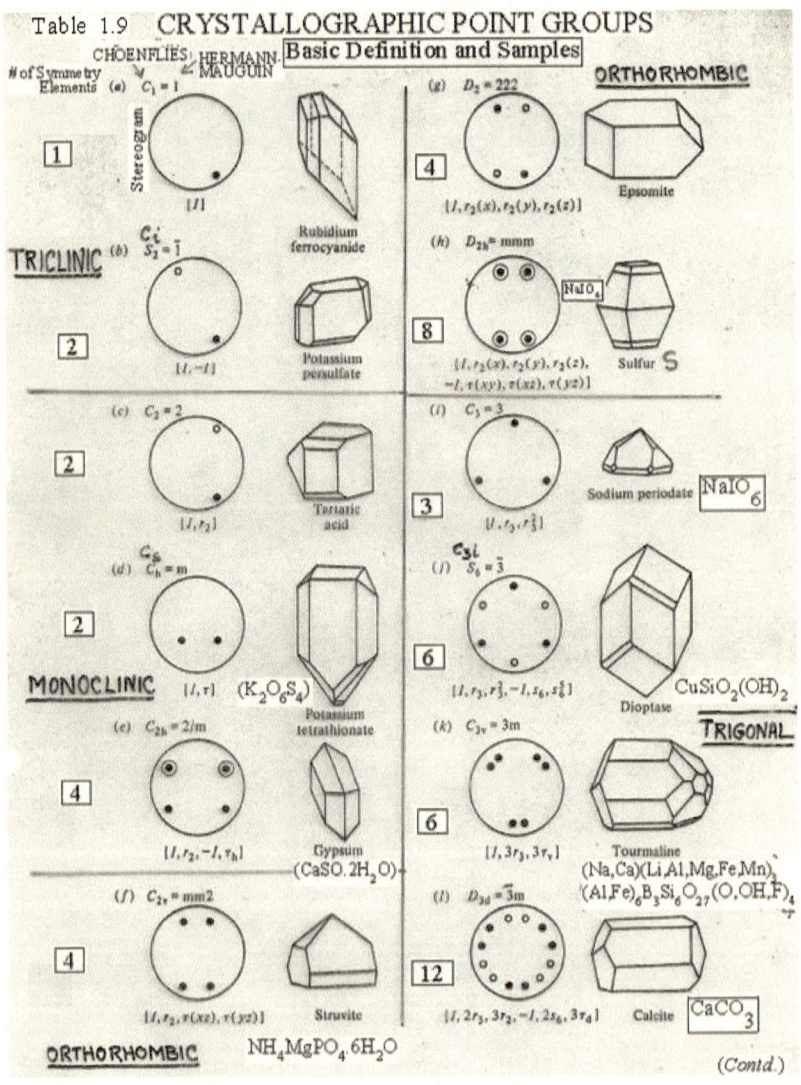

# of Symmetry Elements

CHOENFLIES

HERMANN-MAUGUIN

Stereogram

(a) $C_1 = 1$ — 1 — $[I]$ — Rubidium ferrocyanide

TRICLINIC

(b) C_i $S_2 = \bar{1}$ — 2 — $[I, -I]$ — Potassium persulfate

(c) $C_2 = 2$ — 2 — $[I, r_2]$ — Tartaric acid

(d) C_s $C_h = m$ — 2 — $[I, \tau]$ — $(K_2O_6S_4)$ Potassium tetrathionate

MONOCLINIC

(e) $C_{2h} = 2/m$ — 4 — $[I, r_2, -I, \tau_h]$ — Gypsum $(CaSO.2H_2O)$

(f) $C_{2v} = mm2$ — 4 — $[I, r_2, \tau(xz), \tau(yz)]$ — Struvite $NH_4MgPO_4.6H_2O$

ORTHORHOMBIC

(g) $D_2 = 222$ — 4 — $[I, r_2(x), r_2(y), r_2(z)]$ — Epsomite

ORTHORHOMBIC

(h) $D_{2h} = mmm$ — 8 — $[I, r_2(x), r_2(y), r_2(z), -I, \tau(xy), \tau(xz), \tau(yz)]$ — $NaIO_4$ — Sulfur S

(i) $C_3 = 3$ — 3 — $[I, r_3, r_3^2]$ — Sodium periodate $NaIO_6$

(j) C_{3i} $S_6 = \bar{3}$ — 6 — $[I, r_3, r_3^2, -I, s_6, s_6^5]$ — Dioptase $CuSiO_2(OH)_2$

TRIGONAL

(k) $C_{3v} = 3m$ — 6 — $[I, 3r_3, 3\tau_v]$ — Tourmaline $(Na,Ca)(Li,Al,Mg,Fe,Mn)_3 (Al,Fe)_6B_3Si_6O_{27}(O,OH,F)_4$

(l) $D_{3d} = \bar{3}m$ — 12 — $[I, 2r_3, 3r_2, -I, 2s_6, 3r_4]$ — Calcite $CaCO_3$

(Contd.)

Table 1.10   CRYSTALLOGRAPHIC POINT GROUPS
(Typical Example Crystals)

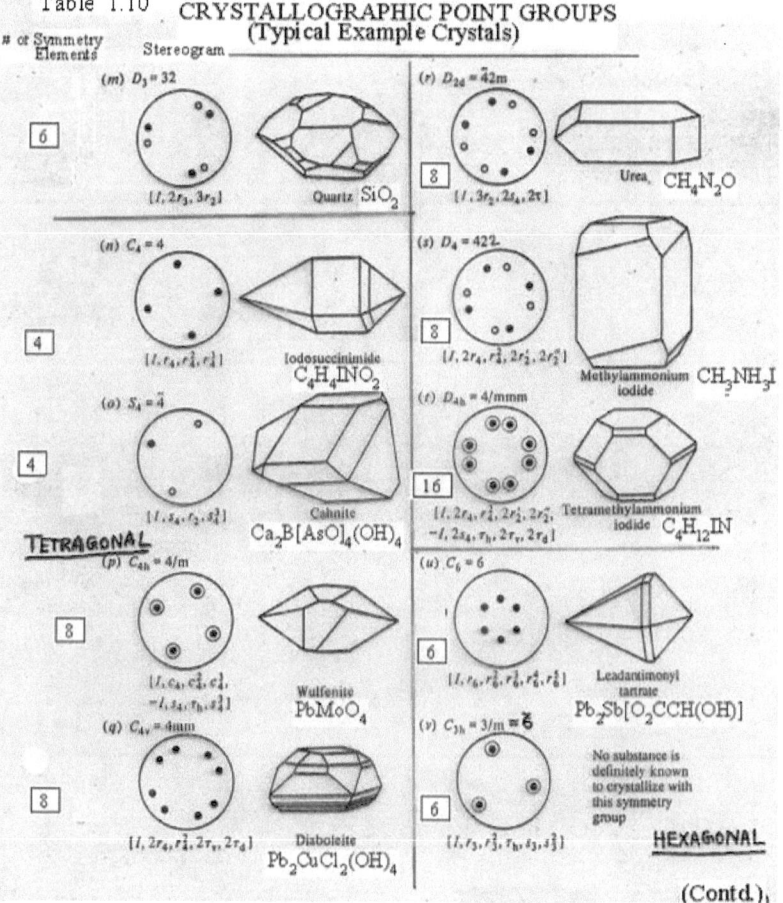

# of Symmetry Elements	Stereogram		

(m) $D_3 = 32$ — 6 — $[I, 2r_3, 3r_2]$ — Quartz SiO_2

(r) $D_{2d} = \bar{4}2m$ — 8 — $[I, 3r_2, 2s_4, 2\tau]$ — Urea CH_4N_2O

(n) $C_4 = 4$ — 4 — $[I, r_4, r_4^{\frac{1}{2}}, r_4^{\frac{3}{4}}]$ — Iodosuccinimide $C_4H_4INO_2$

(s) $D_4 = 422$ — 8 — $[I, 2r_4, r_2^{\frac{1}{2}}, 2r_2^{\prime}, 2r_2^{\prime\prime}]$ — Methylammonium iodide CH_3NH_3I

(o) $S_4 = \bar{4}$ — 4 — $[I, s_4, r_2, s_4^3]$ — Cahnite $Ca_2B[AsO]_4(OH)_4$

(t) $D_{4h} = 4/mmm$ — 16 — $[I, 2r_4, r_2^{\frac{1}{2}}, 2r_2^{\prime}, 2r_2^{\prime\prime}, -I, 2s_4, \tau_h, 2\tau_v, 2\tau_d]$ — Tetramethylammonium iodide $C_4H_{12}IN$

TETRAGONAL

(p) $C_{4h} = 4/m$ — 8 — $[I, c_4, c_4^{\frac{1}{2}}, c_4^{3}, -I, s_4, \tau_h, s_4^{3}]$ — Wulfenite $PbMoO_4$

(u) $C_6 = 6$ — 6 — $[I, r_6, r_6^{\frac{1}{3}}, r_6^{\frac{1}{2}}, r_6^{\frac{2}{3}}, r_6^{\frac{5}{6}}]$ — Leadantimonyl tartrate $Pb_2Sb[O_2CCH(OH)]$

(q) $C_{4v} = 4mm$ — 8 — $[I, 2r_4, r_2^{\frac{1}{2}}, 2\tau_v, 2\tau_d]$ — Diaboleite $Pb_2CuCl_2(OH)_4$

(v) $C_{3h} = 3/m = \bar{6}$ — 6 — $[I, r_3, r_3^{\frac{1}{2}}, \tau_h, s_3, s_3^{\frac{1}{2}}]$ — No substance is definitely known to crystallize with this symmetry group

HEXAGONAL

(Contd.),

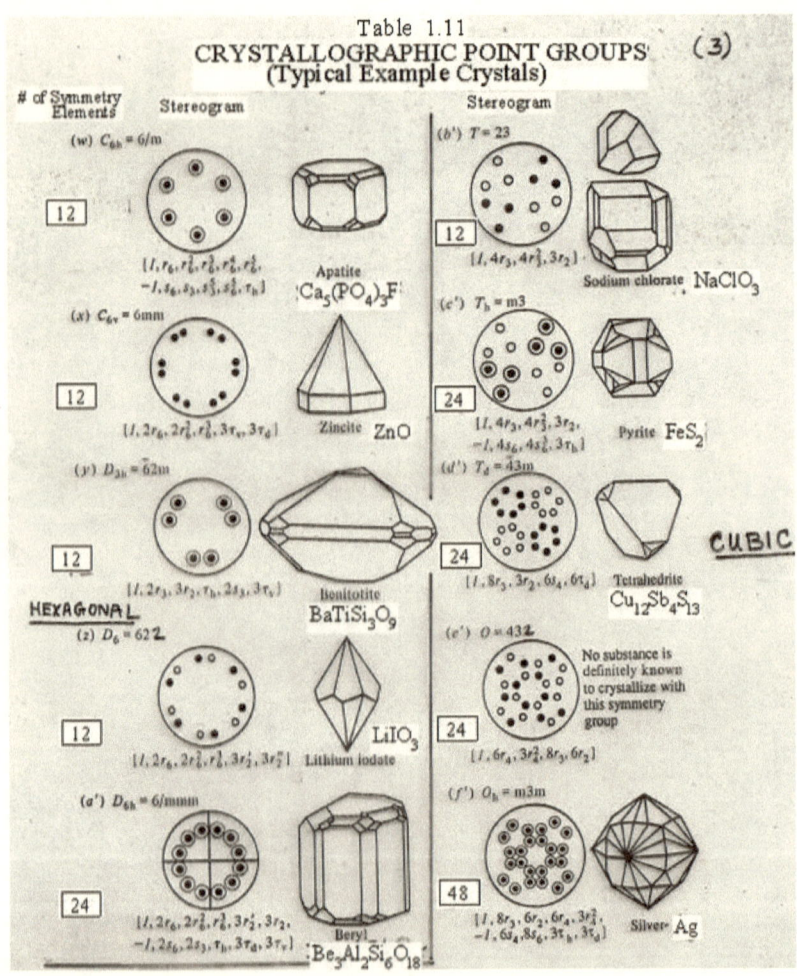

Table 1.11
CRYSTALLOGRAPHIC POINT GROUPS (3)
(Typical Example Crystals)

1.8. Miller Indices

Equation of a plane in space

$$\frac{x}{a}+\frac{y}{b}+\frac{z}{c}=1$$

(1.13)

Intercepts a, b and c

$$\boxed{h = \frac{1}{a}, k = \frac{1}{b}, and \ l = \frac{1}{c}}$$

(1.14)

1. Extend the plane to make it cut the crystal axis system at points (a_1, b_1, c_1)

2. the reciprocals of the intercepts, i.e., $\left(\frac{1}{a_1}, \frac{1}{b_1}, \frac{1}{c_1}\right)$

3. Multiply or divide by the highest common factor

4. Replace negative integers, say $-h$ by $\bar{h}$

5. If the plane is parallel to an axis, it cuts it ∞, and $\frac{1}{\infty} = 0$.

6. Use ordinary braces for single plane, and by double braces to denote a family

of planes.

If a plane has intercepts $\infty, 1, \infty$, this means the plane is represented by $(\frac{1}{\infty} \frac{1}{1} \frac{1}{\infty}) = (010)$.

Generally a plane in a crystal lattice is denoted by $(h\,k\,l)$

and a family of planes by $\{h\,k\,l\}$.

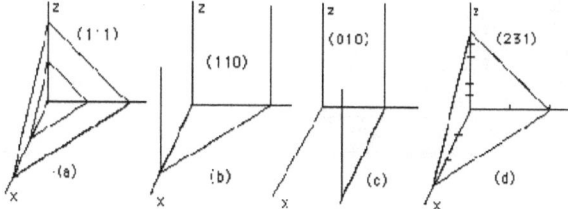

Fig 1.18 **Miller Indices to denote Crystal planes**

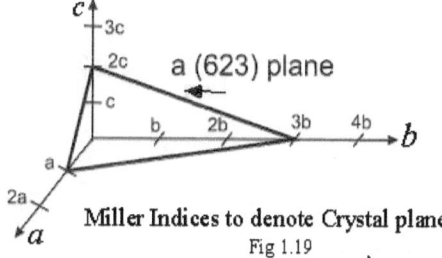

Miller Indices to denote Crystal plane

Fig 1.19

Angle between two planes, (h_1, k_1, ℓ_1) and (h_2, k_2, ℓ_2)

$$\text{Cos}\varphi = \frac{(h_1 h_2 + k_1 k_2 +, \ell_1 \ell_2)}{(h_1^2 + k_1^2 + \ell_1^2)(h_2^2 + k_2^2 + \ell_2^2)}$$

(1.15)

Bravais (Miller-Bravais) indices: A special case arises in hexagonal system. Here the general symbol is $(hkil)$ or $(hk. l)$, This must obey the rule, $(h+k) = -i = \bar{i}$.

1.9 Reciprocal Lattice (RL) :

The inverse scaling between real and reciprocal space is based on Fourier transforms.

Mathematical representation of reciprocal lattice

Reciprocal lattice vector is the inverse in magnitude of the real vector and is normal to the planes separating the original vector.

If $\vec{a}, \vec{b}, \vec{c}$ are linearly independent triad (set

$$\vec{r} = \lambda \vec{a} + \mu \vec{b} + v \vec{c} \neq 0$$
$$\lambda = \mu = v \neq 0$$

(1.16)

which is linearly independent set $V_3(F)$, orthogonal and are non-coplanar. then there exists a reciprocal triad, $\vec{a}*, \vec{b}*, \vec{c}*$ defined by

$$\vec{a}* = [\vec{b} \wedge \vec{c}] / [\vec{a} \, \vec{b} \, \vec{c}]$$
$$\vec{b}* = [\vec{c} \wedge \vec{a}] / [\vec{a} \, \vec{b} \, \vec{c}]$$
$$\vec{c}* = [\vec{a} \wedge \vec{b}] / [\vec{a} \, \vec{b} \, \vec{c}]$$

(1.17)

where $[\vec{a} \, \vec{b} \, \vec{c}] = \vec{a} \cdot (\vec{b} \wedge \vec{c}) = \vec{b} \cdot (\vec{c} \wedge \vec{a}) = \vec{c} \cdot (\vec{a} \wedge \vec{b})$

(1.18)

$$[\vec{a} \, \vec{b} \, \vec{c}] = (\hat{i} \, a_x + \hat{j} \, a_y + \hat{k} a_z) \begin{vmatrix} \hat{i} & \hat{j} & \hat{k} \\ b_x & b_y & b_z \\ c_x & c_y & c_z \end{vmatrix}$$

(1.19)

$$|\vec{a}*| = \frac{1}{d_{100}} = \frac{1}{|a| \cos(\gamma - \pi/2)}$$

(1.20)

Table 1.12

Reciprocal Lattices of Direct CUBIC lattices					
Direct	Lattice constant	Volume	RL	Lattice constant	Volume
1) sc	a	a^3	sc	$\frac{2\pi}{a}$	$(\frac{2\pi}{a})^3$
2) bcc	$\frac{a}{2}$	$\frac{a^3}{2}$	fcc	$\frac{2\pi}{a}(\pm i \pm j)$, $\frac{2\pi}{a}(\pm j \pm k)$ $\frac{2\pi}{a}(\pm i \pm k)$	$2(\frac{2\pi}{a})^3$
3) fcc	$\frac{a}{2}$	$\frac{a^3}{4}$	bcc	$\frac{2\pi}{a}(+i-j+k)$	$4(\frac{2\pi}{a})^3$

1.10 Space Groups

The number of permutations of Bravais lattices with rotation and screw axes, mirror and glide planes, plus points of inversion is finite: there are only 230 unique combinations for three-dimensional symmetry, and these combinations are known as the 230 space groups. Use of powder diffraction for structural studies does not require a knowledge of their derivation, nor does it require you to memorize a list of the 230 combinations. However, you do need to understand some of the properties of space groups. In order to make life simple, the space groups can be classified according to certain symmetry types.

32 Crystal Classes • 14 Bravais Lattices = 230 Space Groups

Table 1.13

Table of Space Group Symbols

1	$P1$	2	$P\text{-}1$	3	$P2$	4	$P2_1$	5	$C2$
6	Pm	7	Pc	8	Cm	9	Cc	10	$P2/m$
11	$P2_1/m$	12	$C2/m$	13	$P2/c$	14	$P2_1/c$	15	$C2/c$
16	$P222$	17	$P222_1$	18	$P2_12_12$	19	$P2_12_12_1$	20	$C222_1$
21	$C222$	22	$F222$	23	$I222$	24	$I2_12_12_1$	25	$Pmm2$
26	$Pmc2_1$	27	$Pcc2$	28	$Pma2$	29	$Pca2_1$	30	$Pnc2$
31	$Pmn2_1$	32	$Pba2$	33	$Pna2_1$	34	$Pnn2$	35	$Cmm2$
36	$Cmc2_1$	37	$Ccc2$	38	$Amm2$	39	$Aem2$	40	$Ama2$

41 $Aea2$	42 $Fmm2$	43 $Fdd2$	44 $Imm2$	45 $Iba2$
46 $Ima2$	47 $Pmmm$	48 $Pnnn$	49 $Pccm$	50 $Pban$
51 $Pmma$	52 $Pnna$	53 $Pmna$	54 $Pcca$	55 $Pbam$
56 $Pccn$	57 $Pbcm$	58 $Pnnm$	59 $Pmmn$	60 $Pbcn$
61 $Pbca$	62 $Pnma$	63 $Cmcm$	64 $Cmce$	65 $Cmmm$
66 $Cccm$	67 $Cmme$	68 $Ccce$	69 $Fmmm$	70 $Fddd$
71 $Immm$	72 $Ibam$	73 $Ibca$	74 $Imma$	75 $P4$
76 $P4_1$	77 $P4_2$	78 $P4_3$	79 $I4$	80 $I4_1$
81 $P\text{-}4$	82 $I\text{-}4$	83 $P4/m$	84 $P4_2/m$	85 $P4/n$
86 $P4_2/n$	87 $I4/m$	88 $I4_1/a$	89 $P422$	90 $P42_12$
91 $P4_122$	92 $P4_12_12$	93 $P4_222$	94 $P4_22_12$	95 $P4_322$
96 $P4_32_12$	97 $I422$	98 $I4_122$	99 $P4mm$	100 $P4bm$
101 $P4_2cm$	102 $P4_2nm$	103 $P4cc$	104 $P4nc$	105 $P4_2mc$
106 $P4_2bc$	107 $I4mm$	108 $I4cm$	109 $I4_1md$	110 $I4_1cd$
111 $P\text{-}42m$	112 $P\text{-}42c$	113 $P\text{-}42_1m$	114 $P\text{-}42_1c$	115 $P\text{-}4m2$
116 $P\text{-}4c2$	117 $P\text{-}4b2$	118 $P\text{-}4n2$	119 $I\text{-}4m2$	120 $I\text{-}4c2$
121 $I\text{-}42m$	122 $I\text{-}42d$	123 $P4/mmm$	124 $P4/mcc$	125 $P4/nbm$
126 $P4/nnc$	127 $P4/mbm$	128 $P4/mnc$	129 $P4/nmm$	130 $P4/ncc$
131 $P4_2/mmc$	132 $P4_2/mcm$	133 $P4_2/nbc$	134 $P4_2/nnm$	135 $P4_2/mbc$
136 $P4_2/mnm$	137 $P4_2/nmc$	138 $P4_2/ncm$	139 $I4/mmm$	140 $I4/mcm$
141 $I4_1/amd$	142 $I4_1/acd$	143 $P3$	144 $P3_1$	145 $P3_2$
146 $R3$	147 $P\text{-}3$	148 $R\text{-}3$	149 $P312$	150 $P321$
151 $P3_112$	152 $P3_121$	153 $P3_212$	154 $P3_221$	155 $R32$
156 $P3m1$	157 $P31m$	158 $P3c1$	159 $P31c$	160 $R3m$
161 $R3c$	162 $P\text{-}31m$	163 $P\text{-}31c$	164 $P\text{-}3m1$	165 $P\text{-}3c1$
166 $R\text{-}3m$	167 $R\text{-}3c$	168 $P6$	169 $P6_1$	170 $P6_5$
171 $P6_2$	172 $P6_4$	173 $P6_3$	174 $P\text{-}6$	175 $P6/m$
176 $P6_3/m$	177 $P622$	178 $P6_122$	179 $P6_522$	180 $P6_222$
181 $P6_422$	182 $P6_322$	183 $P6mm$	184 $P6cc$	185 $P6_3cm$
186 $P6_3mc$	187 $P\text{-}6m2$	188 $P\text{-}6c2$	189 $P\text{-}62m$	190 $P\text{-}62c$
191 $P6/mmm$	192 $P6/mcc$	193 $P6_3/mcm$	194 $P6_3/mmc$	195 $P23$

196 *F*23	197 *I*23	198 *P*2₁3	199 *I*2₁3	200 *Pm*-3
201 *Pn*-3	202 *Fm*-3	203 *Fd*-3	204 *Im*-3	205 *Pa*-3
206 *Ia*-3	207 *P*432	208 *P*4₂32	209 *F*432	210 *F*4₁32
211 *I*432	212 *P*4₃32	213 *P*4₁32	214 *I*4₁32	215 *P*-43*m*
216 *F*-43*m*	217 *I*-43*m*	218 *P*-43*n*	219 *F*-43*c*	220 *I*-43*d*
221 *Pm*-3*m*	222 *Pn*-3*n*	223 *Pm*-3*n*	224 *Pn*-3*m*	225 *Fm*-3*m*
226 *Fm*-3*c*	227 *Fd*-3*m*	228 *Fd*-3*c*	229 *Im*-3*m*	230 *Ia*-3*d*

The initial letter of a space group symbol represents the lattice type which may primitive (*P*), single-face centred (*A*, *B*, or *C*), all-face centred (*F*), body-centred (*I*), or rhomohedrally centred (*R*). (For rhombohedral space groups, a primitive unit cell may also be chosen, but the symbol *R* is still used so as to distinguish these space groups from the primitive trigonal space groups based on hexagonal axes.)

1.11 X-ray Diffraction from Scattering lattices:

1.11.1 Bragg condition

RL structure can be determined by diffraction techniques, Lattice has many 'Bragg Planes' (*hkl*) of atoms in various directions, Each lattice plane will constructively reflect radiation of the proper wavelength λ when incident at the proper angle θ, The interatomic distances of solids are perfectly matched for X-ray wavelengths,

The minimum wavelength of X-rays is

$$\lambda_{min} = \frac{1.24x10^{-6}}{V} Vm = \frac{1.24x10^{-6}}{V(=5x10^{-4})} = 0.25A$$

(1.21)

The spacing or distance between parallel planes of atoms in a cubic crystal, d_{hkl} is

$$d_{hkl} = \frac{a}{\sqrt{h^2+k^2+\ell^2}}$$

(1.22)

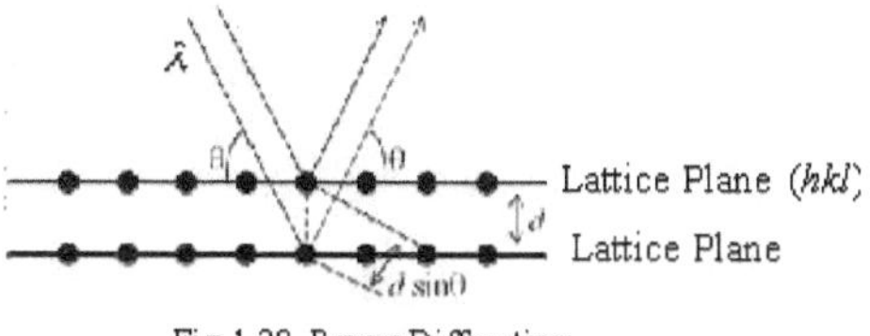

Fig 1.20 Bragg Diffraction

William Bragg diffraction condition

$$2d_{hkl} \, Sin\theta = n\lambda \qquad (1.23)$$

Table 1.14

Wavelengths of Characteristic X-rays of Targets			
Type	Kα	Kβ	Filter
1) Copper	1.542 A	1.392 A	Ni
2) Chromium	2.291 A	2.085 A	V
3) Iron	1.937 A	1.757 A	Mn
4) Cobalt	1.790 A	1.621 A	Fe
5) Molybdenum	0.711 A	0.632 A	Nb or Zr

Scan in with either λ or θ, to satisfy Bragg condition		
Method of diffractiion	λ	θ
!) Laue	variable	fixed
2) Rotating Crystal	fixed	variable
3) Powder	fixed	variable

Two common methods for material analysis and characterization are

1.11.2 Powder X-ray Diffraction (PXRD) Method

Debye-Scherrer powder method in which a rectangular strip of X-ray film is mounted in a cylindrical camera. The sample is a finely ground powder, *eg.*, filled in a thin walled capillary tube, and set exactly on the incident X-ray beam, gives the powder pattern exposed by a W or Cu target X-ray source. There are three types of film mounting. In the Straumanis type, the two ends of the film meet midway between the entry and exit of the rays. The diffraction obtained is between $\theta = 0°$ to $\theta = 90°$

$$\theta = \left(\frac{180}{\pi} \frac{1}{4R}\right) S° \qquad (1.24)$$

The camera is designed to have diameter

$$2R = \left(\frac{180}{\pi}\right) = 57.3 \ mm \qquad (1.25)$$

So $\theta = \frac{S°}{2}$, where S = distance between the arcs of particular to θ. For this camera, S = 1 mm corresponds to $\theta = 1°$

Missing Reflections in the Cubic System

Table 1.15

Missing X-ray Reflections (hkℓ)			
Type of Cubic Structure			
P	BCC(I)	FCC	Diamond
Example CsCl	Vanadium (V)	Coppeer(Cu)	Diamond, Si
Condition All h, k, ℓ	Only (h+k+ℓ) =	h,k and ℓ all	like in FCC
allowed	even or all odd	must be odd	

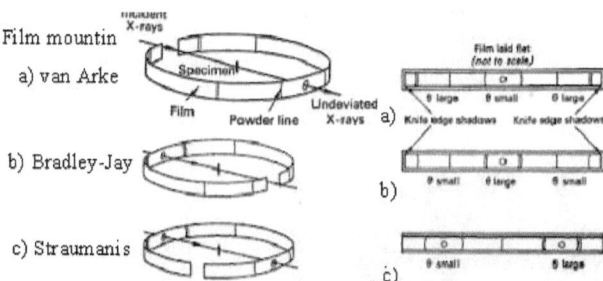

Film mountin

a) van Arke

b) Bradley-Jay

c) Straumanis

Fig 1.21 PXRD Film Mountings

1.11.3.1Laue Transmission method

Polychromatic (λs) radiation is used.

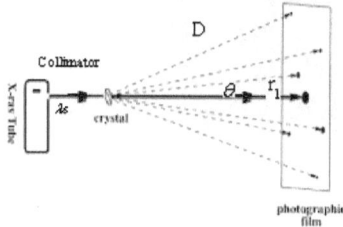

$$\tan 2\theta = \frac{r_1}{D}$$

$$\lambda_{\text{shortest } \lambda \text{ lim}} = \frac{12.4}{V(kV)}$$

Angle is measured from Laue pattern with a <u>Greninger Net</u>.

1.11.3.2 The Laue Back reflection Diffraction method.

A plane X-ray film mounted in a Laue camera gives the Laue diffraction spots of a <u>single crystal</u> on diffraction.

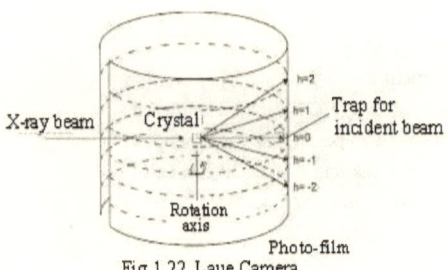

Fig 1.22 Laue Camera

Laue pattern of single crystalline KCl.

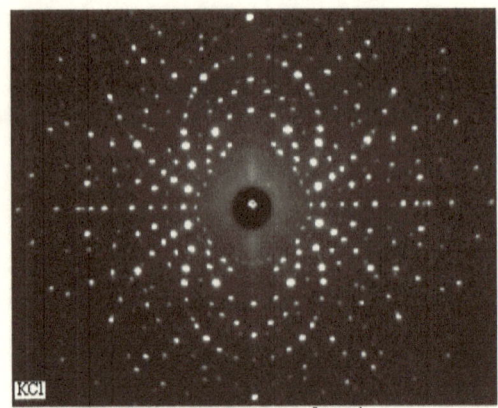

Fig 1.23   Laue Pattern of KCl

$g(x,y)$ = Diffracting aperture details in XY plane,

$U(X,Y)$ = Diffraction pattern in XY plane of Screen.

$$U(\mu,v) = \iint g(x,y) \cdot e^{i(\mu x + vy)} dx dy$$

(1.26)

$U(\mu,v)$ and $g(x,y)$ are functions forming Fourier Transform pair.

Diffraction Pattern in RL space = FT (Crystal structure)

Inverse FT (Diffraction Pattern in RL space) = Crystal structure

$\mu = kX / L$ and $v = kY / L$ are spatial frequencies,

Worked out Example 1.2

Sylvite (KCl) is a cubic crystal with density $\rho = 1.98 \, gcm^{-3}$.

a) Find the distance between adjacent atoms,

b) What is the distance between two neighbouring atoms of the same kind.

Solution: $\boxed{Step \#1}$ Molecular weight of KCl =(39.10+35.45)=74.55 $gmole^{-1}$.

$\boxed{Step \#2}$ Mass of KCl molecule = $(74.55 gmole^{-1})(\dfrac{^r1mole}{6.02x10^{23}g}) = 12.0x10^{-23} g$,

$\boxed{Step \#3}$ # of KCl molecules/unit volume = $(\rho = 1.98 \, gcm^{-3})(\dfrac{^r1mole}{12.0x10^{-23}g})$

$$= 1.60x10^{22} \text{ molecule-}cm^{-3}$$

$\boxed{Step \#4}$ KCl is diatomic. # of atoms /unit volume= $3.20x10^{22} atoms\text{-}cm^{-3}$,

$\boxed{Step \#5}$ Volume of the cubic cell of length (d =interatomic distance) and side 1cm

$= (nd)^3$, with n= number linear density of atoms.

$\therefore (3.20x10^{22} atoms\text{-}cm^{-3})d^3 = 1$

Whence a) $d = 3.14 A^\circ$ .b) $6.28 A^\circ$.

1.11.4   For crystal structure analysis

The Weissenberg photograph is extremely easy to interpret. Generally, the axes - the lines containing spots such as h00 and 0k0 - are readily recognized, and from standard charts (or even sometimes without them) the indices of all the other spots can be read off. Information that might take weeks to acquire by the oscillation method could take only minutes with the Weissenberg method, and it would be more certain. The intensities of the spots can be measured more accurately and weaker spots can be detected. Weissenberg's goniometer had everything to commend it.

1.12   Brillouin Zone construction in 2-D

The *reciprocal lattice* basis vectors span a vector space that is commonly referred to as reciprocal space, or often, **k** space.

Step # 1 Use the real space lattice vectors to find the reciprocal lattice vectors and construct the reciprocal lattice. When constructing Brillouin zones, they are always centred on a reciprocal lattice point.

Step # 2 Draw a line connecting this origin point to one of its nearest neighbours. This line is a reciprocal lattice vector as it connects two points in the reciprocal lattice.

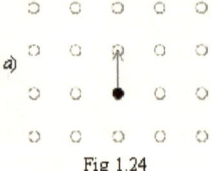
Fig 1.24

Step # 3 Draw on a perpendicular bisector to the first line. This perpendicular bisector is a Bragg plane.

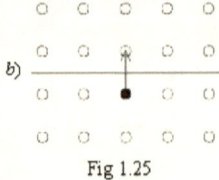
Fig 1.25

Step # 4 Add the Bragg Planes corresponding to the other nearest neighbours.

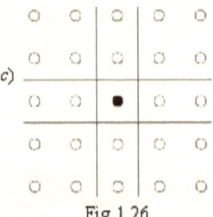
Fig 1.26

The locus of points in reciprocal space that have no Bragg Planes between them and the origin defines the first Brillouin Zone. It is equivalent to the Wigner-Seitz unit cell of the reciprocal lattice. In the picture below the first Zone is shaded.

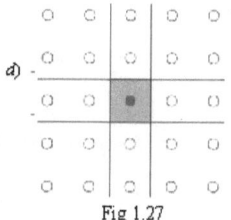

Fig 1.27

Step # 5 Draw on the Bragg Planes corresponding to the next nearest neighbours.

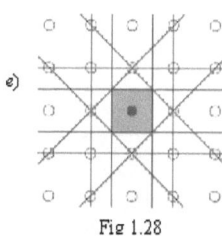

Fig 1.28

The second Brillouin Zone is the region of reciprocal space in which a point has one Bragg Plane between it and the origin. This area is shaded yellow in the picture below. Note that the areas of the first and second Brillouin Zones are the same.

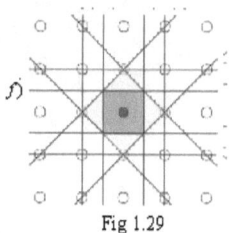

Fig 1.29

It will be worthwhile to state that

1) Crystal structure $\xrightarrow{\text{determines}}$ Diffraction pattern		
2) Shape & size of unit cell $\longleftrightarrow$ Laue spot positions		
3) Atom arrangement within unit cell $\longleftrightarrow$ Line intensities		

Worked out Example 1.2

For glycyl-L-asparagine, $C_6H_{11}N_3O_4$, its orthorhombic unit cell has $a = 4.81$, $b = 12.85$, $c = 13.52 A°$. Find the cell content. Given, $\rho = 1.506 \, gcm^{-3}$ and $M = 189.2 amu$

Solution: $\boxed{Step\#1}$ cell volume $V = abc = 839 A°^3$, $\rho = 1.506 \, gcm^{-3}$, Mol.wt., $M = 189.2 amu$

$\boxed{Step\#2}$ cell content, $= (\dfrac{(V=839 A°^3)(\rho=1.506 \, gcm^{-3})}{(N_A=1.6602x10^{-24} gamu^{-1})(M=189.2 amu)}) \approx 4.03$

$= 4$

---oooOooo---

CRYSTAL GROWTH

1.13 CRYSTAL GROWTH

1.13.1 Introduction

Advancement of solid state science depends on the availability of perfect, defect-free single crystals. Crystal grown at elevated temperatures has certain inherent difficulties. Crystalline imperfections are apt to be present due to thermal vibrations, lattice imperfections, etc. During crystal growth, orderly arrangement of atoms takes place followed by evolution of heat. Entropy decreases. The heat is liberated by the system as a result of crystallization. When in dynamic equilibrium between the crystal and in its present phase energy is at the minimum, and no more growth of crystal will occur. Factors like temperature, pressure, chemical potential or strain in the system enable further growth process.

Single crystals can be obtained through one of the following techniques:

(1) Solid-solid phase transition,

(2) Vapour growth by vapour-solid phase transition,

(3) Melt growth,

(4) Solutions growth, from liquid-solid phase transition, and (5) Gel growth..

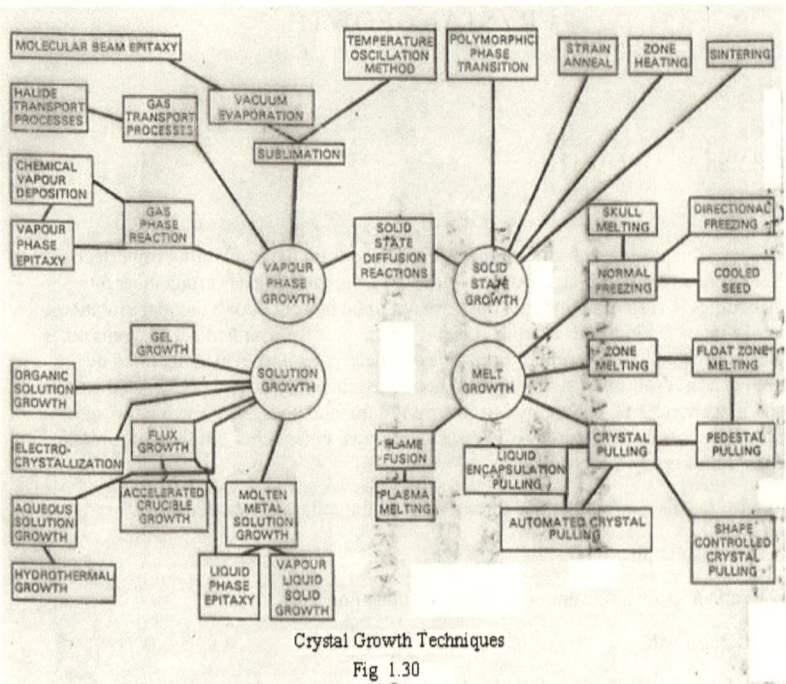

Crystal Growth Techniques

Fig 1.30

1.13.2 Solution Technique

Flux growth: This is a high temperature solution growth of crystals. Here a given high temp a supersaturated solution of the compound is slowly lowered in temp so that saturation of solution and crystal growth takes place. Desired component materials are dissolved in a solvent in a heated crucible forming so-called flux.

1.13.3 Hydrothermal growth

At ambient temp, insoluble compound is made to form solution by increasing temp and simultaneously pressure for crystal formation on lowering temp. Diamond, Calcite, (Quartz) II-VI compounds, etc are crystallized so.

1.13.4 Aqueous solution growth

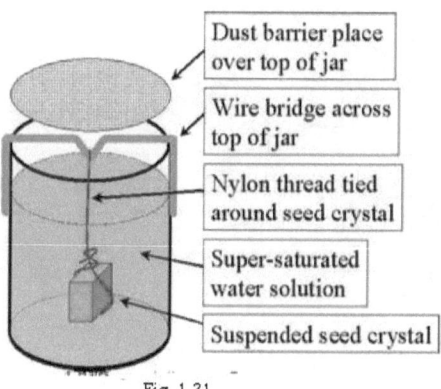

Dust barrier place over top of jar

Wire bridge across top of jar

Nylon thread tied around seed crystal

Super-saturated water solution

Suspended seed crystal

Fig 1.31

1.13.5 Melt Technique

Bridgeman-Stockbarger used two-zone furnace, Flame-fusion method vy Vernuil, crystal pulling method, by Czochralskii, by spontaneous cooling or crystallization on a seed crystal as in Kyrapoulos method.

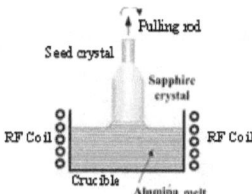

Fig 1.32 Czochralski method
Crystal Pulling

1.13.6 Bridgeman method

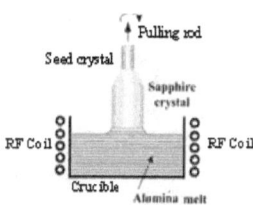

Fig 1.32 Czochralski method
Crystal Pulling

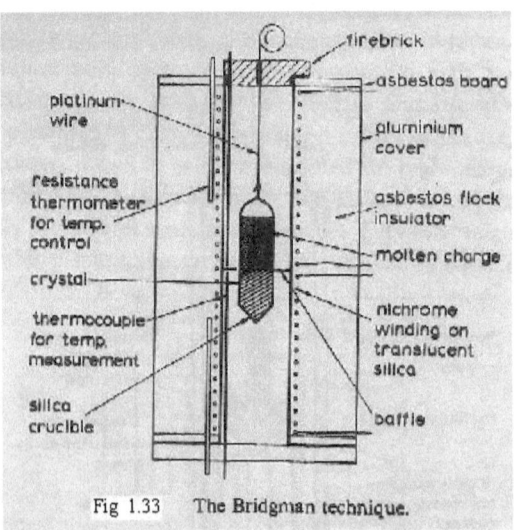

Fig 1.33   The Bridgman technique.

1.13.7 Epitaxial Method

Liquid phase epitaxy requires a vacuum coating unit and a substrate in which thin crystal films are produce.

1.13.8 Electro-crystallization

1.13.9 Crystal growth under micro-gravity conditions.

1.13.10 Gel Growth

Chemical gels like gelatin gel (gelatin powder + water at $50°C$ (add formaldehyde. Tetramethoxysilane (TMS) gel, agar-agar gel *etc*, also are used, A single test tube method or a "U-tube" method can be chosen.

In the 'test tube' method, (Fig 1.34) reactant I is incorporated inside the gel and reactant II is diffused into the gel to interact with reactant I and causing nucleation of crystals and growth.

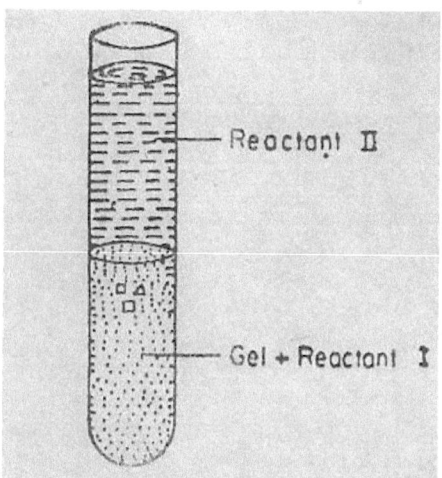

Fig 1.34 Schematic diagram of
test tube apparatus for crystal
growth by reaction

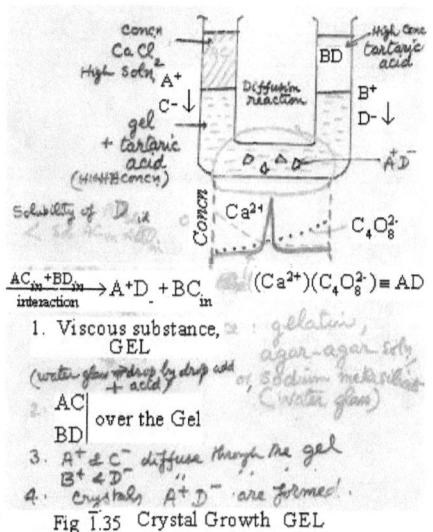

1. Viscous substance, GEL   gelatin,
   agar-agar soln,
   (water glass drop by drop add + acid) or, sodium metasilicate
   AC | over the Gel   (water glass)
   BD |
2.
3. A⁺ & C⁻ diffuse through the gel
   B⁺ & D⁻   "   "   "
4. Crystals A⁺D⁻ are formed.

Fig 1.35 Crystal Growth GEL

In the 'U-tube' method(Fig 1.35, (Fig 1.36)), have the silica gel (mix in 10 cc at pH = 5, titrating with glacial acetic acid, which is to be mixed with 5 cc of 95% Ethanol and allowed to colloidal gel, in 90 hrs.), then add to one arm enough of the reactants, plus ethanol (BD) carefully over the gel top, and to the other arm the nutrient (AC) over the gel. Diffusion takes place A^+ and C^- ions and B^+ and D^- ions. Slowly crystals of AD are formed and grow into perfection.

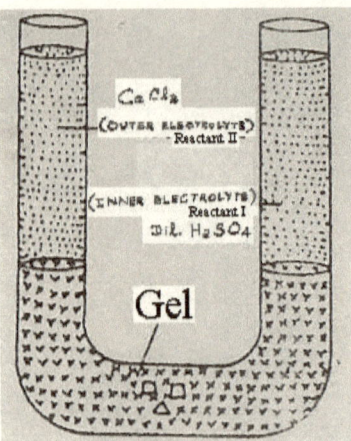

Fig 1.36 Schematic diagram of
U-tube apparatus for crystal
growth by reaction

+^+&+&+&+&+&+&+&+&+^+

REVIEW QUESTIONS

R.Q. 1.1   Write the Miller indices of the planes with intercepts a) $\frac{1}{2},\frac{2}{3},1$, b) $\infty,1,\frac{2}{5}$
c) $\frac{2}{3},\infty,\frac{1}{6}$ d) $\frac{1}{3},\frac{2}{5},\infty$ and e) $\frac{1}{6},\frac{1}{5},\infty$. (Ans: a) 432, b) 025, c) 3012, d) 650, e) 650).

R.Q. 1.2   Draw sketches illustrating a (100) plane, a (110) plane, and a (111) plane in a cubic unit cell.

R.Q. 1.3   A primitive unit cell has $a = 5.00$, $b = 6.00$, $c = 7.00\ A°$,
$\alpha = \beta = \gamma = 90°$. A new unit cell is chosen with edges defining vectors from the origin to the points with coordinates 3,1,0; 1,2,0; and 0,0,1. a) Calculate the volume of the original unit cell, (Ans: $210\ A^{°3}$), b) Calculate the lengths of the edges and angles of the new unit cell (Ans: $a = 16.16\ A°$, $b = 13.00\ A°$, $c = 7.00\ A°$; $\alpha = \beta = 90°$, $\gamma = 45.6°$), c) Calculate the volume of the new unit cell. (Ans: $V = 1050 A^{°3}$), d) How many lattice points are there in the new unit cell? (Ans; 5).

R.Q. 1.4   A unit cell has dimensions $a = 6.00\ A°$, $b = 7.00\ A°$, $c = 8.00\ A°$; $\alpha = \gamma = 90°, \beta = 115.0°$. (a) Calculate the distance between the points 0.200, 0.150, 0.333 and 0.300, 0.050,0.123. (Ans: $2.13\ A°$).

R.Q. 1.5. Differentiate atomic structure, crystal structure and crystal system.

R.Q. 1.6 Show that for a b.c.c structure the unit cell edge length a and atomic radius R are related through $a = \frac{4R}{\sqrt{3}}$.

R.Q. 1.7. NaCl Crystal has a density $\rho = 2.167\ g.cm^{-3}$. What is the separation between neighbouring atoms? (Ans: $0.707a$).

R.Q.1.8. A crystal has a basis of one atom. Given the vectors $\vec{a} = 3\,\hat{i}$, $\vec{b} = 3\hat{j}$, $\vec{c} = 1.5(\hat{i} + \hat{j} + \hat{k})$. A) What is the Bravais lattice type of the crystal?, b) Calculate the volume of the primitive cell and that of its conventional unit cell. (Ans: a) trigonal lattice, since $Cos\alpha = \frac{\vec{b}.\vec{c}}{|\vec{b}||\vec{c}|}, \alpha = 54°$, $\beta = 54°$, $\gamma = 90°$;b) primitive cell $V = 13.5(unit)^3$).

+&+*+&+&+&+&+&+&+&+&*+

PRELIMINARIES – 2A

ATOMIC BONDING

Chapter 2

PRELIMINARIES – 2A

ATOMIC BONDING,

"Remain always strong and steadfast in thy own faith, but eschew all bigotry and intolerance"

Shri Ramakrishna

2.1 ATOMIC BONDING

When the atoms are in equilibrium in a crystal lattice the minimum lattice potential energy

$$U = \frac{A}{R^n} + \frac{B}{R^m}$$

(2.1)

R = Equilibrium separation between two neighboring atoms

Force,
$$F = -\frac{dU}{dR}$$

(2.2)

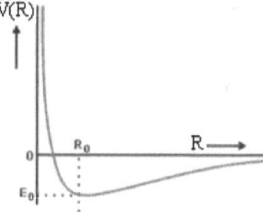

Fig 2.1

In equilibrium, there is minimum pE, when

$$R_{eq} = \left(\frac{mB}{nA}\right)^{1/(m-n)}$$

(2.3)

i.e.,
$$U_{eq} = U_{R=R_{eq}} = -\frac{A}{R_{eq}^n}\left(1-\frac{n}{m}\right)$$

(2.4)

2.2.1 Ionic Bonds,

When one atom takes an electron away from another and the resulting positive and negative ions are attracted to each other, those atoms have formed an **ionic bond**. Ionic bonding is possible only between two unlike atoms, one electro-positive and the other electro-negative.

$$U_{ionic} = \frac{q^+q^-}{|4\pi\varepsilon_o r|} = \frac{e^2 Z^+ Z^-}{|4\pi\varepsilon_o r|} \tag{2.5}$$

$$U_{eq} = -\frac{e^2}{4\pi\varepsilon_o R_{eq}}$$

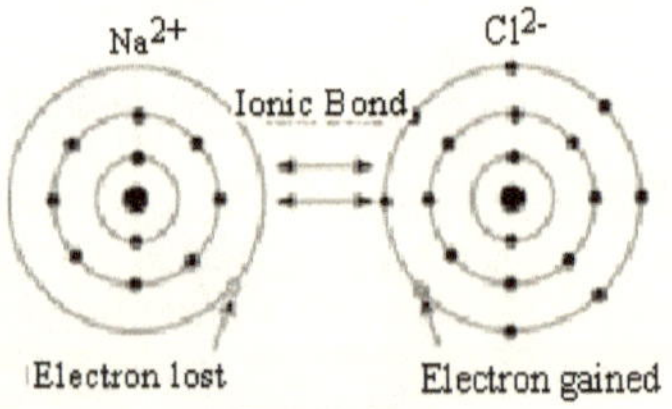

Fig 2.2

or ionic alkali halide crystals, say NaCl, Born and Mayer obtained the lattice (cohesive) energy, i.e., energy released at the time of forming the ionic bond is,

$$U_{eqm} = -\frac{Ae^2 N_A}{4\pi\varepsilon_o R_{eq}}\left(1 - \frac{\rho}{R_{eq}}\right) \tag{2.6}$$

Bulk modulus of a solid,

$$\beta = -V\left(\frac{dp}{dV}\right), \tag{2.7}$$

and compressibility,

$$\left(\frac{1}{K}\right) = \beta = -V\left(\frac{dp}{dV}\right) \tag{2.8}$$

$$\frac{1}{K_o} = \beta = \frac{Ae^2}{18R_{eq}^4}\left(\frac{R_{eq}}{\rho} - 2\right) \tag{2.9}$$

For NaCl, $R_{eq} = 0.2283\,nm$, $\rho = 0.0345\,nm$ (for all alkali halides),

$$U = -\frac{Ae^2N_A}{4\pi\varepsilon_o R_{eq}}\left(1-\frac{1}{n}\right)$$

(2.10)

For a compound with rock salt structure, Coulomb energy

$$U_{ionic} = \frac{e^2Z^+Z^-}{|4\pi\ \varepsilon_o r|}\left(\frac{6}{r}-\frac{12}{|r\sqrt{2}|}+\frac{8}{|r\sqrt{3}|}-\frac{6}{|r\sqrt{4}|}+...\right) = \frac{e^2Z^+Z^-}{|4\pi\ \varepsilon_o r|}NA$$

A = Madulong constant.

2.2.2 **Madulong constant, A**

$$n = 1 + \frac{72\pi\varepsilon_o R_{eq}^4}{Ae^2K_o}$$

(2.11)

Table 2.1

Crystal structure Type	Madelung constant, A	CN
1) Rocksalt	1.748	6
2) CsCl	1.763	8
3) Wurtzite	1.641	4
4) Sphaleraite	1.638	4
5) Fluorite	5.039	8
6) Rutile	4.816	6

2.2.2.1 Dissociation energy, D can be obtains as

$$D = \frac{4A}{5R_{eq}^2}$$

(2.12)

2.2.3 Covalent Bonds,

When two atoms share electrons, they form a **covalent bond**. Covalent bond is formed when two or more outer most electrons in an atom are shared by other atoms, as in chlorine molecule. Or in molecules formed by two non-metallic atoms as in H_2O.

Fig 2.3

2.2.4 Metallic Bonds

Metals have several qualities that are unique, such as the ability to conduct electricity, a low ionization energy, and a low electro-negativity.

A metallic bond is pretty different from covalent and ionic bonds, but the goal is the same: to achieve a lower energy state. Instead of a bond between just two atoms, a **metallic bond** is a sharing of electrons between many atoms of a metal element

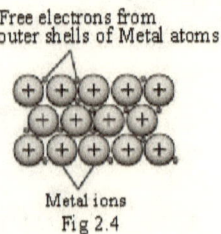

Free electrons from outer shells of Metal atoms

Metal ions

Fig 2.4

Metals engage in a unique type of bonding that provides them with a unique set of properties. Unlike most other non-metallic substances, metals are **malleable** and **ductile** and good conductors of heat and electricity. **Malleable** means a substance can be shaped. It is also ductile..

2.2.5 Van der Waals Bonds,

There is a constant 'sloshing around' of the electrons in the molecule causes rapidly fluctuating dipoles, even in highly symmetric molecule. It even happens in monatomic molecules, say helium (single atom molecule). The polarity of both molecules reverses ,*i.e.*, fluctuating, keeping synchronization, with mutual attraction. The dispersion forces between molecules are much weaker than the covalent bonds within molecules. The shape of the molecules also matter.

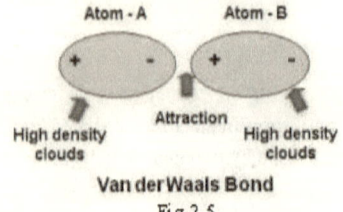

Van der Waals Bond

Fig 2.5

All molecules experience dispersion forces.

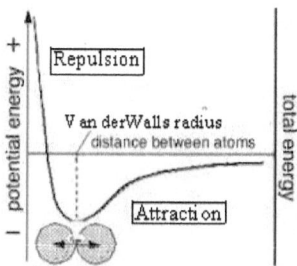

Fig 2.6

$$U_{Lennard-Jones} = 4\varepsilon \left[\left(\frac{\sigma}{r}\right)^{12} - \left(\frac{\sigma}{r}\right)^{6} \right]$$

where ε is the depth of the potential well, σ is the finite distance at which the inter-particle potential is zero, r is the distance between the particles

$$U(\text{Inert gas solids}) < U(\text{Alkali metals}) < U(\text{Diamomd})$$

2.2.6 Hydrogen Bonds,

A chemical bond in which a hydrogen atom that isalready bonded to an atom in a molecul e forms asecond bond with another atom, either in the samemolecule or in a different one. The se cond atom isusually of a type that strongly attracts electrons, suchas nitrogen or oxygen

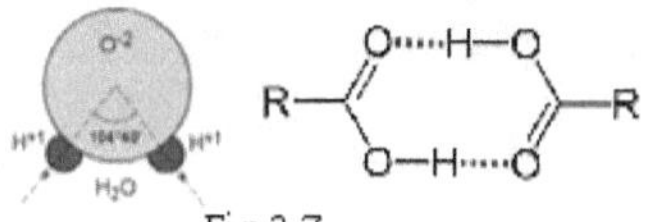

Fig 2.7

Table 2.2

Types of Solieds				
Type	Units present	Characteristic Prioperties	Representative Crystal	Chesive energy (*eV*)
1) Ionic	Anion & Cation	Brittle, Insulator, High Melting Point	NaCl, LiF	184 244
2) Covalent	Atomic	Hard, High MP, Non-conducting	Diamond SiC	170 244
4) Metallic	Anions & electron gas	High conductivity	Fe Na	94 26
5) H-bond	Molecules held by H-bonds	Low MP Insulators	H2O (ice) HF	12 7.0
6)Vander Waals (Molecular)	Molecules Atoms	Soft, Low MP, Vokatile Insulating	Ar	2.0

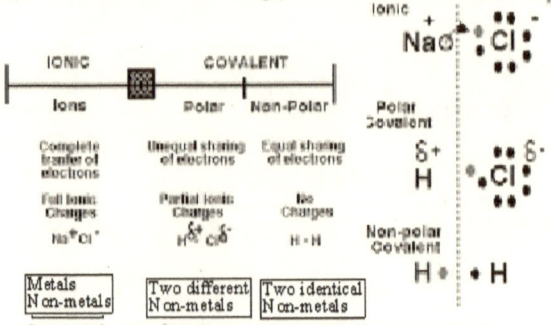

Comparison of Ionic, Polar, and Non-Polar Bonding
Fig 2.8

----oooOooo----

PRELIMINARIES – 2 B

LATTICE VIBRATIONS

PRELIMINARIES – 2 B

LATTICE VIBRATIONS

If I have ever made any valuable discoveries, it has been owing more to patient attention,

than to any other talent - Sir Isaac Newton

2.3   PHONONS - **Lattice Vibrations**

2.3.1   Introduction

Lattice vibrations can explain sound velocity, thermal properties, elastic properties and optical properties of materials. Lattice Vibration is the oscillations of atoms in a solid about the equilibrium position. For a crystal, the equilibrium positions of atoms form a regular lattice. The vibration of these neighboring atoms is not independent of each other. An ideal lattice has <u>harmonic forces</u> between atoms, and normal modes of vibrations are called lattice waves (Born & Huang). Lattice waves range from low frequencies to high frequencies. on the order of $10^{13}\,Hz$ or even higher. However, the wavelengths at extremely high frequencies are of the order of inter-atomic spacing. Due to the shortness of these wavelengths, the motion of the neighboring atoms is uncorrelated; with each atom moving about its average position in three dimensions with average vibrational energy, which is usually $3k_B T$.

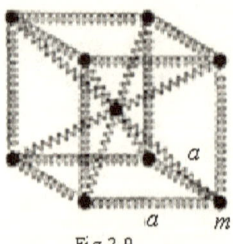

Fig 2.9

Lattice vibrations can also interact with free electrons in a conducting solid which gives rise to electrical resistance.

2.3.2   Monatomic 1-D Lattice.

Lattice dynamics offers two different ways of finding the dispersion relation within the lattice.

2.3.3 Quantum-mechanical approach

It can be used to obtain phonon's dispersion relation, the solution to the Schrodinger equation for the lattice vibrations must be solved.

2.3.4 Semi-classical treatment of lattice vibrations

This treatment gives classical mechanics the use of one additional postulate taken from quantum mechanics, mainly that the energy of lattice vibrations is quantized.

2.3.4.1 Newton's law of mechanics:

Force = Mass x Acceleration

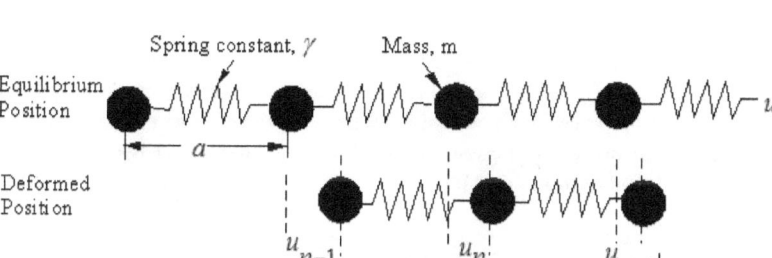

Fig 2.10 **Monatomic Chain**

Position $u(t)$ of an atom of mass m in an instantaneous potential $\varphi(u,t)$ gives

$$\frac{d^2u(t)}{dt^2} = -\frac{1}{m}\nabla\varphi(u,t).$$
(2.13)

This potential energy is the interaction of the atom with the other atoms within the crystal, a = interatomic distance.

2.3.4.2 Dispersion Relation

Solving, the velocity of the lattice wave $v = \frac{\omega}{k}$ plotted against $k = \frac{2\pi}{\lambda}$

Within $\omega = \pm\sqrt{\left(\frac{4\beta}{m}\right)}\,Sin\frac{ka}{2}$
(2.14)

First Brillouin Zone, $viz.,$ $-\frac{2\pi}{a} \leq k \leq +\frac{2\pi}{a}$ is as shown.

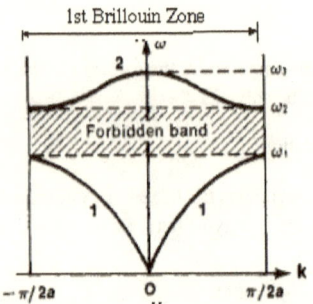

Fig 2.11 Dispersion of Monatomic Chain
within First BZ

2.3.4.3 Optical **Phonon**

Optical phonons are quantized modes of lattice vibrations when two or more charged particles in a primitive cell move in opposite directions with the center of mass at rest. This mode has highest energy for wavelength infinity or $k = 0$ when the two lattices move in opposing direction of each other.

2.3.5.1 Diatomic 1-D Lattice

Diatomic means the lattice with two kinds of atoms with masses m and M. The equations of motion are:

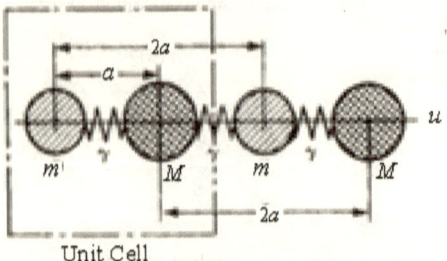

Unit Cell

Fig 2.12 Diatomic Chain

$$m\frac{\partial^2 u_{2n}}{\partial t^2} = \beta(u_{2n+1} - 2u_{2n} + u_{2n-1})$$

(2.15)

$$M\frac{\partial^2 u_{2n+1}}{\partial t^2} = \beta(u_{2n+2} - 2u_{2n+1} + u_{2n})$$

(2.16)

Trial solutions are

$$u_{2n} = A\, e^{i(2nka \pm \omega t)}$$

(2.17)

$$u_{2n+1} = B\, e^{i\{(2n+1)ka \pm \omega t)\}}$$

(2.18)

β = spring constant

The solution of the diatomic lattice is

$$\omega^2 = \beta\left(\frac{1}{m}+\frac{1}{M}\right) \pm \beta\sqrt{\left(\frac{1}{m}+\frac{1}{M}\right)^2 - \frac{4Sin^2ka}{Mm}}$$

(2.19)

2.3.5.2 Dispersion Relation

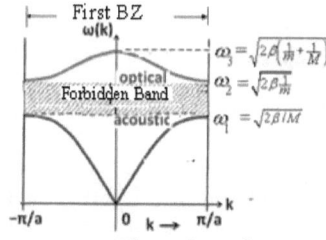

$$\omega_3 = \sqrt{2\beta\left(\frac{1}{m}+\frac{1}{M}\right)}$$

$$\omega_2 = \sqrt{2\beta\frac{1}{m}}$$

$$\omega_1 = \sqrt{2\beta/M}$$

Fig 2.13 **Dispersion of Diatomic Chain**

2.3.5.3 Transverse Optic (TO) mode –

$$TO\ mode\quad k \to 0,\quad \omega = \left(\frac{2\beta}{\mu}\right),$$

(2.20)

$$Effective\ mass,\ \mu = \frac{Mm}{M+m}.$$

(2.21)

In the long-wavelength limit, optical modes interact strongly with electromagnetic radiation in polar crystals, hence the name.

Strong optical absorption is observed (Photons annihilated, phonons created).

$$\omega \to finite\ as\ k \to 0$$

(2.22)

Optical modes arise from folding back the dispersion curve as the lattice periodicity is doubled (*i.e.,* halved in *q*-space).

2.3.5.4 Zone boundary

All modes are standing waves at the zone boundary,

$\frac{\partial \omega}{\partial k} = 0$: a necessary consequence of the lattice periodicity.

In a diatomic chain, the frequency-gap between the acoustic and optical branches depends on the mass difference. In the limit of identical masses the gap tends to zero.

2.3.5.5 Transverse Acoustic (TA) mode,

$$\boxed{\text{TA mode} \quad k \to 0, \quad \omega = \left(\frac{2\beta a^2}{M+m} \right),}$$

corresponds to sound waves in long wave limit, hence the name. $\omega \to 0 \; as \; k \to 0$

2.3.6.1 Origin of Optic and acoustic modes.

Effect of periodicity – of a diatomic chain is the result of that of monatomic

The permitted waves are split into two branches called the optical and acoustical branches. The gap (forbidden band) between the optical and acoustic branch is the region where frequencies are not allowed to propagate. The width of this forbidden band depends on the difference of the masses of the two atoms. If the two masses are equal, the two branches join (become degenerate) at $\frac{\pi}{2a}$. The acoustical branch for the diatomic is similar to that of the monatomic lattice, but the optical branch is different. Pattern of Pattern of displacement of atoms

Acoustic mode

Optic mode

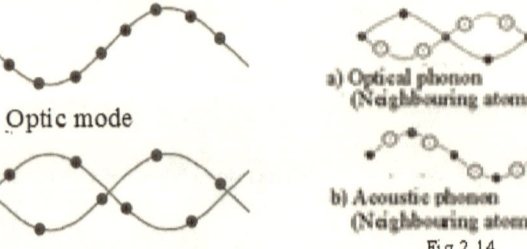

a) Optical phonon
(Neighbouring atoms out of phase)

b) Acoustic phonon
(Neighbouring atoms in phase)

Fig 2.14

Let s = Number of atoms / unit cell,

N = # of $\vec{q}$ s / unit cell of the RL lattice,

$3s$ = # of $\omega(\frac{\vec{q}}{i})$ s for each $\vec{q}$ value,

$3sN$ = total # of modes.

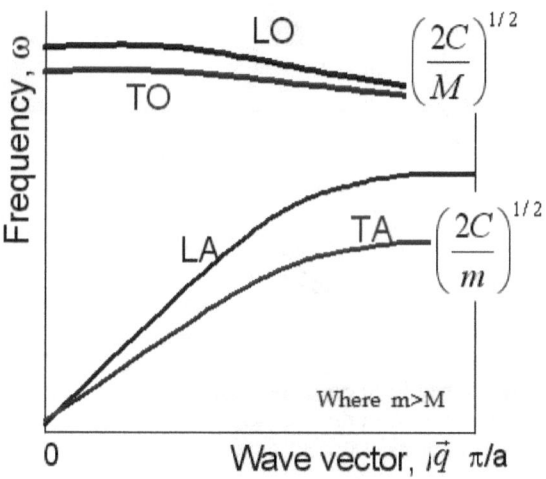

Fig 2.15   Diatomic Chain

i) $\therefore$ # of modes as $\omega(\frac{\vec{q}}{i}) \to 0$, =3

        as $\vec{q} \to 0$ are $\omega(\frac{\vec{q}}{i})\big|_{Acous}$

ii)   # of modes as $\omega(\frac{\vec{q}}{i}) \to$ finite,

        As $\vec{q} \to 0$ are $\omega_i(0) = (3s - 3)$.

        There will be one Optic branch.

In a 3-D cubic lattice,

a) Polarization vector for long wave are $\vec{q} \parallel$ to a crystal direction and are atomic vibrations $\parallel \vec{q}$.

b) For atomic vibrations $\perp \vec{q}$ there arise transverse waves.

   Conventionally, longitudinal modes have highest ω for acoustic mode than transverse acoustic modes.

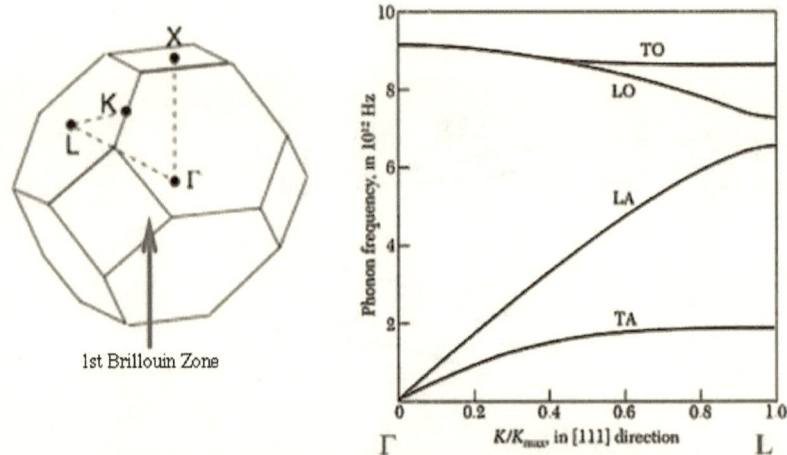

1st Brillouin Zone

Fig 2.16 Measured Phonon dispersion relations
for Ge in Diamond structure (Cold inelastic
scattering of neutrons) (Brockhouse & Iyengar, 1959)

2.3.6.2 Zone boundary modes

$$\boxed{\text{Standing waves} \quad k = \frac{\pi}{2a}, \; \lambda = \frac{2\pi}{k}}$$

(2.23)

Higher energy mode, only *light atoms move*,

Lower energy mode –only *heavier atoms move*.

The difference between the optical and acoustic branch is that the optical branch for the long wavelength limit both atoms in the unit cell move opposite to each other with an increase in the mass amplitude. The acoustical branch for the long wavelength limit, the

2.3.6.3 Phonons in 3-Dimension

In a 3-D crystal, the atoms vibrate in 3-Ds with three vibrational branches, one longitudinal and two transverse. For a 3-D Lattice with N atom per lattice point, there is $3(m-1)$ optical branches, of which $2(m-1)$ are TO phonons and the remaining LO phonons. In a transverse wave, the atomic displacement direction is perpendicular to the direction of the propagated wave. The remaining two transverse waves will overlap if the two vibrational directions are symmetric. In regards to electrons, the phonons are dispersed along different crystallographic direction

Eg., NEON (FCC lattice)

Inelastic neutron scattering results in different crystallographic directions

Many features are explained by the 1-D model:

Dispersion is sinusoidal [Nearest Neighbour (NN) interactions]

All modes are acoustic (monatomic system).

2.3.6.4   NEON- An FCC Monatomic solid.

There are two distinct types of mode:

a) Longitudinal (L), with displacements parallel to the propagation direction,

These generally have higher energy.

b) Transverse (T), with displacements perpendicular to the propagation direction. These generally have lower energy. They are often degenerate in high symmetry directions (not along ($\xi\xi$0)). Minor point (demonstrating that real systems are subtle and interesting, but also implicated):

1) mode along ($\xi\xi$0) has 2 Fourier components, suggesting next- NN (Next Nearest-Neighbour) interactions.

2) In fact there are only NN interaction.

The effect is due to the FCC structure.

Nearest Neighbour interactions from atom, A (in plane I) join to atom C (in plane II) and to atom B (in plane III) thus linking nearest- and next-nearest-planes.. Phonons in 3-D lattice, Diatomic solid eg., NaCl has sodium chloride structure! two interpenetrating f.c.c. lattices.

Main points: The 1-D model gives several insights, as before. There are: Optical and acoustic modes (labels O and A); Longitudinal and transverse modes (L and T). Dispersion along (00ξ) is simplest and most like our 1-D model. ($\xi\xi$0) planes contain, alternately, Na atoms and Cl atoms (other directions have Na and Cl mixed).

NaCl Phonons

Note the energy scale. The highest energy optical modes are ~8 *THz* (*i.e.* approximately 30 *meV*). Note the higher phonon energies than in Neon. The strong, polar bonds in the alkali halides are stronger and stiffer than the weak, van-der-Waals bonding in Neon.

<u>Minor point</u>:

Modes with same symmetry cannot cross, hence the avoided crossing between acoustic and optical modes in (00ξ) and $(\xi\xi 0)$ directions.

Ignore the detail for present purposes

&&&&&&&&&&

PRELIMINARIES – 2C

Specific heats of solids

PRELIMINARIES – 2C

Specific heats of solids

Anyone who has not been shocked by quantum physics has not understood it Niels Bohr

2.4.1   Specific heats of solids

Performing a *normal mode analysis* of the oscillations, one gets $3N$ independent modes of oscillation of the solid. Each mode has its own particular oscillation frequency, and its own particular pattern of atomic displacements. Any general oscillation can be written as a linear combination of these *normal modes*. Let q_i be the (appropriately normalized) amplitude of the i^{st} normal mode, and p_i the momentum conjugate to this coordinate. In *normal mode coordinates*, the total energy of the lattice vibrations takes the particularly simple form

$$E = \frac{1}{2} \sum_{i=1-3N} (p_i^2 + \omega_i^2 q_i^2)$$

(2.24)

ω_i = Frequency of normal mode, lattice modes are non-localized.

$$\Delta E = \hbar \omega$$

(2.25)

is the reason for lattice vibrations are more closely spaced than vibrational energy levels of vibrations of gaseous molecules. Lattice modes if obey classically, as per equi-partition of energy, mean value per mole,

$$\widehat{E} = 3Nk_B T$$

(2.26)

Molar heat capacity at constant volume, C_V

$$C_V = \frac{1}{V} \left(\frac{\partial \widehat{E}}{\partial T} \right)_V = 3R$$

(2.27)

2.4.2   Dulong and Petite's law

It follows that the C_V is for solids, stated in equation (2.27) gives a value of 24.9 $J\ mole^{-1}\ C^{-1}$. In fact, at room temperature most solids (in particular, metals) have heat capacities which lie remarkably close to this value. This fact was discovered experimentally by Dulong and Petite at the beginning of the 19th Century, and was used to make some of the first crude estimates of the molecular weights of solids (one can obtain an estimate of the molecular weight). The C_V is somewhat less than the constant pressure value, C_p but not by much, because solids are fairly incompressible.

It is essentially a high temperature limit.

2.4.3 Einstein's approximation,

All vibrate at the same frequency.

$$C_V = -\frac{3N_A \hbar\omega}{k_B T^2}\left[\frac{\hbar\omega \, e^{\beta\hbar\omega}}{[e^{\beta\hbar\omega}-1]^2}\right]$$
(2.28)

2.4.4 Einstein Model (1907)

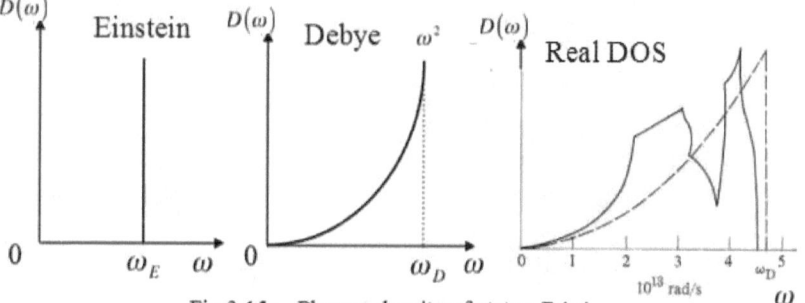

Fig 2.15a   Phonon density of states, $D(\omega)$

Einstein temperature $\boxed{\Theta_E = \frac{\hbar\omega}{k_B}}$
(2.29)

When $T \ll \Theta_E$

$$C_V = 3R\left(\frac{\Theta_E}{T}\right)^2\left[\frac{e^{\Theta_E/T}}{[e^{\Theta_E/T}-1]^2}\right]$$
(2.30)

$$C_V \sim 3R\left(\frac{\Theta_E}{T}\right)^2 e^{-\Theta_E/T}$$
(2.31)

In this model the specific heat approaches zero exponentially as $T \to 0$. Experimentally at low temperatures is more like

$$\boxed{C_V \propto T^3}.$$

(2.32)

2.4.5 Debye approach.(1912)

In this model, choosing the total number of normal modes as, $3N$ define Debye frequency

$$\boxed{\omega_D = c\left(6\pi^2 \frac{N}{V}\right)^{1/3}}$$

(2.33)

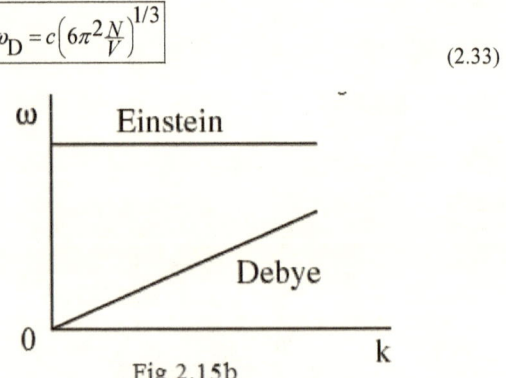

Fig 2.15b

leading to

$$\boxed{C_V \sim 3Rf_D(\beta\hbar\omega_D) = 3Rf_D\left(\frac{\Theta_D}{T}\right)}$$

(2.34)

$$f_D(y) = \frac{3}{y^3} \int_0^y \frac{e^x}{(e^x-1)^2} x^4 dx.$$

(2.35)

In asymptotic limit $T \geq \Theta_D$., for small y,

$$f_D(y) \to \frac{3}{y^3} \int_0^y x^2 dx = 1.$$

(2.36)

In the low temperature limit,

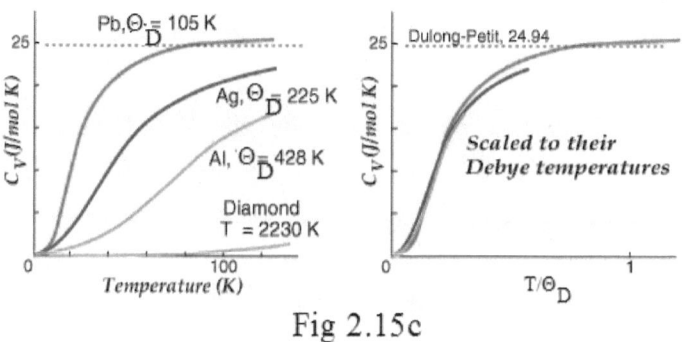

Fig 2.15c

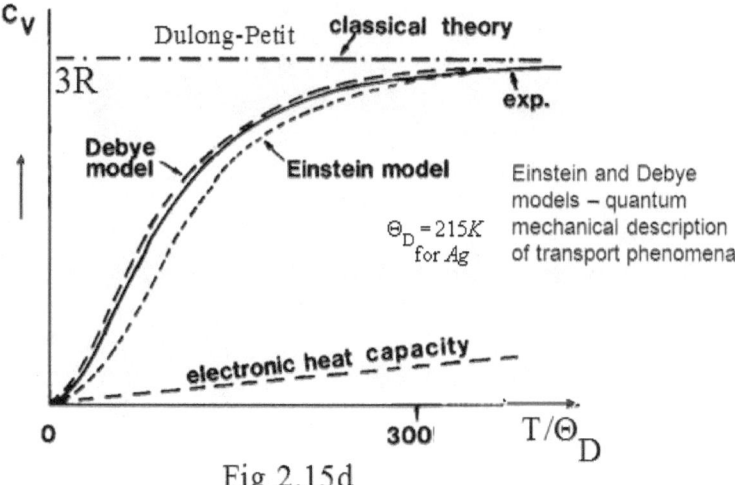

Fig 2.15d

The Debye theory has seen valid with experiment for very low temperatures for non-metals.

For metals (Fig 2.15d) electron specific heat C_{vel} becomes significant at low temperatures, and has to be combined with the phonon value above in the Einstein-Debye heat capacity.

For metals:

$$C_V = C_{vphonon} + C_{vel}$$

(2.37)

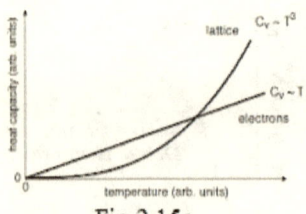

Fig 2.15e

For non-metals:

$$\boxed{C_V = C_{vphonon}}.$$

(2.38)

----oo0oo---

PRELIMINARIES – 2D

THERMAL EXPANSION OF SOLIDS

PRELIMINARIES – 2D

THERMAL EXPANSION OF SOLIDS

The whole of science is nothing more than a refinement of everyday thinking- A. Einstein

2.5   THERMAL EXPANSION OF SOLIDS

Ideal crystal has its each atom vibrate in a harmonic (parabolic) potential well'

$$V(x) = \frac{1}{2} kx^2$$

(2.39)

This means as temperature is increased the amplitude of vibration increases but the equilibrium position x_o does not change with T.

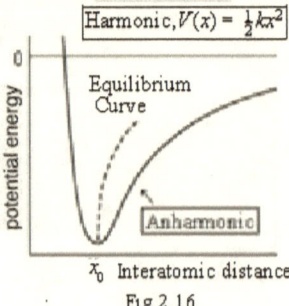

Fig 2.16

For a harmonic solid thermal expansion is ZERO .

Anharmonicity in the potential causes the solid to expand on increase on T (Fig 2.16).)

2.10.1   Coefficient of linear thermal expansion, α

$$\alpha = \frac{1}{L} \frac{\Delta \ell}{T_1 - T_2}$$

(2.40)

For isotropic (Cubic) solids, there is only one expansion coefficient, α

For anisotropic solids, uniaxial crystals have two different values, whereas for biaxial crystals there are three different coefficients of expansion.

2.10.2 Volume expansion

$$\beta = \sum_{i=1,2,3} \alpha_i$$

(2.41)

1) Thermal expansion of a crystal is a structure sensitive property of the crystal,
2) α is related intimately to the normal modes of vibration of crystalline lattice,
3) α is quite essential to convert the experimentally determined molar specific heat C_P value to C_V value required by theorists for theory of specific heats.

$$C_P - C_V = \frac{\beta^2 VT}{\chi_T J}$$

(2.42)

χ_T = compressibility of the solid, V = molar volume of the solid,

J = Joules Mechanical equivalent of heat.

Thermal expansion is T (or $\theta°C$) dependent.

$$\alpha_\theta = A + B\theta + C\theta^2$$

(2.43)

For non-metals: $\boxed{\alpha = \alpha_{phonon}}$

(2.44)

For Metals: $\boxed{\alpha = \alpha_{lattice\ phonon} + \alpha_{electronic}}$

(2.45)

2.10.3 Gruneisen's Rule

γ = Gruneisen constant of a solid is

$$\frac{3\alpha}{\chi_T} = \gamma \frac{C_V}{V}$$

(2.46)

2.10.4 Thermal expansion is a second-rank tensor, $[\alpha_{ik}]$

$$[\alpha_{ik}] = \begin{bmatrix} \alpha_{11} & \alpha_{12} & \alpha_{13} \\ \alpha_{21} & \alpha_{22} & \alpha_{23} \\ \alpha_{31} & \alpha_{32} & \alpha_{33} \end{bmatrix}$$

(2.47)

In terms of the principal axes, the tensor simplifies to

Table 2.3 *Thermal Expansion of Crystals*

Crystal System	Axial relations	No. of cons-tants	Thermal expansion tensor referred to axes in the conventional orientation		
1. Triclinic	$a \neq b \neq c$ $\alpha \neq \beta \neq \gamma$	6	α_{11} α_{21} α_{31}	α_{21} α_{22} α_{32}	α_{31} α_{32} α_{33}
2. Monoclinic	$a \neq b \neq c$ $\alpha = \beta = 90°$ $\gamma \neq 90°$	4	α_{11} 0 α_{31}	0 α_{22} 0	α_{31} 0 α_{33}
3. Orthorhombic	$a \neq b \neq c$ $\alpha = \beta = \gamma = 90°$	3	$\alpha_{11} = \alpha_1$ 0 0	0 $\alpha_{22} = \alpha_2$ 0	0 0 $\alpha_{33} = \alpha_3$
4. Tetragonal	$a = b \neq c$ $\alpha = \beta = \gamma = 90°$				
5a. Hexagonal (*Trigonal*)	$a = b = c$ $\alpha = \beta = \gamma \neq 90°$, $< 120°$	2	$\alpha_{11} = \alpha_1$ 0 0	0 $\alpha_{22} = \alpha_1$ 0	0 0 $\alpha_{33} = \alpha_3$
5b. Hexagonal	$a = b \neq c$ $\alpha = \beta = 90°$, $\gamma = 120°$				
6. Cubic and Isotropic	$a = b = c$ $\alpha = \beta = \gamma = 90°$	1	$\alpha_{11} = \alpha$ 0 0	0 $\alpha_{22} = \alpha$ 0	0 0 $\alpha_{33} = \alpha$

(Some authors distinguish, in different ways, an equivalence of the hexagonal system, calling it "*trigonal*" or "*rhombohedral*". M.J. Buerger points out in his book on "*Contemporary Crystallography*" that such practices lead to serious inconsistencies and should be avoided).

$$[\alpha_{ii}] = \begin{bmatrix} \alpha_1 & 0 & 0 \\ 0 & \alpha_2 & 0 \\ 0 & 0 & \alpha_3 \end{bmatrix}$$

(2.49)

The volume coefficient of thermal expansion,

$$\beta = \alpha_1 + \alpha_2 + \alpha_3$$

(2.50)

where the principal coefficients are in general different from

$$\alpha_{11}, \ \alpha_{22}, \ \alpha_{33}.$$

2.10.5 Reduction of the observed data in terms of the principal axes of the expansion ellipsoid.

In order to fix the orientation of the thermal expansion ellipsoid with respect to the crystallographic axes, (i) find the thermal expansion tensor, and (ii) determine the principal expansion coefficients (Devanarayanan, 1969).

(i) To find α_{11}, α_{22} and α_{33} by Least Squares Method.

If θ_1, θ_2 and θ_3 are known angles of the axes X, Z and W, respectively, with respect to the Z-axis, the transformation matrix $[\theta]$ is

$$[\theta] = \begin{bmatrix} Sin^2\theta_1 & Sin2\theta_1 & Cos^2\theta_1 \\ Sin^2\theta_2 & Sin2\theta_2 & Cos^2\theta_2 \\ Sin^2\theta_3 & Sin2\theta_3 & Cos^2\theta_3 \end{bmatrix}$$

Substitute for known values of θ_1, θ_2 and θ_3 and find the transpose of $[\theta]$, viz., $[\theta]_t$

Next find the determinant $\Delta = \det[\theta]_t[\theta]$, and the inverse of $[\theta]_t[\theta]$, which is $([\theta]_t[\theta])^{-1}$.

Therefore, the transformation matrix $[T]$ required is

$$[T] = ([\theta]_t[\theta])^{-1}.[\theta]_t$$

Now, $\begin{bmatrix} \alpha_{11} \\ \alpha_{22} \\ \alpha_{33} \end{bmatrix} = [T]. \begin{bmatrix} \text{Value along X} = \alpha_X \\ \text{Value along Z} = \alpha_Z \\ \text{Value along W} = \alpha_W \end{bmatrix}$

Thus one gets the tensor components of the expansion ellipsoid in terms of α_{11}, α_{31} and α_{33} by means of least square fitting of observed data.

$$\begin{vmatrix} \alpha_{11} & 0 & \alpha_{31} \\ 0 & \alpha_{22} & 0 \\ \alpha_{31} & 0 & \alpha_{33} \end{vmatrix}$$

(ii) Fixing of Ellipsoid orientation related to the Crystallographic axes (Devanarayanan, 1969).

The principal coefficient of expansion ellipsoid α_1, α_3 and angle φ which relates the one of these principal axes with a crystal axis in the (010) plane is obtained as follows:

$$\tan 2\varphi = \frac{2 |\alpha_{31}|}{\alpha_{33}-\alpha_{11}}$$

If α_3 is inclined at angle φ measured counter-clockwise to the c-axis, the radius of the Mohr circle r is expressed as

$$\alpha_1 = \tfrac{1}{2}(\alpha_{11}+\alpha_{33})-r,$$

$$\alpha_2 = \tfrac{1}{2}(\alpha_{11}+\alpha_{33})+r,$$

whence one gets
$$\begin{vmatrix} \alpha_1 & 0 & 0 \\ 0 & \alpha_{22} & 0 \\ 0 & 0 & \alpha_{33} \end{vmatrix}.$$

2.10.6 Thermal Expansion and Phase Transitions in Crystals.

It is well known that there exist two types of phase transitions in solids. They are: a) "First order" (also known as the first kind) and b) the "second order" (or variously known as the 'continuous' transitions, Curie Points, λ-points, transitions of higher order). The first order transitions are characterized by discontinuous change in energy, volume and crystal structure and singularities in the first-order derivatives of the Free energy, G On the other hand, the second-order phase changes are accompanied by continuous change in energy, and volume with singularities in the second-order derivatives of G. Sometimes one distinguishes the onset of rotation of molecular groups within a relatively narrow temperature range, ΔT, as a 'third order' phase transition (or homomorphous transition).

The four state functions are:

U = Internal energy of solid, = potential energy + kinetic energy,   (2.51)a

H = Enthalpy, = $U + PV$, $\hspace{4cm}$ (2,51)b

F = Helmholtz Function = $U - TS$, $\hspace{3.5cm}$ (2.51)c

G = Gibbs Function = $H - TS$ $\hspace{4cm}$ (2.52)

Thus *first-order derivatives*

$$\frac{\partial G}{\partial T}\bigg)_P = -S \text{, the entropy} \hspace{3cm} (2.53)$$

$$\frac{\partial G}{\partial P}\bigg)_T = -V \text{, volume.} \hspace{3cm} (2.54)$$

Leading to the Clausius-Clapeyron relation, *viz.*,

$$\frac{dP}{dT} = \frac{L}{T(V_2 - V_1)}. \hspace{3cm} (2.55)$$

where L = latent heat involved in the transition,

V_2 and V_1 are volumes of the crystal before and after the phase change.

Second-order derivatives are

Specific heat, $\quad C_V = \frac{\partial U}{\partial T}\bigg)_P = T \frac{\partial S}{\partial T}\bigg)_V$ $\hspace{2.5cm}$ (2.56)

$$C_P = \frac{\partial H}{\partial T}\bigg)_P = T \frac{\partial S}{\partial T}\bigg)_{P'} \hspace{2.5cm} (2.57)$$

Thermal expansion, $\quad \beta = \frac{1}{V} \frac{\partial V}{\partial T}\bigg)_P$ $\hspace{2.5cm}$ (2.58)

Elastic modulus, $\chi_T = -\frac{1}{V} \frac{\partial V}{\partial P}\bigg)_T$ $\hspace{2.5cm}$ (2.59)

All these show discontinuously in a second-order phase transition.

Abrupt changes in these thermodynamic quantities are related by Ehrenfest equation,

$$\frac{dP}{dT} = \frac{1}{TV} \frac{C_{P2} - C_{P1}}{\beta_2 - \beta_1)} \hspace{3cm} (2.60)$$

$$\frac{dP}{dT} = \frac{1}{TV} \frac{\Delta C_P}{\Delta \beta} = \frac{\Delta \beta}{\Delta \chi_T} \hspace{3cm} (2.61)$$

In a λ-point (T_λ), both $\Delta\beta = \infty$ and $\Delta C_P = \infty$, causing $\frac{dP}{dT}$ =indeterminate parameter. So Pippard arrived at the following relation to λ-transition.

$$C_P = CVT_\lambda\beta + \text{constant}, C_0 \qquad\qquad (2.62)$$

where $C = \frac{\partial P}{\partial T}\Big)_\lambda$, and $C_0 = CT_\lambda \frac{\partial S}{\partial P}\Big)_\lambda$. $\qquad\qquad (2.63)$

Pippard's $C_P - \beta$ relation, above has been found valid in the case of NH_4Cl, below $T_\lambda = 242K$, also for dielectric, magnetic and ferroelectric transitions.

$$NH_4Cl: \quad \boxed{NaCl} \xleftarrow{\quad 436.1K \quad} \boxed{CsCl} \xleftarrow{\quad 242.6K \quad} \boxed{CsCl}$$

Other phase transitions, with no thermodynamic properties known, are such as

$$ZnS: \quad \boxed{Wurtzite} \xleftarrow{\quad 1293K \quad} \boxed{Sphalerite}$$

$$SiO_2: \quad \boxed{crystobalite} \xleftarrow{\quad 1743K \quad} \boxed{tridymite}$$

2.10.7. Thermal Expansion and Ferroelectricity

Crystals belonging to certain classes are known to possess structural phase transitions below which they exhibit ferroelectric phenomenon. Ferroelectric crystals can be for convenience classified as, (i) displacive or hard type, and (ii) soft type. Perovskite type, *e.g.*, $BaTiO_3$, crystals are 'hard' type, whereas a representative of the 'soft' type is Rochelle salt and are generally hydrogen bonded.

Any kind of phase transition in a crystal should give rise to anomalous behaviour in thermal expansion. A study of the thermal expansion behaviour of ferroelectric crystals is of great importance for getting an insight into the nature of the phase transition. Thermal expansion measurements made continuously and precisely in a temperature range can not only reveal the nature of phase transition in a crystal in that range of T but also direction of maximum dielectric anomaly in a ferroelectric crystal, thereby identify the possible orientation of the ferroelectric axis (Devanarayanan, 1968, Krishnan *et al.*, 1970).

2.11 Physical Properties and Tensors

2.11.1 Rank Rwo Tensors

Number of components = 9, of which 6 are independent

These are Symmetric tensors

1) Dielectric constant, ε: $\quad D_i = \varepsilon_{ij}.E_j$

E_i = Electric field, P_i =Polarization

t = mechanical stress, r = strain

$$\varepsilon_{ij} = \begin{pmatrix} \varepsilon_{11} & \varepsilon_{12} & \varepsilon_{13} \\ \varepsilon_{21} & \varepsilon_{22} & \varepsilon_{23} \\ \varepsilon_{31} & \varepsilon_{32} & \varepsilon_{33} \end{pmatrix}$$

2) Magnetic permeability, μ: $J_i = \mu_o \, \chi_{ij}.H_j$; and $B_i = \mu_{ij}H_j$

3) Susceptibility, χ: $\quad P_i = \chi_o \, \chi_{ij}.E_j$

4) Electric conductivity, σ: $\quad j_i = \sigma_{ij}.E_j$

5) Thermal expansion, α: $\quad r_{ij} = \alpha_{ij}.\Delta T$

6) Piezo-electric coefficient, d: $\quad \Delta Q = T.d_{ij}.t_{ij}$

2.11.2 Rank Three Tensors

Number of components = 27, of which 17 are independent

1) Piezoelectric modulii $d_{ijk} = d_{ikj} \equiv d_{ij}$

$$d_{ij} = \begin{pmatrix} d_{11} & d_{12} & d_{13} & d_{14} & d_{15} & d_{16} \\ d_{21} & d_{22} & d_{23} & d_{24} & d_{25} & d_{26} \\ d_{31} & d_{32} & d_{33} & d_{34} & d_{35} & d_{36} \end{pmatrix}$$

E_i = Electric field, $\qquad P_i$ =Polarization

t = mechanical stress, r = strain

$$r_{ij} = \begin{pmatrix} r_{11} & r_{12} & r_{13} \\ r_{21} & r_{22} & r_{23} \\ r_{31} & r_{32} & r_{33} \\ r_{41} & r_{42} & r_{43} \\ r_{51} & r_{52} & r_{53} \\ r_{61} & r_{62} & r_{63} \end{pmatrix}$$

$$P_i = d_{ijk} \cdot t_{jk}$$

$$P_i = e_{ijk} \cdot r_{jk}$$

$$E_i = -g_{ijk} \cdot t_{jk}$$

2)   EM coupling factor, $k = \dfrac{d_{ij}}{\sqrt{e_i^2 \, s_{ij}^2 / 4\pi}}$

3)   Electro-optic coefficient, $\Delta n_{ij} = r_{ijk} \cdot E_k$

2.11.3 Higher rank Tensors (4th rank)

Number of components $= 81$, of which 36 reduced $\rightarrow 21$ are independent

1) Elastic compliance, $s \equiv s_{ijkl}$

$$= \begin{pmatrix} s_{11} & s_{12} & s_{13} & s_{14} & s_{15} & s_{16} \\ s_{12} & s_{22} & & & & \\ s_{13} & & s_{33} & & & \\ s_{14} & & & s_{44} & & \\ s_{15} & & & & s_{55} & \\ s_{16} & & & & & s_{66} \end{pmatrix}$$

Hookes law requires $(i, j, l = 1, 2, 3)$

$$r_{ij} = s_{ijkl} \cdot t_{kl}$$

2) Elastic stiffness coefficient, $c \equiv c_{ijkl}$: $t_{ij} = c_{ijkl} \cdot r_{kl}$

t and r tensors are both symmetric.

$$s_{ijkl} = s_{jikl}; \quad s_{ijkl} = s_{ijlk}$$

$$c_{ijkl} = c_{jikl}; \quad c_{ijkl} = c_{ijlk}$$

causing 36 reduced $\rightarrow 21$ independent,

$$s_{ij} = (-1)^{i+j} \Delta c_{ij} / \Delta^c; \quad c_{ij} = (-1)^{i+j} \Delta s_{ij} / \Delta^s$$

3) Piezo-optic (Photo-elastic) coefficients π:

$$\Delta n_{ij} = \pi_{ijkl} \cdot t_{kl}$$

4) Elasto-optic coefficients, p:

$$\Delta n_{ij} = p_{ijkl} \cdot t_{kl}$$

5) Electrostriction

$$r_{ij} = R_{ijmn} \cdot E_m \cdot E_n$$

2.11.4 Rank 5 Tensor

1) Electro-elastic

$$\Delta c_{ijkl} = g_{mijkl} \cdot E_m$$

2.11.5 Adiabatic piezo-electric Equations

$$D = dT + \varepsilon^r E \quad \text{(Direct)}$$

$$S = s^E T + dE \quad \text{(Converse)}$$

$$D = \varepsilon^S T + eS \quad \text{(Direct)}$$

$$T = c^E S - eE \quad \text{(Converse)}$$

$$E = \beta^T D - gT \quad \text{Direct)}$$

$$E = \beta^S D - hS \quad \text{(Direct)}$$

Table   Tensor Components

Tensor order	Quantity	# of components Voigt Laval Wooster		
		Voigt	Laval	Wooster
1	E, D	3	3	3
2	S, T, ε, β	6	9	9
3	d, e, g, h	18	27	27
4	s, c	21	45	39

$$c = s^{-1}$$
$$\beta = \varepsilon^{-1}$$
$$d = e^{S}E = \varepsilon^{r}g$$
$$e = dc^{E} = \varepsilon^{S}$$
$$g = hs^{D} = \beta^{T}d$$
$$h = gc^{D} = \beta^{S}e$$

----oo0oo----

PRELIMINARIES – 2E

CRYSTAL DEFECTS

PRELIMINARIES – 2E

CRYSTAL DEFECTS

"It is difficult to discuss the beginning of the universe without mentioning the concept of God. My work on the origin of the universe is on the borderline between science and religion, but I try to stay on the scientific side of the border. It is quite possible that God acts in ways that cannot be described by scientific laws, but in that case, one would just have to go by personal belief."- Stephen Hawking

2.11   CRYSTAL DEFECTS

A perfect crystal, with every atom of the same type in the correct position, does not exist

2.11.1 Introduction

Materials are often stronger when they have defects.

Crystal defects are results of thermal dynamic equilibrium contributed also by the increase in entropy *(S)* term of the Gibb's Free energy.

All real solids are impure. A very high purity material, say 99.9999% pure (called 6N – six nines) contains $\sim 6\,x10^{16}$ impurities per cm^3.

Impurities are often added to materials to improve the properties. For instance, carbon added in small amounts to iron makes steel, which is stronger than iron. Boron impurities added to silicon drastically change its electrical properties.

Solid solutions are made of a host, the solvent or matrix) which dissolves the solute (minor component). The ability to dissolve is called *solubility*. Solid solutions are:

- homogeneous
- maintain crystal structure
- contain randomly dispersed impurities (substitutional or interstitial)

Factors for high solubility

- Similar atomic size (to within 15%)
- Similar crystal structure
- Similar electro-negativity (otherwise a compound is formed)

- Similar valence

Composition can be expressed in weight percent, useful when making the solution, and in atomic percent, useful when trying to understand the material at the atomic level

Table 2.3 lists the different type of imperfections in solids'

Table 2. 4

Imperfections in crystalline solids	
Classifications	Types
1) Point Defects (Zero-dimensions)	a) Vacancy (Scottky defect) (Colour centres) b) Interstitial c) Self-Interstitial d) Substitutional e) Frenkel (Colour centres)
2) Line Defects (1- Dimensions) Dislocations	a) Edge b) Screw
3) Plane Defects (2-Dimensional) External surface	
3) Volume Defects (3-Dimensions)	a) Grain boundary b) Crystal Twin c) Twist d) Stacking faults e) Voids f) Precipitates

2.11.2 Point Defects (Zero-Dimensional)

Point defects are where an atom is missing or is in an irregular place in the lattice structure. Point defects include.
Simplest Point defects are:
i) Vacancy, ii) Interstitial impurity atom, iii) Self-interstitial atom, iv) Substittution impurity atom, v) Frenkel defect

a) Vacancy or Schottky Defect

Vacancies are empty spaces where an atom should be, but is missing. They are common, especially at high temperatures when atoms are frequently and randomly change their positions leaving behind empty lattice sites. In most cases diffusion (mass transport by atomic motion) can only occur because of vacancies.

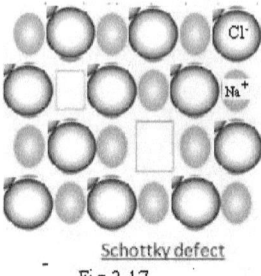

Schottky defect

Fig 2.17

A vacancy is a lattice position that is vacant because the atom is missing. It is created when the solid is formed. There are other ways of making a vacancy, but they also occur naturally as a result of thermal vibrations.

An interstitial is an atom that occupies a place outside the normal lattice position. It may be the same type of atom as the others (self interstitial) or an impurity atom (Fig 2.17). In the case of vacancies and interstitials, there is a change in the coordination of atoms around the defect. This means that the forces are not balanced in the same way as for other atoms in the solid, which results in lattice distortion around the defect.

The number of vacancies formed by thermal agitation follows the law:

$$N_V = N_A e^{-Q_V/k_B T}$$

where N_A is the total number of atoms in the solid, Q_V is the energy required to form a vacancy, k_B is Boltzmann constant, and T the temperature in Kelvin (note, not in °C).

When Q_V is given in J, $k_B = 1.3805x10^{-23} JK^{-1}$/ atom.

When using eV as the unit of energy, $k_B = 8.62 x10^{-5} eVK^{-1}$/ atom.

Note that

$$k_B T (T = 300K) = 0.025 \ eV$$

is much smaller than typical vacancy formation energies. For instance,

$$Q_v(Cu) = 0.9 \ eV \ / \ atom.$$

This means that

$$\frac{N_V}{N_A}(T = 300K) = e^{-36} = 2.3 \, x10^{-16}$$

an insignificant number. Thus, a high temperature is needed to have a high *thermal* concentration of vacancies. Even so,

$$\frac{N_V}{N_A}(T = \text{M.P. of Cu}) = 1 \, x10^{-4}.$$

2.11.2 b) Interstitial impurity

These atoms are much smaller than the atoms in the bulk matrix. Interstitial impurity atoms fit into the open space between the bulk atoms of the lattice structure. An example of interstitial impurity atoms is the C carbon atoms that are added to Fe iron to make steel. C atoms, with a radius of 0.071 *nm*, fit nicely in the open spaces between the larger (0.124 *nm*) Fe atoms.

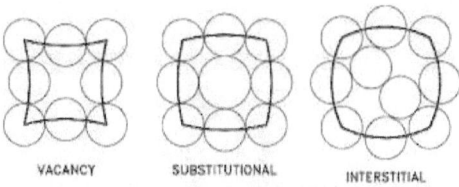

VACANCY   SUBSTITUTIONAL   INTERSTITIAL

Fig 2.18   Lattice defects

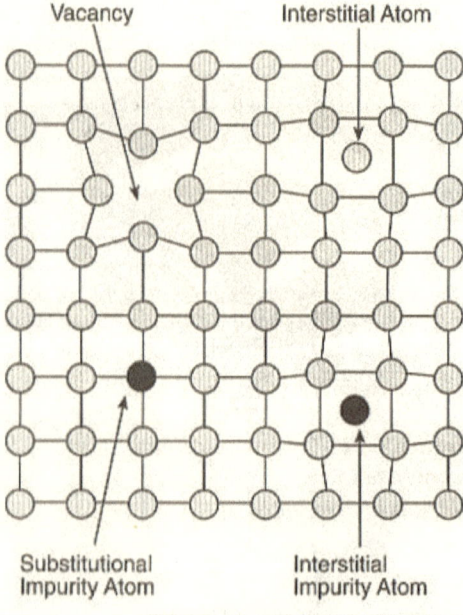

Vacancy

Interstitial Atom

Substitutional
Impurity Atom

Interstitial
Impurity Atom

Fig 2.19 Various types of point defects within a crystal.

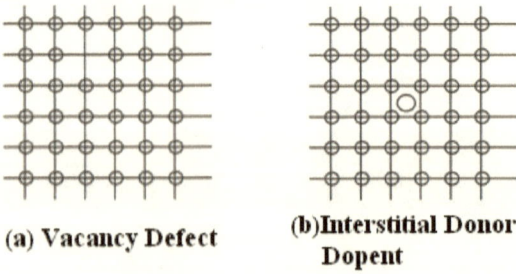

(a) Vacancy Defect

(b)Interstitial Donor Dopent

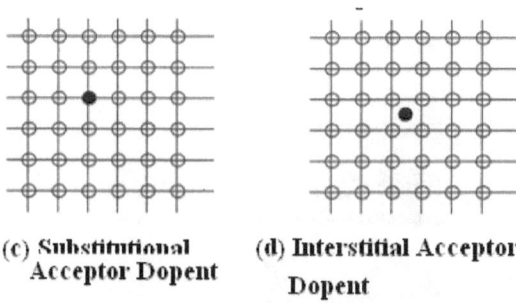

(c) Substitutional Acceptor Dopent   **(d) Interstitial Acceptor Dopent**

Fig 2.20 Point defects in a crystal.

2.11.3 Other Miscellaneous

2.11.3 1 Line defects (Dislocations)(One-Dimensional)

This weakens the structure along a one-dimensional space, and the defects type and density affects the mechanical properties of the solids

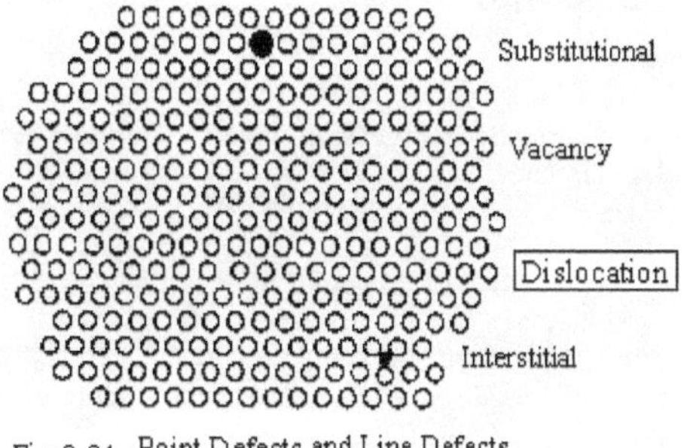

Fig 2.21 Point Defects and Line Defects

This type includes a) Edge dislocation, b) Screw dislocation. In crystals mostly both edge and screw dislocations are mixed. Dislocations are abrupt changes in the regular ordering of atoms, along a line (dislocation line) in the solid. They occur in high density and are very important in mechanical properties of material.

Mathematically, skip or Burgers vector characterizes displacement of atoms around dislocation. (Fig 2.22). If one imagines going around the dislocation line, and exactly going back as many atoms in each direction as one has gone forward, one will not come back to the same atom where started. The Burgers vector points from start atom to the end atom of the journey (This "journey" is called Burgers circuit in dislocation theory).

The Burgers vector in metals points in a close packed direction.

Edge dislocations occur when an extra plane is inserted. The dislocation line is at the end of the plane. In an edge dislocation, the Burgers vector is perpendicular to the dislocation line.

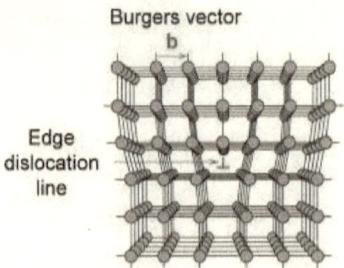

Fig 2.22 Edge dislocation

Screw dislocations result when displacing planes relative to each other through shear. In this case, the Burgers vector is parallel to the dislocation line.

2.11.3.2. Plane Defect (2-D) (External surface)

2.11.3.3 Volume Defects (3-D)

This type comprises various types like Grain boundary, crystal twin, Twist, stacking fault, voids and precipitates.

2.11.3.4 Colour Centres

These are imperfections in crystals that cause colour (defects that cause colour by absorption of light).

A crystal defect thatabsorbs light in a spectral region inwhich the crystal itself does not absorb light. Due to defects, metal oxides may also act as semiconductors, because there are many different types of electron traps. Electrons in defect region only absorb light at certain range of wavelength. The colours seen are due to lights not absorbed.

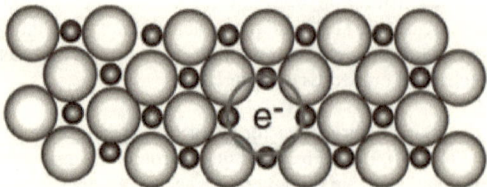

Fig 2.23 Colour Centre

%&%&%&%&%&%&

Chapter 3

DIELECTRIC PROPERTIES

Chapter 3

DIELECTRIC PROPERTIES

"In science one tries to tell people, in such a way as to be understood by everyone, something that no one ever knew before. But in poetry, it's the exact opposite" - PAM Dirac

3.1   INTRODUCTION

Dielectric materials are a special class of substances that, under almost all conditions are insulators. They have the interesting and useful property that their electrons, ions, or molecules may be polarized under the influence of an external electric field. When such materials are place in between the plates of a capacitor (Condenser), they increase the total capacity of these devices. This application constitutes one of the most important applications of these materials.

A small subset of this class of materials consists of crystals that are spontaneously polarized. Their behaviour parallels that of ferromagnetic materials and are known as ferroelectric crystals. Ferroelectrics constitute a category of materials within the class of pyroelectrics, materials that change their polarizations with temperature.

All of the ferroelectric materials and some of the other dielectric materials are piezoelectric, which can change their polarizations and dimensions under the influence of external mechanical forces or electric fields. These materials can serve as electromechanical transducers. Ferroelectric crystals and quartz (which is not ferroelectric) are used both a s frequency generators and as detectors of electromagnetic radiation and ultrasonic waves.

The largest engineering application of dielectric materials is for purposes of electrical insulation. When a voltage is applied across a dielectric substance, there is a limit beyond which the dielectric breaks down and material destroyed. This limiting voltage is called the dielectric strength of the material, and its magnitude is listed in Table 3.1 for typical matter.

Table3.1

Dielectric Strength

1) Air gap	$30\ kV\ cm^{-1}$	at 1 *atm* pressure
2) Paper	$10\text{-}15\ kV\ cm^{-1}$	
3) Oil	$> 100\ kV\ cm^{-1}$	
4) Mica	$26\ kV\ cm^{-1}$	

3.2 POLARIZATION OF ATOMS AND MOLECULES

To examine the behaviour of isolated ions or molecules is useful for eliminating any Neighbour-Neighbour (NN) reactions.

3.2.1 Dipole Moments, $\vec{p}$

Electric dipoles are formed when a dielectric is inserted between the charged plates of a capacitor, and the electric dipole moment, $\vec{p}$ is given by

$$\boxed{\vec{p} = +q\vec{d}}\,, \tag{3.1}$$

where the positive charge $+q$ separated from a negative charge $-q$ of the same magnitude by the distance d (Fig 3.1),

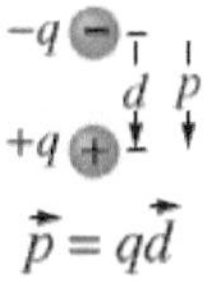

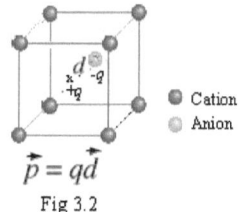

$$\vec{p} = q\vec{d}$$

Fig 3.1

For example, in the diagram below (Fig 3.2), the centre of positive charge from the 8 cations shown is at X, while the centre of negative charge is located some distance d away on the anion.

○ Cation
○ Anion

$$\vec{p} = q\vec{d}$$

Fig 3.2

This view of dipole moment is more useful, since it can be applied over a large area containing many charges in order to find the net dipole moment of the material, and can also be used in situations where it is inappropriate to consider the charges as belonging to discrete particles – *e.g.* in the case of the electron cloud that surrounds the nucleus in an atom, which must be described by a wavefunction. The dipole moments of typical cases have been presented in Table 3.2. It is known that the magnitude of the dipole moment varies and depends on the nature of bonding between the two ions.

Table3.2

Dipole moments, $\vec{p} = q\vec{d}$
$(1D = 1Debye = 3.33 \, x10^{-30} Cm)$

Atom pair	p
1) p_{C-H}	0.4 D
2) p_{O-H}	1.6 D
3) p_{N-H}	1.7 D
4) p_{C-O}	0.7 D
5) p_{N-O}	3.2 D
6) p_{N-C}	1.7 D

$$O^- \text{----} H^+$$
$$p=1.6D$$

Typical tri-atomic molecules are represented in Fig 3.3.

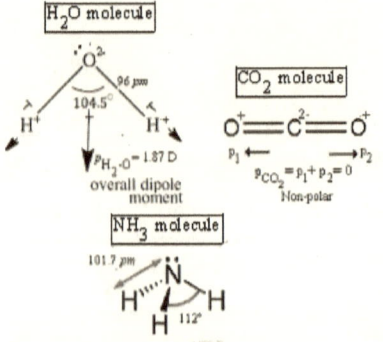

Fig 3.3 Typical tri-atomic polar molecules

3.2.2   To find $\vec{p}_{NH_3}$

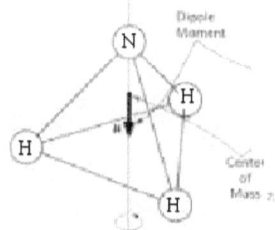

Fig 3.4

$\vec{p}_{NH} = 1.7$ D, $Sin56° = x/1.7$, $Sin60° = x/$ hori.comp (Diagram)

$\therefore h = 1.7 \dfrac{Sin56°}{Sin60°} = 1.63$, giving the vertical component,

$\therefore v = \sqrt{(1.7^2 - 1.63^2)} = 0.5$.

$\vec{p}_{NH_3} = 3v = 3x0.5 = 1.5$ D

Because of different centres of positive and negative charges react differently with the EM field, the stronger the $\vec{p}$, the greater is the bonding between the structures! ε and n are related to $\vec{p}$.

3.2.3 Polarization $\vec{P}$:

The polarization of a material is simply the total dipole moment for a unit volume.

$$P = \frac{\sum_i \vec{p}_i}{V},$$ (3.2)

where V = volume of the sample.

3.2.4 Polarization Mechanisms:

3.2.4.1 Electronic $\vec{p}_{el}$:

Electron displacement around atoms; deformation of electron shell with respect to the nucleus of its atom (Fig 3.5) contributes this polarization, $\vec{p}_{el}$, up to $10^{16} Hz$

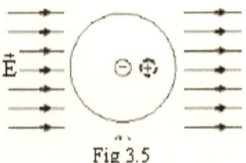

Fig 3.5

3.2.4.2 Ionic, $\vec{p}_{ion}$:

Ion displacement within the crystal contributes to $\vec{p}_{ion}$, up to $10^{13} Hz$. For example in sodium chloride crystal (Fig 3.6),

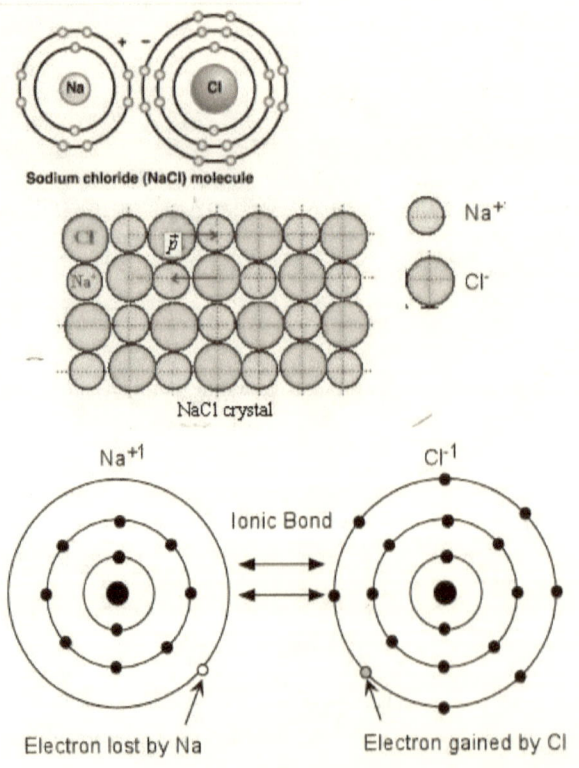

Fig 3.6

In an external electric field Na^+ ions and Cl^- ions are displaced in opposite directions to give polarization.

3.2.4.4 Orientation (Dipolar) polarization, $\vec{P}_{dip}$:

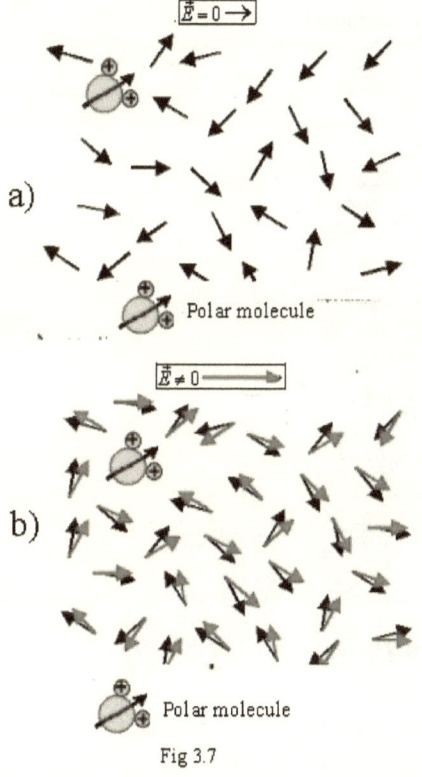

Fig 3.7

The dipole moment P may be (i) permanent dipole moment, P_0 and (ii) induced dipole moment, P_i.

$$\vec{P}_i = [\frac{P^2}{3k_BT} + \alpha] \vec{E} \qquad (3.3)$$

α = molecular polarizability.

Thus the electric polarization $\vec{P}$ can be defined quantitatively in two ways, *viz.*,

a) $\quad \vec{P} = \vec{D} - \varepsilon_0 \vec{E} = (\varepsilon - \varepsilon_0)\vec{E}$,   (3.4)

b) $\quad \vec{P} = (\varepsilon_r - 1)\varepsilon_0 \vec{E}$.   (3.5)

where $\chi = (\varepsilon_r - 1) =$ electric susceptibility

$\therefore \quad \boxed{\vec{P} = \chi\, \varepsilon_0 \vec{E}}$ (3.6)

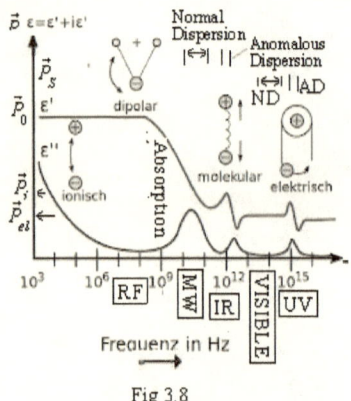

Fig 3.8

3.3.1 Parallel Plate Capacitor

A capacitor is meant to store charges. It is a device having two parallel conducting plates with a dielectric medium in between. The capacitance C is defined by

$$\boxed{Q = CV}$$ (3.9)

where $Q =$ charge(C) on each plate, area A (m^2) and $V =$ voltage (V) applied between the plates. $1F = 1CV^{-1}$.

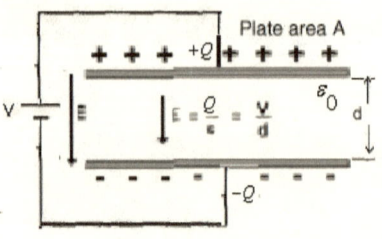

Fig 3.9

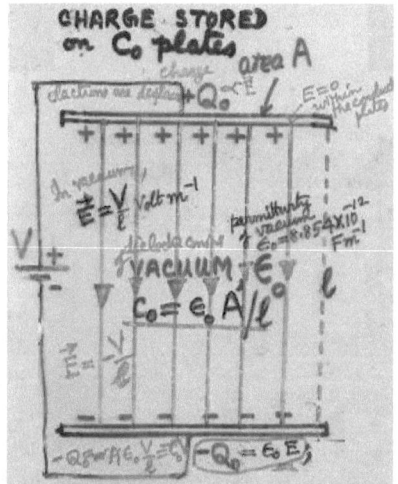

Fig 3.10   Charge stored in Capacitor in vacuun

$$C = \frac{\varepsilon A}{d} = \frac{k\varepsilon_0 A}{d}$$ (3.10)

where κ = dielectric constant of the dielectric medium which changes the value of C to C'..

$$\kappa = \frac{C'}{C} \text{ and } \kappa = \frac{\varepsilon}{\varepsilon_0}.$$ (3.11)

$$d \ll A$$

ε = dielectric permittivity,

ε_0 = vacuum permittivity.

$$\kappa_S = n^2$$ (3.12)

κ_S = static dielectric constant, n = refractive index of the dielectric.

Actually, $\kappa_S = n^2$ is valid only for non-polar materials and for dielectric constant of any dielectric only at optical (very high) frequencies.

When a homogeneous electric field strength E (Nm^{-3}), between the capacitor plates when a vacuum (free space) is present between them is

$$\boxed{E = 4\pi q = D} \tag{3.13}$$

where   q =surface charge density,

D = flux density, the electric induction or the <u>electric displacement</u>.

The potential difference V existing between the plates with free space is

$$V=E\,d \tag{3.14}$$

The capacitance of the assembly

$$C = \frac{4q}{V} = QV \tag{3.15}$$

3.3.2 Dielectric Capacitor

If a dielectric is placed between the plates of a capacitor its electrical effect is to increase the capacitance of the assembly. This is the result of a polarization of the molecules of the intervening material. Substances composed of polar molecules, those with a permanent electric dipole moment, have random molecular orientations, in the presence of an external electric field. The field between the charged plates aligns, or polarizes, the molecular dipoles of the intervening material. This tendency of molecular alignment to become parallel to the field called ORIENTATION POLARAIZATION. If the intervening, insulating substance is a liquid or a gas, as is sometimes the case, the polar molecules can move much more freely and become aligned more readily than in most solids. The molecules of the insulating material become predominantly arrayed such that the negative poles at one of its surface are adjacent to the positive poles on the surface of the positive plate. The dipolar molecules in the interior in the dielectric material array themselves in a -, +, -, +, order until, at the opposite surface of the material, the positive molecular poles are adjacent to the negative plate. A layer of negative charge thus forms on the surface of the dielectric adjacent to the surface of the positive pole and one of positive charge is created adjacent to the surface of the negative plate. The result is the effective cancellation of some of the charges on each of the plates and the consequent reduction of the electric field strength. This increases the ability of the plates to hold more charges; their capacity is increased. In cases in which the molecules are not polar (those with symmetric charge distributions) polarization may be induced by an applied field to produce results similar to those just noted. This occurs in solids in which the molecules or ions are not free to rotate.

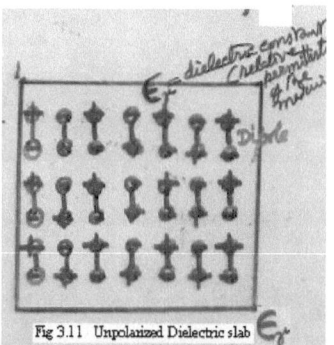

Fig 3.11 Unpolarized Dielectric slab

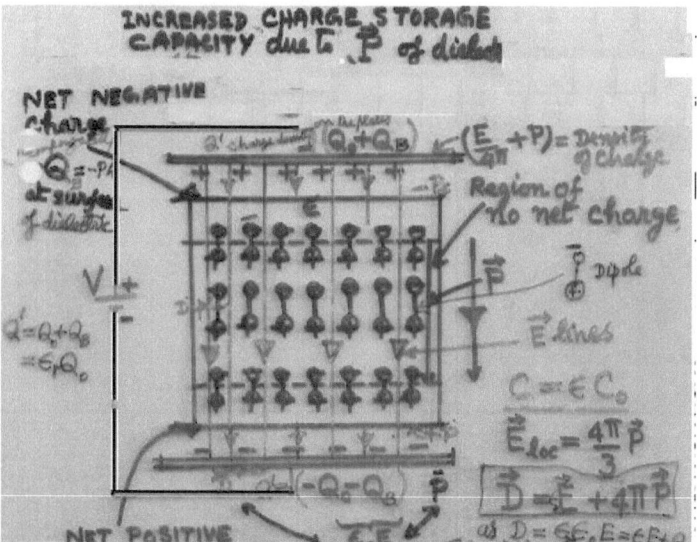

Fig 3.12 Increased Charge storage Capacity due to Polarization P of the Dielectric

It should be noted that orientation polarization is temperature dependent. The polarization decreases with increasing temperature. Non-polar materials are less affected by temperature. In this case, the change in properties is largely a result of changes in such factors as intermolecular or interionic distances, bond angles, and / or shifts in electronic charge distribution around the nuclei of molecular or ionic component.

The reduction in the field strength caused by the presence of the dielectric material between the plates is expressed as

$$E_\varepsilon = \varepsilon E = D \qquad (3.16)$$

$$\therefore \varepsilon = \frac{D}{E} = \frac{E_\varepsilon}{E} \qquad (3.17)$$

The reduced surface charge density q' as a result of the presence of the dielectric in between the plates is

$$q' = \frac{E}{4\pi} \qquad (3.18)$$

This reduction $(q - q')$ is the <u>polarization</u> p. This causes an increase in the capacitance

$$p = (q - q') = \frac{E_\varepsilon}{4\pi} - \frac{E_D}{4\pi} \qquad (3.19)$$

$(- \frac{E_D}{4\pi})$ is the <u>depolarization</u> factor of the plates by the dielectric. This is expressed by the equation

$$p = (\frac{E_\varepsilon}{4\pi} - \frac{E_D}{4\pi}) = \frac{E_\varepsilon}{4\pi}\{1 - \frac{1}{\varepsilon}\}$$

i.e., $$\boxed{p = \frac{E_\varepsilon}{4\pi\varepsilon}(\varepsilon - 1)} \qquad (3.20)$$

i.e., $$\boxed{p = \frac{E}{4\pi\varepsilon}(\varepsilon - 1)}$$

$$\therefore \varepsilon = \frac{4\pi P}{E} + 1 \qquad (3.21)$$

Or $$E_\varepsilon = D = (4\pi P + E) \qquad (3.22)$$

The relationships given include those basic properties of dielectric materials of greatest use here, and useful for future discussion.

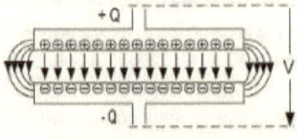

Fig 3.13(a) in Vacuum

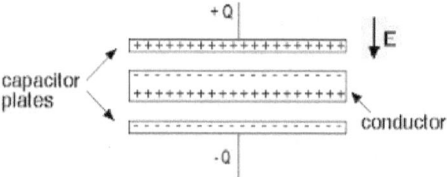

Fig 3.13 (b) Conductor

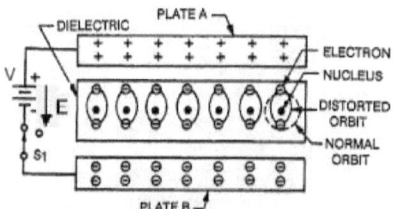

Distortion of Electron Orbits in a Dielectric.

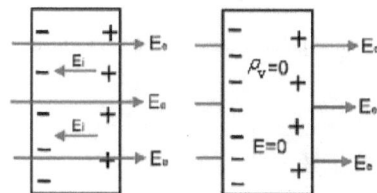

(a) Conducting slab   (b) Dielectric slab

Fig 3.14 Dielectric capacitor

Fig 3.14 Dielectric capacitor

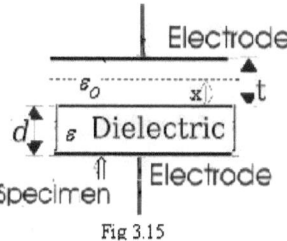

Fig 3.15

In Fig 3.15, A is the area of electrode. d is the thickness of the specimen. t is the gap between the electrode and specimen (here this gap is filled by compressed gas or air).

C_{spec} is the capacitance of specimen, C_t is capacitance due to spacing between electrode and specimen. C_{eff} is the effective combination of C_{spec} and C_t.

$$C_{eff} = \frac{C_{spec}\, C_t}{C_{spec} + C_t}$$ (3.23)

From Fig above, as two capacitance are connected in series,

$$= \frac{\varepsilon_r \varepsilon_0 A}{\varepsilon_r t + d}$$ (3.24)

ε_0 is permittivity of free space, ε_r is relative permittivity, when we remove specimen and the spacing readjusted to have same value of capacitance, the expression for capacitance reduces to

$$= \frac{\varepsilon_0 A}{t + d - x}$$ (3.25)

On equating (1) and (2), we will get the final expression for of ε_r as:

$$\varepsilon_r = \frac{d}{d - x}$$ (3.26)

In a solid dielectric the static polarization $\vec{P}_{St}$ is a net value of all the above components.

$$\boxed{\vec{P}_{St} = \vec{P}_{el} + \vec{P}_{ion} + \vec{P}_{dip} + \vec{P}_{s}}$$ (3.27)

$$\boxed{\vec{P} = \varepsilon_0 \chi_{el} \vec{E}}$$ (3.28)

ε_0 =dielectric permittivity of free space, $\varepsilon_0 = 8.8542 \times 10^{-12}\, Fm^{-1}$ (3.29)

χ =electric susceptibility

$$\chi_{el} = \varepsilon - 1$$ (3.30)

Dielectric displacement

$$\boxed{\vec{D} = \varepsilon_0 \vec{E} + \vec{P}}$$ (3.31)

$$\chi_{el} = 0, \text{for vacuum}$$

$$\left.\begin{array}{l} \varepsilon_S = 81 \\ \varepsilon_0 = 1.77 \end{array}\right\} \text{for water}$$

Total electric field (local field) $E_{\varepsilon Loc}$

$$\vec{E}_{\varepsilon Loc} = E_0 + E_1 + E_2 = (E + \frac{P}{\varepsilon_0}) - \frac{P}{\varepsilon_0} + \frac{P}{3\varepsilon_0} + = (E + \frac{P}{3\varepsilon_0}) \qquad (3.32)$$

3.4 LOCAL FIELD, $E_{\varepsilon Loc}$;

The electric field at any molecular site in the interior of a dielectric situated between the plates of a charged condenser actually experiences, called the local field (or polarizing field), is known to be larger than the applied field E. This is related to the polarization which occurs within and on the surfaces of the dielectric. The actual field acting on the molecule is, therefore, called the LOCAL FIELD, $E_{\varepsilon loc}$;

$$\boxed{\vec{E}_{\varepsilon Loc} = \underset{\text{External}}{\overrightarrow{\vec{E}_0}} + \underset{\text{Depolarization}}{\overrightarrow{\vec{E}_1}} + \underset{\text{Lorentz}}{\overrightarrow{\vec{E}_2}} + \underset{\substack{\text{Atoms inside} \\ \text{a cavity}}}{\overrightarrow{\vec{E}_3}}} \qquad (3.33)$$

Lorentz was the first to calculate it in the following way.

3.4.1 When a dielectric medium ε_0 is under the influence of an applied electric field,

the dielectric displacement is

$$D = \varepsilon_0 E + P \qquad (3.34)$$

But as a result of polarization of the dielectric medium

$$D = \varepsilon_0 E_0 \qquad (3.35)$$

So one gets the value of the external field E_0 is,

$$\boxed{E_0 = E + \frac{P}{\varepsilon_0}} \qquad (3.36)$$

3.4.2 $\vec{E}_1$ = depolarization field, and has direction opposite to that of the external field.

N = depolarization factor, depends on the geometrical shape of dielectric (Table 3.3).

$$\boxed{\vec{E}_1 = -\frac{N\,P}{\varepsilon_0}}$$ (3.37)

where $N = N_x + N_y + N_z$

Table 3.3

Shape	Axis	N	
		CGS	SI
1) Sphere	any	$\frac{4\pi}{3}$	$\frac{1}{3}$
2) Thin slab	$\perp$	4π	1
3) Thin slab	‖	0	0
4) Long circular cylinder	Longitudinal	0	0
5) Long circular cylinder	Transverse	2π	$\frac{1}{2}$

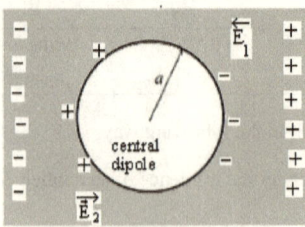

Fig 3.16

3.4.3 Lorentz field, $\vec{E}_2$

The polarization charge on the surface of the Lorentz sphere (Fig 3.16) is calculated by considering the spherical cavity (Fig.3.17).

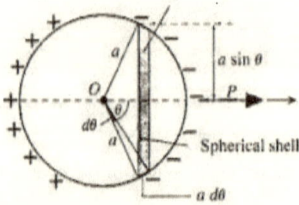

Fig 3.17 Dielectric

θ = polar angle that an elemental area dA of the annular shell subtending at O with the polarization direction

$(P \| z) \ P\cos\theta dA,$

$dA = 2\pi(aSin\theta)ad\theta = 2\pi a^2 Sin\theta d\theta .$

The force dF due to the surface charge on the charge q at O is give by the Coulomb's law,

$$dF = \frac{q(P\cos\theta dA)}{4\pi\varepsilon_0 a^2}$$

$$\vec{F}_z = +\frac{q\,P}{2\varepsilon_0}\int_0^\pi Cos^2\theta\ Sin\theta d\theta ,$$

$$\vec{F}_z = -\frac{q\,P}{2\varepsilon_0}\int_{+1}^{-1} x^2 dx$$

where $x = -Sin\theta d\theta$, giving

$$\vec{F}_z = -\frac{q\,P}{2\varepsilon_0}\left|\frac{x^3}{3}\right|_{+1}^{-1} = -\frac{q\,P}{3\varepsilon_0} .$$

Lorentz force, $\boxed{\vec{E}_2 = \frac{\vec{F}_z}{q} = \frac{P}{3\varepsilon_0}}$ (3.38)

3.4.4 Field of Dipoles inside Cavity, $\vec{E}_3$

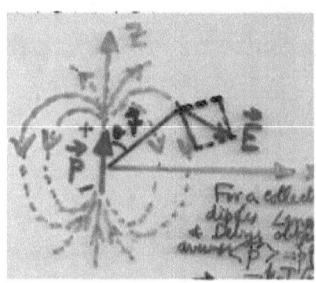

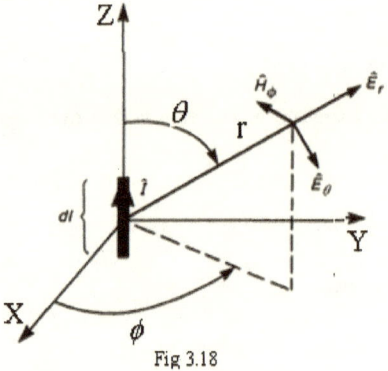

Fig 3.18

Fig 3.18

$\vec{E}_3$ is the contribution solely depends on the structure of the dielectric crystal.

The electric field due to a dipole is shown schematically in Fig 3.17.

It is known that the $\vec{E}_{dip} = \dfrac{P}{4\pi\varepsilon_0 r^3}[2\hat{r}\,Cos\theta + \hat{\theta}\,Sin\theta]$

i.e., $\qquad \vec{E}_{dip} = \dfrac{1}{4\pi\varepsilon_0 r^3}[3(\vec{p}.\vec{r})\vec{r} - \vec{p}r^2]$ $\hfill (3.39)$

Considering the effect of all the dipoles, parallel to one another along their axix as z, treated as point dipoles, within the cavity, $\vec{E}_3(z) = \sum_i \dfrac{(3\vec{p}_i z_i^2 - \vec{p}_i r_i^2)}{r_i^5}$

Assuming cubic symmetry for the lattice, one gets

$$\sum_{i=x,y,z} \frac{r_i^2}{r_i^5} = 3 \sum_{i=x,y,z} \frac{z_i^2}{r_i^5}$$

whence for a cubic lattice

$$\boxed{\vec{E}_3 = \sum_{i=x,y,z} \frac{(3\vec{p}_i z_i^2 - \vec{p}_i r_i^2)}{r_i^5} = 0}$$ $\hfill (3.40)$

Table 3.4

Lattice Symmetry	$\vec{E}_3(z)$
1) Cubic	0
2) Non-cubic	$\sum_i \dfrac{(3\vec{p}_i z_i^2 - \vec{p}_i r_i^2)}{r_i^5}$

3.4.5 Net electric field (Local Field) $E_{\varepsilon Loc}$ at the dielectric Cavity centre of a

cubic dielectric:

$$\vec{E}(\vec{r}) = \sum_{i=1} \left(\frac{3(\vec{p}_i \cdot \vec{r}_i)\vec{r}_i - r_i^2 \vec{p}_i}{r_i^5} \right) = \vec{E}_1 + \vec{E}_2 + \vec{E}_3 \qquad (3.41)$$

Space average field inside the dielectric,

$$\vec{E} = \vec{E}_1 + \vec{E}_0 \qquad (3.42)$$

$\vec{E}$ occurs in Maxwell's Equations.

$$E_{\varepsilon Loc} = E_0 + E_1 + E_2 + E_3$$

$$= E_0 + E_1 + E_2 \qquad (3.43)$$

$$E_{\varepsilon Loc} = (E + \frac{P}{\varepsilon_0}) - \frac{P}{\varepsilon_0} + \frac{P}{3\varepsilon_0} + 0, \text{ giving}$$

$$\boxed{E_{\varepsilon Loc} = E + \frac{P}{3\varepsilon_0}} \qquad (3.44)$$

3.5 Polarizability

The dipole moment induced in the molecule by the local field is given by

$$\boxed{P_{mol} = \alpha E_{\varepsilon loc}} ; \qquad (3.45)$$

where P_{mol} represents its dipole moment and α is called the POLARIZABILITY of the molecule.

For dielectrics containing N molecules per unit volume (density of dipoles per unit volume), the total dipole moment, or polarization, is

$$\vec{P} = \sum_i ex_i = N\vec{p}_{mol} = N\alpha \, E_{\varepsilon loc} \; ; \tag{3.46}$$

3.5.1 An Isolated atom or ion

In the absence of an external field, the centroid of its electric charge distribution is at its nucleus. The positive and negative charges are considered to act as though they are superimposed. The atom or ion has no net dipole moment in this situation.

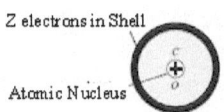

Fig 3.19 A Neutral atom

The application of a homogeneous, static electric field can cause the centre of the surrounding electric charge to become offset from its initial central position (Fig 3.19). This elastic displacement is in a direction opposite to the applied field in a manner analogous to that shown in Fig. 3.20.

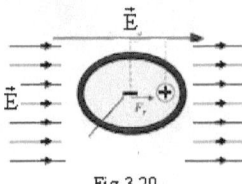

Fig 3.20

The attractive forces between the nucleus and the electrons counteract the displacement of the electrons induced y the field. An equilibrium is reached between these two effects and the resulting offset between the centres of the two charges results in a dipole moment. This induced dipole moment is given approximately by

$$\vec{p} = \alpha_{elec} \, \vec{E} \tag{3.47}$$

where α_{elec} is the electronic polarizability of the atom or ion..

In the case of a neutral atom in an electric field.

$$\vec{P}_{elec} = \alpha_{elec}\vec{E} = Ze\,\vec{x} = r^3 \, \vec{E} \tag{3.48}$$

i.e., displacement of the electron and nucleus, $\vec{x} = 10^{-15} \, \vec{E}$

3.5.2 In Alternating Electric fields

In alternating electric fields the electronic polarizability is essentially constant up to UV frequencies. According to an empirical relationship (by J.C. Slater and N.H. Frank), for each electron in an outer level,

$$\alpha_{elec} = r^3 = \frac{a\ (n^2 a_o)}{(Z - S)},$$

(3.49)

a_o = Bohr radius, $a_o = 0.053\ nm$, n = principal quantum number of the electron in the highest filled level.

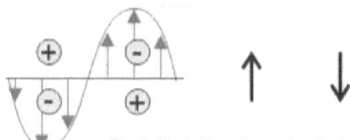

Fig 3.21 Induced atomic dipole

. Z = atomic number. The coefficient A and the screening constant S are given in Table 3.5.

Table 3.5

Shielded electron	Shielding electron						
	1s	2s	2p	3s	3p	n	A
1s	0.35	0.00	0.00	0.00	0.00	1	4.5
2s	0.85	0.35	0.35	0.00	0.00	2	1.1
2p	0.85	0.35	0.35	0.00	0.00	3, etc	0.65
3s	1.00	0.85	0.85	0.35	0.35		
3p	1.00	0.85	0.85	0.35	0.35		

3.6.1   The CLAUSIUS-MOSSOTTI RELATION

What a dielectric 'equation of state' actually looks like?

The field at the molecule due to the surface charges on the sphere is $\vec{E} = \frac{p}{3\varepsilon_o}$.

The electric field at a distance r from a dipole $\vec{p}$ is

$$\vec{E} = -\frac{1}{4\pi\varepsilon_o}\left[\frac{\vec{p}}{r^3} - \frac{3(\vec{p}.\vec{r})\ \vec{r}}{r^5}\right]$$

(3.50)

$$\varepsilon_0 = 8.8542x10^{-12}\ Fm^{-1}$$

The net electric field seen by an individual molecule

$$\boxed{\vec{E}_{loc} = \vec{E}_0 + \frac{P}{3\varepsilon_0}}$$

(3.51)

$$\boxed{\varepsilon = \frac{\varepsilon_0 E + P}{\varepsilon_0 E} = 1 + \chi}$$

(3.52)

$$\boxed{\chi = \frac{P}{\varepsilon_0 E} = \varepsilon - 1}$$

(3.53)

To derive the Clausius-Mossotti relation between ε and α :

N_i = number of atoms in unit volume

α_i = polarizability of each atom,

For cubics, $\vec{E}_{loc} = \vec{E}_{loc}(i)$

$$\vec{P} = \sum_i N_i \alpha_i \vec{E}_{loc}(i)$$

(3.54)

$$\vec{D} = \vec{E} + 4\pi \vec{P}$$

$$\varepsilon_r = \frac{\vec{D}}{\vec{E}} = 1 + 4\pi \frac{\vec{P}}{\vec{E}} = 1 + 4\pi\chi$$

$$\chi = \frac{\vec{P}}{\vec{E}} = \frac{\sum_i N_i \alpha_i \vec{E}_{loc}(i)}{\vec{E}_{loc} - \frac{4\pi}{3}\sum_i N_i \alpha_i \vec{E}_{loc}(i)} = \sum_i \frac{N_i \alpha_i}{1 - \frac{4\pi}{3}\sum_i N_i \alpha_i}$$

Solving for $\sum_i N_i \alpha_i$,

$$\boxed{\frac{\varepsilon_r - 1}{\varepsilon_r - 2} = \frac{4\pi}{3}\sum_i N_i \alpha_i}$$

(3.55)

For electrons, $\varepsilon_r \to \varepsilon_r(\omega)$, at $\vec{E} \propto e^{-j\omega t}$

$$\boxed{\varepsilon_r = \frac{1 + \frac{8\pi}{3}\sum_i N_i \alpha_i}{1 - \frac{4\pi}{3}\sum_i N_i \alpha_i}}$$

(3.56)

which is the Clausius-Mossotti relation.

3.6.1.1 For non-polar substances, $\alpha_i = \alpha_{el}(i) + \alpha_a(i)$, then $\sum_i N_i \alpha_i = \frac{3}{4\pi}$.

and $\vec{P}$ = finite. $\varepsilon_0 \to \infty$, means <u>catastrophe</u>.

The temperature at which happens is the <u>critical temperature</u>, T_0.

Putting, $\frac{8\pi}{3}\sum_i N_i\alpha_i = (1-3s)$, and $s \ll 1$ near T_0,

If $s \simeq \dfrac{T-T_0}{\xi}$, with ξ=constant;

$$\boxed{\varepsilon \cong \frac{\xi}{T-T_0}} \tag{3.57}$$

3.6.1.2 <u>Debye Relaxation</u>

$$\varepsilon^* \equiv \varepsilon^*(\omega) = \varepsilon_\infty + \frac{\Delta\varepsilon}{1+j\omega\tau} \tag{3.58}$$

$\Delta\varepsilon = \varepsilon_s - \varepsilon_\infty$

ε_s = static (low frequency) dielectric constant of the medium,

ε_∞ = high frequency dielectric constant,

τ = characteristic (Debye) <u>relaxation time</u> of the medium.

At optical frequencies, since $\varepsilon = n^2$,

$$\frac{n^2-1}{n^2+2} = \frac{4\pi}{3}\sum_j N_j\alpha_j(electr) \tag{3.59}$$

The Clausius-Mossotti relation shows that for ac fields, $\varepsilon_r(\omega)$ becomes a complex variable.

$$\varepsilon_r(\omega) \equiv \varepsilon_r^* = \varepsilon_r' - j\varepsilon_r''$$

3.6.2 COLE-COLE RELATION: COMPLEX DIELECTRIC CONSTANT, ε^*

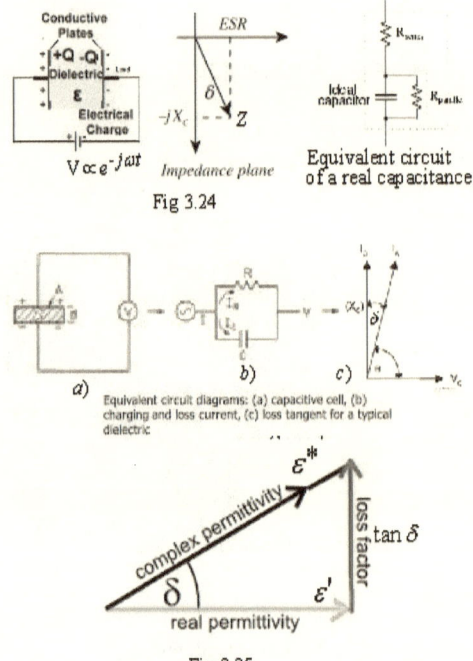

Fig 3.24

Equivalent circuit diagrams: (a) capacitive cell, (b) charging and loss current, (c) loss tangent for a typical dielectric

Fig 3.25

Dielectric loss angle = δ

Lag phase angle = θ

$$\delta = 90° - \theta$$

Loss tangent, $\tan \delta = \dfrac{\varepsilon_r{}''}{\varepsilon_r{}'}$

$\tan \delta = D$. Dissipation factor, $\tan \delta = \dfrac{1}{Q}$, Q = quality factor.

$$\tan \delta = D = \frac{1}{Q} = \frac{\text{Energy loss /cycle}}{\text{Energy stored /cycle}} \tag{3.60}$$

Power factor = $Cos\theta$

Under ac conditions, because $\vec{E}$ and $\vec{P}$ lag by δ,

Dielectric constant $\varepsilon_r \to \varepsilon_r{}^*(\omega)$ is the complex dielectric constant.

$$\boxed{\varepsilon_r(\omega) \equiv \varepsilon_r^* = \varepsilon_r' - j\varepsilon_r''} \tag{3.61}$$

Power dissipation factor, $D \propto V^2\omega\, \varepsilon_r'' = \dfrac{V^2}{8\pi}\varepsilon_r' \tan\delta$

As $\omega = 0$, $\varepsilon_s \equiv \varepsilon(0)$, and $\varepsilon_0 \equiv \varepsilon_\infty$.

The Cole-Cole relation (1941) is

$$\boxed{(\varepsilon' - \frac{\varepsilon_s + \varepsilon_0}{2}) + \varepsilon'' = (\frac{\varepsilon_s - \varepsilon_0}{2})^2} \tag{3.62}$$

For more details one may refer Seaife & Seaife,(1971).

3.6.2.1 Cole-Cole diagram

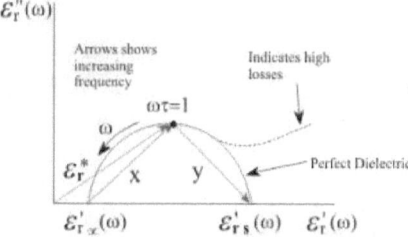

Fig 3.26 Impedance plots in complex plane

Table 3.6

Dielectric Constants at 20°C	
Material	Dielectric Constant
Vacuum	1
Glass	5-10
Mica	3-6
Mylar	3.1
Neoprene	6.70
Plexiglas	3.40
Polyethylene	2.25
Polyvinyl chloride	3.18
Teflon	2.1
Germanium	16
Strontiun titanate	310
Titanium dioxide	173 $\perp$
(rutile)	86 $\parallel$
Water	80.4
Glycerin	42.5
Liquid (-78°C) ammonia	25
Benzene	2.284
Air(1 atm)	1.00059
Air(100 atm)	1.0548

The real part of the Clausius-Mossotti factor is a determining factor for the dielectro-phoretic force on a particle, whereas the imaginary part is a determining factor for the electro-rotational torque on the particle.

3.7.   Experimental Method of measurement of Dielectric constant

A Tansformer Bridge circuit (Robinson & Hollis Hallett, 1966) was used to measure dielectric constant of NaCl, KCl and KBr in the temperature range 4.2 K to 300K.

A.C. Schering bridge is one of the very important and useful methods of measurement of capacitance and dielectric loss of a capacitor. In fact it is a device that can compare a very imperfect capacitor C2 with a loss-free standard capacitor. The imperfect capacitor is equivalent to loss-free pure capacitor in series with a resistor.

Schering Bridge (1920) is a 4-arm bridge circuit with 'screened' arms. (Fig 3.27).

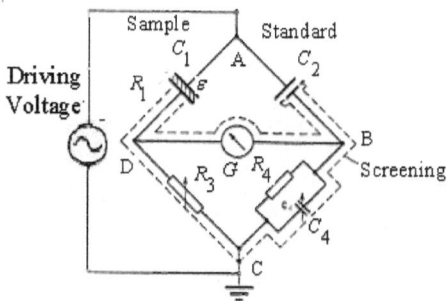

Fig 3.27 Schering Bridge

This A.C/bridge is used to measure to the Capacitance of the capacitor C_1, dissipation factor and measurement of relative permittivity, ε_r. Let us consider the circuit of Schering bridge as shown below:

G = A standard CRO or vibration galvanometer.

The driving voltage is supplied from a standard frequency generator which had a variable amplitude and frequency

C_1 = the unknown capacitance whose value is to be determined with series electrical resistance R_1.

C_2 = a standard laboratory capacitor, of value, say, $0.3\mu F$

C_4 = a standard laboratory variable capacitor, in the range $10^{-1} - 10^{-4} \mu F$.

R_4 = a pure resistor (*i.e.* non inductive in nature).

and R_3 a variable non inductive resistor ($0\Omega - 10k\Omega$) connected in parallel with variable capacitor C_4.

All the components were connected by screened co-axial leads. All of the component terminals were threaded axles (apart from the variable capacitor) with a hole through for wires to be placed in. The axles then had a screw-down cap to hold any connected wires secure. If short screened leads are used and all of the screens are connect to earth points, the noise current can be minimized. (all the bridge elements had metallic casing connected to an *earth terminal* which provided a good earth connection)

Now the supply is given to the bridge between the points A and C The detector G is connected between B and D.

The resistor R_4 was held constant while R_3 and C_4 were adjusted (alternately) until a minimum signal was observed on the CRO

Theory: From the theory of ac bridges we have at balance condition one gets

$$\left(R_1 + \frac{1}{j\omega C_1} \right) \left(\frac{R_4}{1 + j\omega C_4 R_4} \right) = \frac{R_3}{j\omega C_2} \tag{3.63}$$

C_s and C_p are series and parallel circuit values

$$C_s(\text{of } C_4) = \frac{C_2 R_4}{R_3}$$

$$C_p(\text{of } C_1) = \frac{C_2 R_4}{R_3} Cos^2 \delta$$

$$\text{Power factor}(p.f) = R_1 C_2 \tag{3.64}$$

At balance condition,

$$\tan \delta = \omega C_1 R_1 = \left(\omega C_2 \frac{R_4}{R_3} \right) \left(\frac{R_3 C_4}{C_2} \right) \tag{3.65}$$

$$= \omega C_4 R_4$$

$$\tan \delta = \frac{\omega C_4 R_4}{10^6}, \text{ with } C_4 \text{ in } \mu F.$$

Equating the real and imaginary parts of the balance equation above,

$$R_1 = \frac{R_3 C_4}{C_2} \tag{3.66}$$

$$C_1 = C_2 \frac{R_4}{R_3} \tag{3.67}$$

$$\varepsilon = \frac{C_p(\text{Dielectric})}{C_p(\text{Vacuum})} \tag{3.68}$$

$$\text{Or } \varepsilon_r = \frac{C_1 d}{\varepsilon_0 A}$$

One may use a Q-meter or a "Twin-T circuit" instead of the Schering Bridge.
A "3-Terminal dielectric capacitor" is perhaps the best sample cell (Fredericks, Phys. Rev., B4, 911,1971).

High Voltage Schering bridge

A simple Schering bridge (which uses low voltages) is used for measuring dissipation factor, Capacitance and measurement of other properties of insulating materials like insulating oil etc. What is the need of high Voltage Schering bridge? For the measurement of small Capacitance it is necessary to apply high Voltage and high frequency as compare to low Voltage which suffers many disadvantages.

(a) The high Voltage supply is obtained from a transformer 50 Hz and the detector in this bridge is a vibration galvanometer. (b) The impedances of arms AB and AD are very large therefore this circuit draws low current hence power loss is low but due to this low current one requires a very sensitive detector to detect this low current. (c) The fixed standard capacitor C_2 has compressed gas which works as dielectric therefore dissipation factor can be taken as zero for compressed air. Earthed screens are placed between high and low arms of the bridge to prevent errors caused due to inter capacitance.

4.8. SUMMARY

1) Dielectrics are electrical insulators that support charge.
2) The properties of dielectrics are due to polarization.
3) There are three main mechanisms by which polarization arises on the microscopic scale: electronic (distortion of the electron cloud in an atom), ionic (movement of ions) and orientational (rotation of permanent dipoles).
4) A capacitor is a device that stores charge, usually with the aid of a dielectric material. Its capacitance is defined by $Q = CV$.
5) The dielectric constant κ indicates the ability of the dielectric to polarize. It can be defined as the ratio of the dielectric's permittivity to the permittivity of a vacuum.
6) Each of the polarization mechanisms has a characteristic relaxation or resonance frequency. In an alternating field, at each of these (materials dependent) frequencies, the dielectric constant will sharply drop.
7) The dielectric constant is also affected by structure, as this affects the ability of the material to polarize.
8) Polar dielectrics show a decrease in the dielectric constant as temperature increases.
9) Dielectric loss is the absorption of energy by movement of charges in an alternating field, and is particularly high around the relaxation and resonance frequencies of the polarization mechanisms.
10) Sufficiently high electric fields can cause a material to undergo dielectric breakdown and become conducting.

REVIEW QUESTIONS

1. A Ca^{2+} cation and an O^{2-} anion are separated by a distance of 2.4 Å. Calculate the resultant dipole moment. (Charge on an electron = $1.6\,x10^{-19}C$)

2. Consider a capacitor in a computer power supply, possessing a capacitance of 2200 μF. If a voltage of 10 V is applied to this capacitor, what will the charge on the positive plate be? (2 sig figs)

3. Given the molecular dipole moment of H_2O is $6,0\,x10^{30}Cm$, $\rho_{water} = 1\,x10^3\,kgm^{-3}$, $N_A = 6.0225x10^{23}\,mol^{-1}$. Find out the polarization of a drop of water with radius 1.0 mm. (Ans: $8.4\,x10^{-10}Cm^{-2}$).

4. In the case of molecular CO_2, $\rho_{CO_2} = 1.977\,x10^3\,kgm^{-3}$ and electric susceptibility $\chi = 985\,x10^{-6}Cm^{-2}$. Calculate the total polarizability of CO_2. $\varepsilon_0 = 8.8542x10^{-12}\,Fm^{-1}$ (Ans: $\alpha = 3.23\,x10^{-40}\,Fm^2$)'

5. When an atom of oxygen is placed under the influence of an electric field, it becomes a dipole with moment $5.0\,x10^{-23}Cm$, with its electron charge density shifting from the nucleus a distance of $4.0\,x10^{-17}m$. Given, $\varepsilon_0 = 8.8542x10^{-12}\,Fm^{-1}$, what will be the polarizabilty of the atom. (Ans: $\alpha = 1.9\,x10^{-47}\,Fm^2$).

6. In a laboratory test on a Bakelite sample at $20kV$, 50 Hz, having a standard capacitor of 106 pF, balance was obtained with a capacitor of $0.35\mu F$ with a non-inductive resistance of 318Ω, the non-inductive resistance in the remaining arm of the bridge being 130Ω. Determine the capacitance, $p.f$, and equivalent series resistance of the specimen. (Ans: $C_2 = 106pF(318\Omega/130\Omega) = 259.3pF$; $p.f. = [130\Omega(0.35\mu F/106pF)](259.3pF) = 0.035$; $130\Omega(0.35\mu F/106pF) = 0.429\,M\Omega$)

7. Consider a lossy capacitor is tested with a Schering Bridge circuit. Balance obtained with a capacitor under test in one arm, the succeeding arms being a non-inductor resistance of 100Ω, a non-reactive resistor 309Ω in parallel with a pure capacitor $0.5\mu F$, and a standard capacitor of $109\mu\mu F$. The supply frequency is $50Hz$, Calculate from the equivalent equation of balance, the equivalent series capacitance, and power factor (p.f) at 50 Hz) for the capacitor under test. (Ans: $C_3 = 109pF(309/100) = 336.8pF$, $p.f = C_3R_4 = 314x0.5\mu F309 = 0.0485$)

&&*&*&*&*&*&*&*

Chapter 4

FERROELECTRICITY IN CRYSTALS

Chapter 4

FERROELECTRICITY IN CRYSTALS

4.1.   INTRODUCTION

Crystals fall under 32 classes (point groups). Of these 11 classes have centre of symmetry, so that they are called 'centric' and non-polar. The rest of the 21 classes do not posses centre of symmetry meaning 'acentric' or non-centric class. Except the point crystals of point group 432, remaining 20 can exhibit 'piezo-electricity'. Among these piezo-electric crystals, the following have a unique polar axis (*i.e.,* polar crystals) having point groups: 1, 2, 3, 4, 6, m, mm2, 3m, 4mm, and 6mm.

4.1.1   Electrostriction

Electrostriction occurs in all dielectrics, amorphous, crystalline, centric or non-centric,

$$\boxed{(\text{Strain})_{\text{electrostr}} \propto (\text{Electricfield, } \vec{E})^2}$$

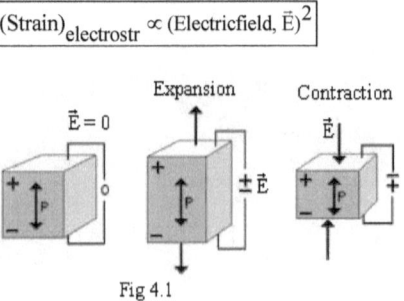

Fig 4.1

In the Fig. it is seen that the sign of the deformation is the same and is independent of the polarity of $\vec{E}$. It is weak.

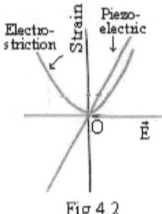

Fig 4.2

4.1.2 Piezoelectricity

This phenomenon was discovered by the brothers Pierre and Jacques Curie in 1880 that some crystals when compressed in some particular directions, show positive and negative electric charges on certain portions of their surfaces, and disappear when the pressure is withdrawn (W.G. Cady, 1946, 1964).

$$\boxed{\text{(Polarization)}_{\text{Piezoelec}}, P_i \propto (\text{Mechanical stress, t}_{jk})} \text{ within Hooke's law region.}$$

P_i = polarization generated along i-axis.

$$\boxed{P_i = d_{ijk} t_{jk}} \quad \text{for direct piezoelectricity}$$

$$\boxed{r_{ij} = d_{kij} \vec{E}_k} \quad \text{for converse effect)}$$

$\vec{E}_k$ = electric field (charges) appearing along k-axis

It is reversible, and observable only in non-centric crystals, *i.e.*, solely dependent on the symmetric properties of the crystals.

Piezoelectricity represented by a (3x3) tensor, has 27 components.

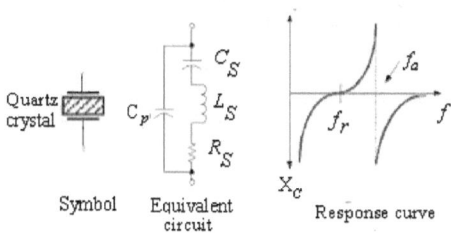

Symbol   Equivalent   Response curve
circuit

Fig 4.3

$L_S \equiv$ mass, $C_S \equiv$ elastic compliance, $R \equiv$ mechanical damping

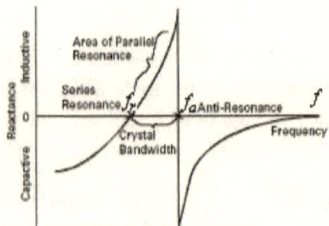

Plot of reactance vs frequency of a quartz crystal

Fig 4.4

Series resonant frequency, $f_s = (\frac{1}{2\pi})\sqrt{\frac{1}{L_s C_s}}$

P_i =polarization, E_i = electric field,

t = mechanical stress , r = mechanical strain.

$d_{ijk} \equiv d_{ikj} = d_{ij}$ = piezoelectric modulii,

The piezoelectric constants are d, j, e, h.

$$d = (\frac{\partial r}{\partial E})_t, \quad P_i = d_{ijk}\,t_{jk}; \quad g = (-\frac{\partial E}{\partial t})_D; \quad E_i = -g_{ijk}\,t_{jk};$$

$$e = (-\frac{\partial t}{\partial E})_r; \quad P_i = e_{ijk}r_{jk}; \quad h = (-\frac{\partial t}{\partial D})_r; \quad E_i = -h_{ijk}\,r_{jk}$$

r, t, E, D are strain, Stress, field, displacement.

$$d_{ijk} \equiv d_{ikj};$$

$\therefore d_{ij}$ is used. $i = 1 - 3$ and $j = 1 - 6$.

Thus there only <u>18 independent components</u>.

$$\therefore d_{ijk} \equiv d_{ikj} = d_{ij} \text{ is } \begin{pmatrix} d_{11} & d_{12} & d_{13} & d_{14} & d_{15} & d_{16} \\ d_{21} & d_{21} & d_{23} & d_{24} & d_{25} & d_{26} \\ d_{31} & d_{32} & d_{33} & d_{34} & d_{35} & d_{36} \end{pmatrix}.$$

Electro-mechanical coupling factor, $k = \dfrac{d_{ij}}{\sqrt{\varepsilon_i^t s_{jj}^2/4\pi}}$

In matrix notation of the third rank tensor property will be simplified.

The matrix equation is.

$$\begin{pmatrix} P_1 \\ P_2 \\ P_3 \end{pmatrix} = \begin{pmatrix} d_{11} & d_{12} & d_{13} & d_{14} & d_{15} & d_{16} \\ d_{21} & d_{21} & d_{23} & d_{24} & d_{25} & d_{26} \\ d_{31} & d_{32} & d_{33} & d_{34} & d_{35} & d_{36} \end{pmatrix} \begin{pmatrix} t_1 \\ t_2 \\ t_3 \\ t_4 \\ t_5 \\ t_6 \end{pmatrix}$$

where (d_{ijk}) are the piezoelectric elements of the piezoelectric tensor $[d_{ijk}]$

<u>Giebe-Scheibe circuit</u> is used for a quick test for piezoelectricity in a crystal.

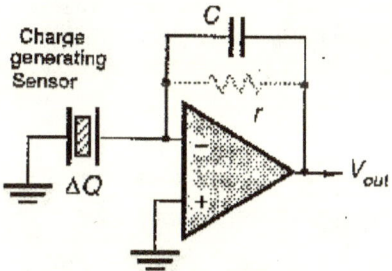

Fig 4.5 Giebe-Scheibe circuit

A qualitative measurement circuit is used

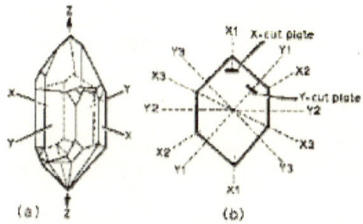

Fig 4.6 a) Quartz crystal and b) cuts

4.1.3 Pyroelectricity

Crystals possessing unique polar axis means that they have polarization in absence of an external electric field. This polarization is called spontaneous polarization,

$\vec{P}_s$. It has been found that Anharmonicity in lattice vibrations cause $\vec{P}_s \cdot \vec{P}_s$ is temperature dependent.

$$\boxed{P_i = \gamma_i \Delta T}$$

So these polar crystals are called popularly 'pyroelectrics' . This spontaneous polarization appears to have either a very high or low value of coercive force. In the former case the $\vec{P}_s$ may not be changed for its alignment by applying a maximum electric field without causing damages to the dielectric. It becomes clear that all pyro-electric crystals are piezo-electric , whereas the reverse is not true. Under uniform heating a pyro-electric crystal can develop electric charge. Pyro-electricity was known about 300 B.C. when Theophrastus had published for the first time the effect. It was in the dark later till in 1703 when it was re-discovered in Tourmaline [(Na,Ca)(Mg,Al)$_6${B$_3$Al$_3$Si$_6$(O,H)$_{30}$}, when it was brought from Sri Lanka to Holland. Piezo-electricity was discovered by the Curie brothers in 1880, and the foundation was laid for the Physics of Crystals.

$$\boxed{\lambda = \frac{dP_i}{dT}}$$

where λ = pyroelectric coefficient. It is possible to detect a charge of $10^{-16} C$ with an electrometer, $\Delta T = 10^{-6} °C$ can be measured by pyroelectric effect.

RL Byer and CB Roundy (1972) (Ferroelectrics, 3, 333) have described a direct method to measure λ .

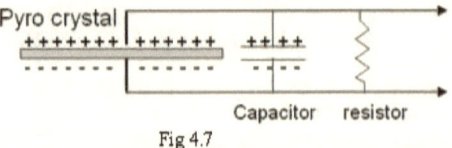

Fig 4.7

A thin specimen in the form of disc may be heated by incident radiation from a hot body, and pyroelectric current measured. A blackened tip of a soldering iron can act as heating source.:

W =Incident radiation intensity,

A = specimen area

z = specimen thickness

c = specific heat of specimen

ρ = density of specimen

Heat balance equation is $WA = z\rho cA \frac{dT}{dt}$

$$Q = A\lambda \, dT$$

Current (initial) = $I = \frac{dQ}{dt}$

$$\boxed{\lambda = I \, z\rho \, c / WA}$$

Table 4.1

Material	Pyroelectric coefficients λ at 20°C ($10^5 \ C \ m^{-2} {}^\circ C^{-1}$)
1) LaTaO$_3$	19
2) Ba$_{0.27}$Sr$_{0.73}$Nb$_2$O$_6$	280
3) Bi$_4$Ti$_3$O$_{12}$	12
4) GASH	14
5) PbTiO$_3$	40
6) TGS	55
7) TGSe	300
8) LiNaSO$_4$	0.8
9) KLiSO$_4$	1.5
10) Tartaric acid	3
11) PVDF	2.7

$[0.75Pb(Mg_{1/3},Nb_{2/3})O_3 + 0.25PbTiO_3]$ has $\lambda \approx 1300 \mu Cm^{-2}K^{-1}$.

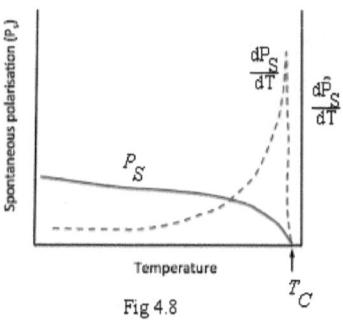

Fig 4.8

$$
\begin{pmatrix} \varepsilon_{11} \\ \varepsilon_{22} \\ \varepsilon_{33} \\ 2\varepsilon_{23} \\ 2\varepsilon_{13} \\ 2\varepsilon_{12} \end{pmatrix} = \begin{pmatrix} 0 & 0 & d_{13} \\ 0 & 0 & d_{31} \\ 0 & 0 & d_{33} \\ d_{14} & d_{15} & 0 \\ d_{15} & -d_{14} & 0 \\ 0 & 0 & 0 \end{pmatrix} \begin{pmatrix} E_1 \\ E_2 \\ E_3 \end{pmatrix}
$$

Table 4.2

Crystal Structure	Point Groups	Centro-Symmetric	Non-centrosymmetric	
			Piezoelectric	Pyroelectric
1) Triclinic	1, $\bar{1}$	$\bar{1}$	1	1
2) Monoclinic	2, m, 2/m	2/m	2, m	2, m
3) Ortho-rhombic	222, mm2, mmm	mmm	222, mm2	mm2
4) Tetragonal	4, $\bar{4}$, 4/m, 422, 4mm, $\bar{4}$2m, 4/mmm	4/m, 4/mmm	4, $\bar{4}$, 422, 4mm, $\bar{4}$2m	4, 4mm
5) Trigonal	3, 3, 32, 3m, 3m	3, 3m	3, 32, 3m	3, 3m
6) Hexagonal	6, $\bar{6}$, 6/m, 622, 6mm, $\bar{6}$m2, 6/mmm	6/m, 6/mmm	6, $\bar{6}$, 622, 6mm, $\bar{6}$m2	6, 6mm
7) Cubic	23, m3, 432, $\bar{4}$3m, m3m	m3, m3m	23, $\bar{4}$3m	_

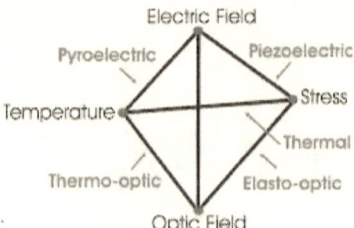

Fig 4.9 Application of Temperature, Stress, Electric field or EM photons in a Crystal and the resulting phenomena

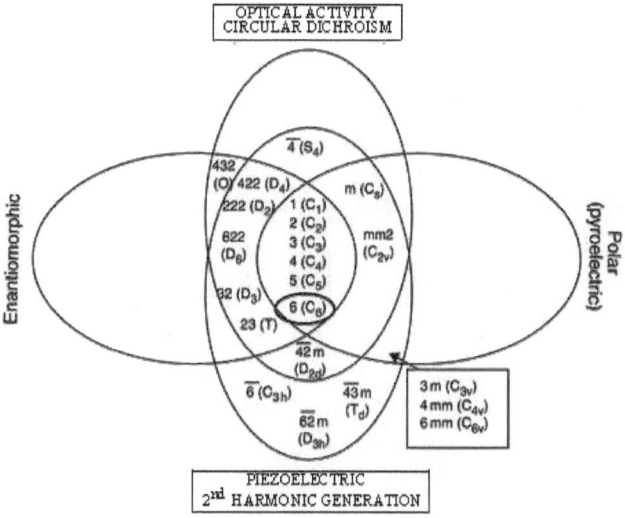

Fig 4.10

Keithley electrometer Model 616 may be used to measure the current.

4.2 FERROELECTRICITY

Ferroelectricity was discovered in 1921 in Czechoslovakia by Joseph Valasek. He was studying the dielectric properties of Rochelle Salt (Ellie Seignette of La Rochelle in France synthesized this material in 1655). Rochelle salt was also known as Seignette Salt. Ferroelectricity is an International term coined to indicate its close analogy with ferromagnetism. It is popularly referred to by Russian scientists as "Seignette electricity".

4.2.1 Definition

The phenomenon of ferroelectricity is said to be exhibited by crystals in which spontaneous polarization $\vec{P}_S$ can be reversed by means of an externally applied electric field without causing damage to the material. Thus a pyro-electric material whose $\vec{P}_S$ can be reversed by an external influence is called a ferroelectric material. All ferroelectrics are pyroelectrics, whereas all pyroelectrics are not ferroelectrics. Crystal exhibiting ferroelectricity should have a permanent electric dipole moments aligned along a unique polar axis, and which can be reversed by an externally applied electric field.

4.2.1 Various Criteria

Let us now understand the different criteria under which the different crystalline properties stated above can be predicted.

1) Piezo-electricity: it is solely determined by the symmetry properties, *viz.*, if the crystal is centro-symmetric or non-centric.

2) Pyroelectricity: this can be predicted as soon as the crystal structure of the crystal is determined.

3) Ferroel3ectricity: Reversal of spontaneous polarization $\vec{P}_s$ is an additional criterion that a pyrolectric crystal must have, and this cannot be predicted from crystal structure. Reversibility of $\vec{P}_s$ is a consequence of the fact that the polar structure, of a ferroelectric, is a slightly distorted one of the non-polar structure of the material. Further a crystal is said to be ferroelectric when it has two or more states in the absence of an electric field, and which can be shifted from one to another of these states by an applied electric field, here any of the two states are identical or enantiomorhous in the crystal structure and different in the electric polarization vector at zero field ('state' mentioned here is not conceptually the same as 'state in thermodynamics or in quantum mechanics'). This definition was due to K. Aizu.

4.3   BEHAVIOUR OF FERROELECTRIC CRYSTALS.

A ferroelectric material found to exhibit the following PROPERTIES.

1) The static dielectric constant, represented by ε_s (or $\varepsilon(0)$ or ε_0), at micro-wave and low frequencies ($10^2 - 10^4\,Hz$) of a ferroelectric is higher than that of the normal dielectric, *i.e.*, non-ferroelectric.

2) It has a unique axis called the 'ferroelectric axis' along which the spontaneous polarization $\vec{P}_s$ is aligned.

3) The static dielectric constant is much anomalously higher along the ferroelectric axis.

4) At room temperature only a small fraction of the polar crystals can show ferroelectricity. So in most of the ferroelectric crystals a phase transition temperature, *viz.*, the Curie point T_C exists at which the crystal undergoes a crystallographic (*i.e.*, structural) phase transition. Therefore ferroelectricity appears at a temperature $T < T_C$. This means a ferroelectric crystal is invariably polymorphic substance, so that over a temperature range a phase with reversible $\vec{P}_s$ will be present.

5) Since the atomic dipole moments produced by means of an external electric field can be only small the energy difference between the bi-polar state must be fairly low. This means the polar state stable at a temperature is only a slightly distorted non-polar state. Thus raising the temperature of the crystal one gets a slightly less stable non-polar state, called the paraelectric phase. Thus the ferroelectric phase at $T < T_C$; and

the paraelectric phase occurs at $T > T_C$. In other words,

$$\boxed{\text{Modification with } \vec{P}_S = 0 \xleftarrow{\;\;T_C\;\;} \text{Modifications with } \vec{P}_S \neq 0}. \quad T_C \text{ varies from 10 K to}$$

1500 K in various ferroelectric crystals.

6) $\vec{P}_s$ is a vector, and $\vec{P}_s$ is parallel to the ferroelectric (FE) axis. $\left|\vec{P}_s\right| \approx 0.1\mu C/cm^2$ to $\left|\vec{P}_s\right| \approx 100\mu C/cm^2$. This is measureable experimentally.

7) There appears a dielectric hysteresis between the net macroscopic polarization and the electric field. This is popularly referred to as the <u>D-E</u> hysteresis or <u>P-E hysteresis loop</u>. This provides parameters like $\left|\vec{P}_s\right|$ and E_C.

8) There is a coercive electric field E_C.

9) The magnitude of the dielectric constant increases as temperature approaches T_C, (the dielectric constant changes anomalously peak value at T_C). In general, in the paraelectric phase, the static dielectric constant $\varepsilon(0)$ obeys the Curie-Weiss Law, viz.,

$$\boxed{\varepsilon(0) - 1 = \frac{C}{T - T_0}}$$

where C = the Curie-Weiss constant has the range $C \sim 10^3$ to $10^5 K$.

T_0 = Curie-Weiss temperature, which is always $\underline{T_0 < T_C}$.

10) At constant temperature and stress the electric field strength is expressed by Taylor series as

$$E = \frac{1}{\chi} P + \xi\, P^3 + \varsigma\, P^5$$

where ξ and ς are measures of dielectric non-linearity. In the paraelectric phase,

$$\chi = k_0(\varepsilon_0 - 1)$$

where k_0 = permittivity of free space.

$$\boxed{\varepsilon_0 = \varepsilon_\infty + \frac{C}{T - T_0}}$$

11) Unit cell dimensions, P_s, ε, E, etc., change discontinuously in a second-order phase transition, and $T_0 \approx T_C$ In first-order transition, energy, volume, etc., change discontinuously, whereas ε, specific heat C_v, thermal expansion coefficient α, etc., .change continuously. Further $T_0 < T_C$ and $T_C - T_0 \approx 10K$.

12) There is a domain structure of a ferroelectric crystal that can be delineated by external means.

13) Non-linear properties of polarization and strain against electric field and mechanical stresses are characteristic of a ferroelectric substance.

14) Evidence of ferroelectricity in a crystal need not rest on electrical means as mentioned above. The ultimate test of ferroelectric transition is to measure the frequency of a soft optical phonon mode in the crystal as a function of temperature, using neutron inelastic coherent scattering experiment or using Raman spectroscopic methods.

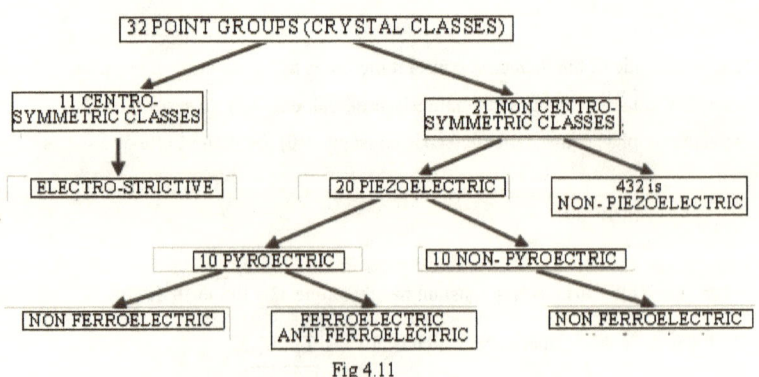

Fig 4.11

4.4.1 DISCOVERY OF FERROLECTRIC CRYSTALS

The different ferroelectric substances can be classified under different headings without listing all of them.

1) <u>Rochelle Salt</u>: This is the Sodium potassium tartrate tetrahydrate having the chemical structural formula, $NaKC_4H_4O_6.4H_2O$. It was the first ferroelectric material to be discovered, in 1921, in Czechoslovakia, by Joseph Valasek. It has two Curie points, the upper one at $24\,°C$ and the lower at $-18\,°C$. The transitions can be illustrated as follows:

Orthorhombic	$T_C = 24°C$	Monoclinic	$T_C = -18°C$	Orthorhombic
Space Group $P2_12_12_1$	$\longleftrightarrow$	Space Group $P2_1$	$\longleftrightarrow$	Space Group $P2_12_12$

The deuterated (heavy) Rochelle salt is also found to be ferroelectric, between the two Curie temperatures, $+35\,°C$ and $-22\,°C$.

2) <u>KDP (Potassium Dihydrogen phosphate)</u>: Its chemical formula is KH_2PO_4

Busch and Scherrer discovered this to be ferroelectric in 1935. It has the Curie point at $T_c = -150°C$.

Tetragonal	$T_C = -150°C$	Orthorhombic
Point Group, D_{2d} - $\bar{1}42d$	$\longleftrightarrow$	Point Group Fdd2

The deuterated crystal KD_2PO_4 is also a ferroelectric. These crystals come under the group XH_2YO_4 family, where $X = K, Cs, Rb$, and $Y = P$ and As, and were discovered only after Barium titanate to be ferroelectric. Till World War II these were the only two families of ferroelectrics materials to be known.

3) <u>Barium Titanate</u>, It has the structural formula $BaTiO_3$.

During 1944-1946, von Hippel *et al.* and Wul independently, reported ferroelectricity in barium titanate. This is a typical crystal having <u>perovskite</u> structure. It is known to have three phase transitions, with the Curies temperature at $T_c = 120°C$.

Cubic	$T_C = 120°C$	Tetragonal	$5°C$	Orthorhombic	$-90°C$	Trigonal
Pm3m	$\longleftrightarrow$	P4mm	$\longleftrightarrow$	Amm2	$\longleftrightarrow$	R3m

This is the first room temperature ferroelectric. $BaTiO_3$ is ferroelectric in all the three phases below 120 °C Compared to the two families above it does not have hydrogen in its structure. Soon other perovskite crystals like $KNbO_3$, $KTaO_3$, $PbTiO_3$. *etc.*, were found to be ferroelectric.

4) <u>Lithium Niobate</u>: $LiNbO_3$ was found to be ferroelectric in 1949.

Rhombohedral	$T_C = 1210°C$	Rhombohedral
3m	$\longleftrightarrow$	3m

5) <u>Pyrochlore type family</u>: In 1952 a typical pyrochlore material Cadmium pyro-niobate, $Cd_2Nb_2O_7$, was found to be ferroelectric, and was studied as follows.

Cubic	$T_C = -88°C$	Tetragonal
m3m	$\longrightarrow$	4mm

6) .<u>Tungsten bronze type</u>: In 1953 Lead meta-niobate, $Pb_2Nb_2O_6$, was found to exhibit ferroelectric properties, with $T_c = 570°C$.

Tetragonal	$T_C = 570°C$	Orthorhombic
4/mmm	$\longleftrightarrow$	mm2

$KSr_2Nb\,O_5$ ($T_c = 156°C$), $NaB_2Nb_5O_{15}$ ($T_c = 560°C$), $A_6^{1+}B_4^{1+}C_7^{5+}D_3^{5-}O_{30}$ -type,

7) <u>GASH</u> family: Guanidinium Aluminium Sulphate hexa-hydrate, $C(NH_2)_3Al(SO_4)_2.6H_2O$, was found to exhibit ferroelectricity at all temperatures. It is rhombohedral at all temperatures, and has <u>no Curie point</u>.

Hexagonal	$T_C = None$
Point Group, P31m	$\longleftrightarrow$

8) ALUM family: Methylammonium Aluminium Sulphate Dodecahydrate, (MASH) $CH_3(NH_3)Al(SO_4)_2.12H_2O$ has been reported with $T_c = -96°C$,

Cubic	$T_C = -96C$	Monoclinic
$P2_13$	$\longleftarrow$	$P2_1$

9) Triglycine Sulphate (TGS) family: Another very popular ferroelectric, discovered in 1956, was TGS, with formula $(CH_2NH_2COOH)_3$.

Monoclinic	$T_C = +49.4 °C$	Monoclinic
$2_1/m$	$\longleftarrow$	$P2_1$

Colemanite, $(C_2B_6O_{11}.5H_2O$, with $T_c = -7°C$, was also discovered in 1956.

Monoclinic	$T_C \approx -7 °C$	Monoclinic
$P2_1/a$	$\longleftarrow$	2

10) Complex Organic crystals: Thio-urea $SC(NH_2)_2$ was reported in 1956. Calcium Strontium propionate, Ammonium monochloro-acetate, *etc.*, were also reported soon.

11) Langbeinite (Double salt) family: The representative member of this is $(NH_4)_2Cd_2(SO_4)_3$ found in 1956.

12) Sodium nitrite, $NaNO_2$:, found in 1958.

PE	163.8°C	Orthorhombic	−105.0°C	
Immm	$\longleftarrow$ T_C	Im2m	$\longleftarrow$ T	

13) Ammonium bisulphate, $NH_4H(SO_4)$: was found to be ferroelectric in 1958.

Monoclinic	$T_C = -3C$	Monoclinic
$P2_1/c$	$\longleftarrow$	m

14) Boracites family: $Mg_3B_7O_{13}Cl$ in 1957.

	265°C	Tetragonal
Pca	$\longleftarrow$ T_C	$F\bar{4}3c$

Nickel boracites in 1966,

Paraelectric PE, PM	130 K	Paraelectric PE, AFM	64 K	FE, FM
Pca	$\longleftarrow$ T_C	$\bar{4}3c$	$\longleftarrow$ T_C	P ca

15) Potassium Nitrate KNO_3 family: in 1958. $NaNH_4SO_4.2H_2O$ has $T_c = 101K$.

Leconite, in 1959; Lithium Selenite in 1959;
KFCT (Potassium Ferro-cyanide) in 1959.

Monoclinic, PE	−24.5°C	Monoclinic
$C2_1/c$	$\longleftarrow$ T_C	Cc

16) Sodium salt of DNA: In 1960 the first biomolecularmaterial was reported to have ferroelectric behaviour.

17) Barium Magnesium Fluoride $BaMgF_4$ family : in 1960 it was found to have no T_c in its only orthorhombic phase.

18) Antimony sulpho-iodide SbSI family: Found FE in 1962.

PE		Orthorhombic		Monoclinic
Pnam	$\xrightarrow{\sim-15.7°C}{T_C}$	Pna2₁	$\xrightarrow{-36.6°C}{T}$	2

$$\boxed{\dfrac{PE}{Pnam} \underset{T_C}{\xleftrightarrow{\sim-15.7°C}} \dfrac{Orthorhombic}{Pna2_1} \underset{T}{\xleftrightarrow{-36.6°C}} \dfrac{Monoclinic}{2}}.$$

19) Manganites $YMnO_3$ family: Discovered FE in Yttrium manganite in 1963.

$$\boxed{\dfrac{PE}{6_3cm} \underset{T_C}{\xleftrightarrow{640°C}} \dfrac{FE}{P6_3/mcm}}.$$

20) Molybdate family: $Gd_2(MO_4)_3$ was found to be FE in 1966, with $T_c = 150°C$.

21) IV – VI compounds family: Germanium Telluride GeTe, a metallic diatomic solid having $T_c = 400°C$,was reporte to be FE in 1966.

$$\boxed{\dfrac{PE}{Cubic} \underset{T_C}{\xleftrightarrow{400°C}} \dfrac{FE}{Rhombohedral}}$$

22) Liquid crystal:. P-azoxy-phenetole is FE, $T_c = 168°C$.

23) Layer-structure Oxides family: Bi_2WO_6 was found to be ferroelectric in 1972, with $T_c = 935°C$. $SrBi_2Ta_2O_9$ ($T_c = 310°C$) in 1961, $Bi_4Ti_3O_{12}$ $T_c = 675°C$ in 1961.

24) Solid solutions of compounds: Perovskite type $Pb_3(Mg,Ta_2)O_6$ ($T_c = -98°C$), $Pb_5Ge_3O_{11}$, halides like Rubidium lithium sulphate, $Li_2Ti_2O_7$ discovered in 1974, has the highest $T_c = 1500°C$. Of the 390 ferroelectric substances discovered till 1976.

4.4.2 PEROVSKITE STRUCTURE, Barium Titanate

The perovskite type ferroelectrics have the general composition ABO_3 . Examples are: $BaTiO_3$, $PbTiO_3$, $KbTiO_3$, *etc*.

4.4.2.1 Paraelectric Phase:

The unit cell is shown I Fig and a three-dimensional network of corner sharing octahedra of O_2^{2-} ions.

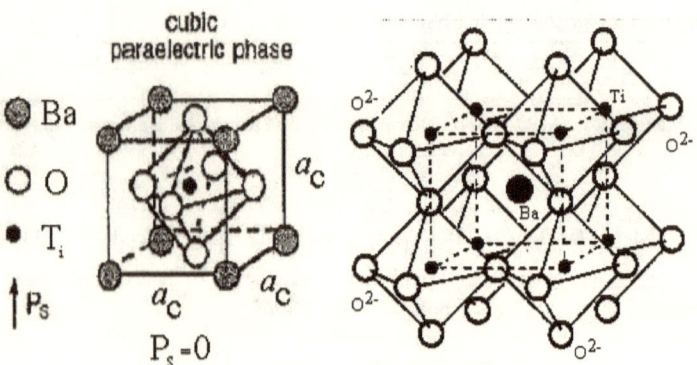

Fig 4.12   Perovskite Structure, $BaTiO_3$

Space group is O_h^1, *i.e.*, Pm3m, , $Z = 1$.

with Ba ion as placed in the origin,

Ba atom   at $0, 0, 0$
Ti   " $\frac{1}{2}, \frac{1}{2}, \frac{1}{2}$
O   " $\frac{1}{2}, \frac{1}{2}, 0;\ 0, \frac{1}{2}, \frac{1}{2};\ \frac{1}{2}, 0, \frac{1}{2}$

$T_C = 120°C$

4.4.2.2 Ferroelectric phase, $T < T_C (= 120°C)$

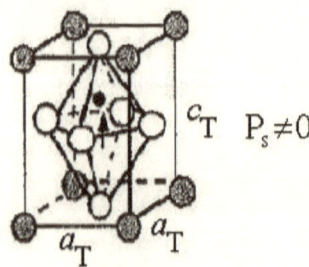

Fig 4.13   **tetragonal ferroelectric phase**

Tetragonal, Space group C_{4v}^1, *i.e.*, P4mm, $a_{atRT} = 3.9920 A°$, $c_{atRT} = 4.0361 A°$, $Z = 1$.

The ions are displaced from their original highly symmetric sites along the polar axis, c, along z-axis, to

$$Ba^{2+} \text{ ion} \quad \text{at} \quad 0,0,0$$
$$Ti^{4+} \quad " \quad \tfrac{1}{2},\tfrac{1}{2},\tfrac{1}{2}+\delta z_{Ti}$$
$$O_I^{2-} \quad " \quad \tfrac{1}{2},\tfrac{1}{2}, \delta z_{O_I};$$
$$2O_{II}^{2-} \quad " \quad \tfrac{1}{2},0,\tfrac{1}{2}+\delta z_{O_{II}}; \ 0,\tfrac{1}{2},\tfrac{1}{2}+\delta z_{O_{II}}$$

At room temperature, $BaTiO_3$, has (W. Kanzig, 1951) $\delta z_{Ti} = 0.014$ units $= 0.5\ nm$, $\delta z_{O_I} = -0.032$, assuming $\delta z_{O_{II}} = 0$. Later Frazer $et\ al,,$(1955) by neutron diffraction found; $\delta z_{Ti} = 0.014$ units $= +0.5\ nm,$, $\delta z_{O_I} = -0.023$ units $= -0.9\ nm$, $\delta z_{O_{II}} = -0.020$ units $= -0.5\ nm$.

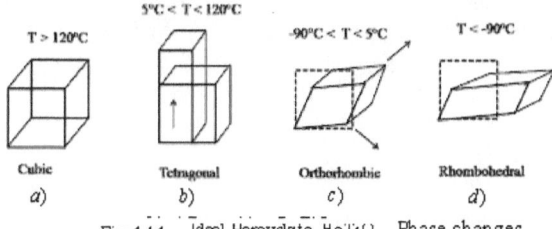

T > 120°C   -90°C < T < 5°C   T < -90°C

Cubic   Tetragonal   Orthorhombic   Rhombohedral
a)   b)   c)   d)

Fig 4.14   Ideal Perovskite, $BaTiO_3$ Phase changes

Worked out Problem 4.1

Consider perovskite crystal of $BaTiO_3$. According to X-ray measurements, Ba^{2+} and Ti^{4+} ions were moved by $x = 0.1 A°$ with respect to the O^{2-} ions. The cell centre at $T < T_C$ is $Z = 1$ formula unit. Find a) the dipole moment / unit cell and b) the polarization in the unit cell.

Solution: $\boxed{Step\ \#\ 1}$ Given perovskite structure meaning cubic $BaTiO_3$. Positive and negative ions are separated at $x = 0.1 A°$. Cell dimension $a = 4 A°$.

$\boxed{Step\ \#\ 2}$ unit cell volume $= V = a^3$, $= (a = 4 A°)^3 = 64\ x10^{-24} cm^3$,

$\boxed{Step\ \#\ 3}$ Dipole moment / unit cell. One Ba^{2+} and one Ti^{4+} ion together has $6+$ charge units, and three O^{2-} ions together has $6-$ units of charge.

Dipole moment / unit cell $= 6.\ e.\ x = 6\ (4.8\ x10^{-10} esu)(0.1\ x10^{-8} cm)$

$$= 3\, x10^{-18} esu.cm\, .$$

$\boxed{Step\ \#\ 4}$ Polarization $P = (3\, x10^{-18} esu.cm)/(64\, x10^{-24} cm^3) = (3\, /64)(10^6 esu.cm^{-2})$

$$= (3\, /64)(10^6 esu.cm^{-2})(3\, x10^9\, esu.C^{-1}) = 15.6\, x10^{-6} Ccm^{-2}$$
$$= 15.6\ \mu Ccm^{-2}$$

4.5.1   Dielectric constant of Ferroelectrics:

Usually the dielectric constant of a ferroelectric is determined based on the measurements on capacitance. A Schering bridge is used (Chapter 3) for measuring the dielectric loss, $\tan \delta$ Typical variation of ε with temperature is shown for the representative crystal $BaTiO_3$. In Fig.4.15.

Above $T > T_c$ the Curie-Weiss law is followed. The Curie constant, $C = 1.5\, x10^5 K$. $T_0 = 150°C$, $\varepsilon_a(0) = 1920$, and $\varepsilon_c(0) = 168$ at room temperature. The frequencies involved in the measurement of dielectric constant are in the range 0 to microwave frequencies. $Li_{0.1}NaBi_{1.8}Nb_5O_{15}$ is a tungsten-bronze ferroelectric type of 1974, and it has the highest values, $\varepsilon_{33}^T = 1.28\, x10^5$ at T_c, and $\varepsilon_{33}^T = 710$ at RT (room temperature), so far measured..

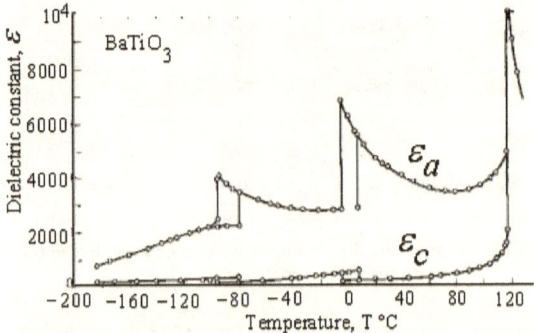

Fig 4.15 ε versus T in $BaTiO_3$ as Phase changes

4.5.2   Spontaneous Polarization, P_s, measurement

The most popular technique is using the Sawyer-Tower circuit of 1930. (Modified, Fig 4.16).

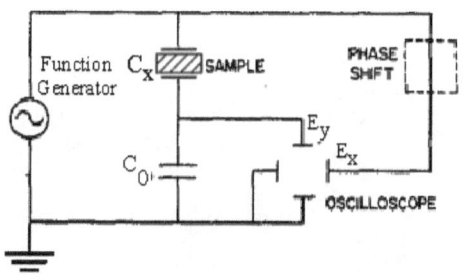

Fig 4.16 Sawyer-Tower Circuit

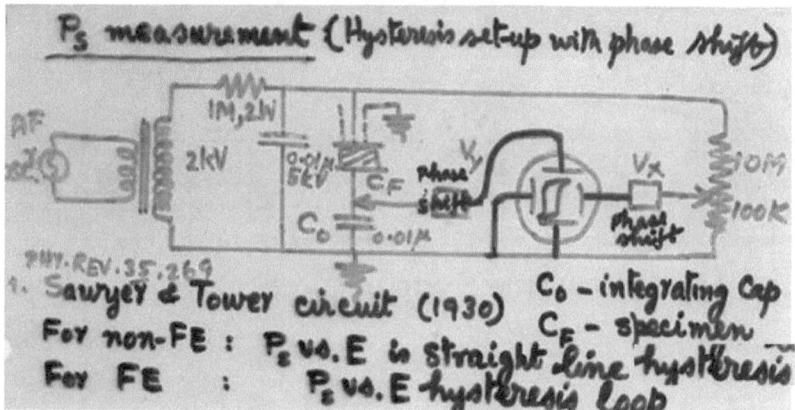

Fig 4.17 Sawyer-Tower Circuit

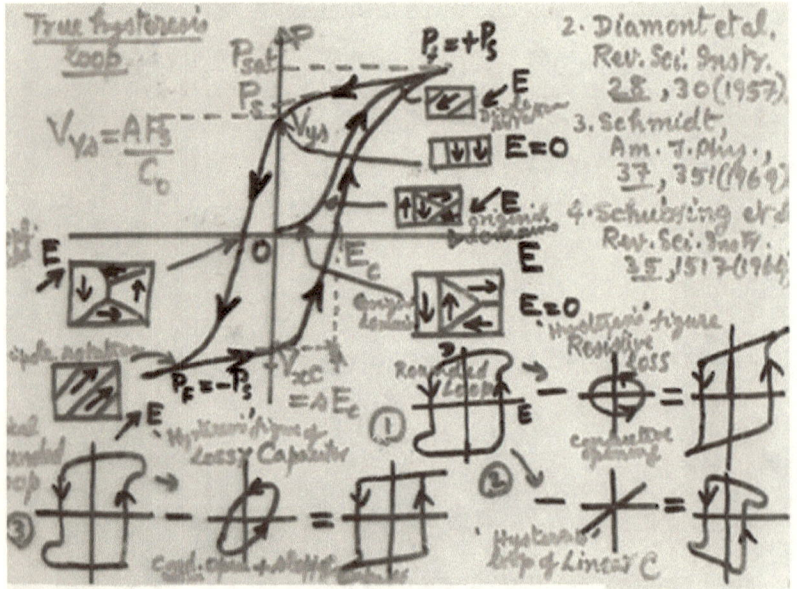

Fig 4.18   a) True Hysteresis loop

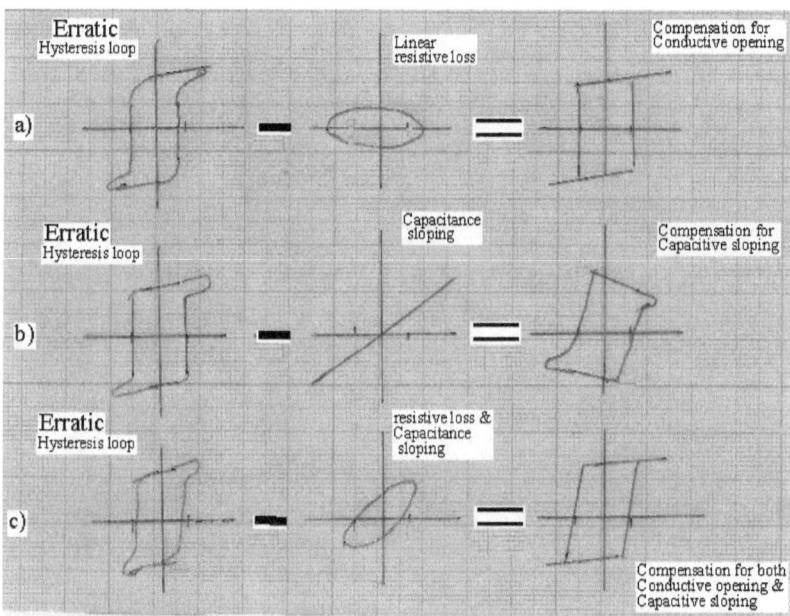

Fig. 4.19 b) Compensating Erratic Hysteresis Loop

Wayne-Kerr Capacitor Bridge can be used for the measurement of dielectric constant at various frequencies.

Dias and Das-Gupta had made use of a setup given in Fig 4,20.

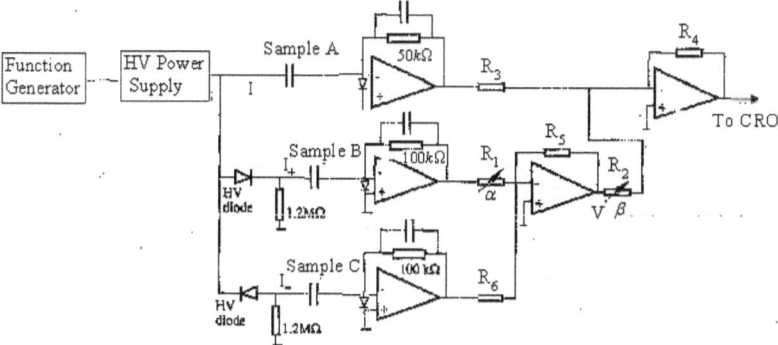

Fig 4.20   Schematic Setup for Hysteresis loop Measurement
(Dias and Das-Gupta, 1993)

In this setup, hysteresis loop of the sample is obtained with correction to the resistive loss and cpacitive opening. Though the circuit is self-explanatory, a brief description is probably useful. The sample is immersed in silicone oil to avoid arcing, and use of a hot plate can change the temperature. The function generator provides the desired voltage signal of suitable shape, frequency and amplitude. A HV power supply (say, TREK 610C) can amplify the signal in a 1:1000 ratio. This voltage is applied to the three identical Sample capacitors, A, B and C. When a high voltage ac is applied, the polarization of the sample A is switched repeatedly, while sample B is positively polarized, C will be negatively polarized, or the *vice-versa*.

The currents, I_+, I_- are shown in Fig 4.20. are the resistive and capacitive current. polarization currents, and I includes polarization I_{Pol}, as well as both resistive, capacitive currents.

$$I_{Pol} = I - \beta(I_+ - \alpha I_-)$$

α and β represent variables related to the thickness and area of the samples.

The procedure to use the setup of Fig 4.20 one may refer to the paper by Dias & Das-Gupta.

A simple relaxation equation for polarization reversal is

$$I_{Pol} = \frac{dP}{dt} = v(P_S - P) = v_0 \, e^{(-E_\alpha/E)(P_S - P)}$$

E_α = Activation field, v_0 =a rate constant independent of the applied field E.

When $E = E_C$, the coercive field, the polarization current becomes

$$I_0 = v_0 \, P_S \, e^{(-E_\alpha/E_C)}$$

E_α may be obtained as the slope of the plot of $\log(I_0)$ and $(1/E_C)$, while the ordinate of the vertical axis gives $v_0 P_S$.

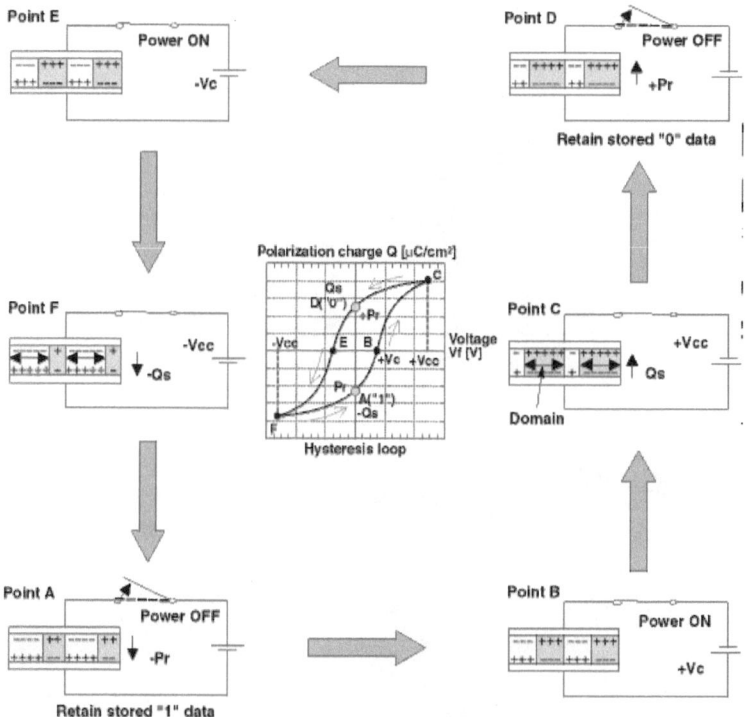

Fig 4.19   Ferroelectric P-E hysteresis Cycle

The principle is applying an alternated voltage and relating the stored charge in the capacitor with the instantaneous voltage. A large value integrating capacitor, C_0, is place in series with the sample, C_x the applied voltage is displayed as the horizontal deflection, of a voltage across C_0 .(*i.e.*, as a measure

of the charge stored in the test sample), is conventionally displayed as the vertical deflection of a Cathode Ray Oscilloscope (CRO). A phase shifting network included as shown (Fig 4.17 and Fig 4.20) enables getting the ferroelectric hysteresis loop. Its function is to compensate for the conductive opening, *viz.*, the d.c. resistance of the dielectric material, and the capacitance sloping due to the lossy capacitance formed by the free space. A typical D-E hysteresis loop is as shown in Fig.4.18. P_S and E_C are measured from the hysteresis, whereas the dielectric constant,

$$\varepsilon = (\partial D/\partial E)_{at\ E=0},$$

is also obtainable. Only few ferroelectric substances give "<u>square</u>" hysteresis loops (Fig 4.21). All the other FE materials and ceramics provide '<u>round</u>' shaped hysteresis loops (Fig 4.19).

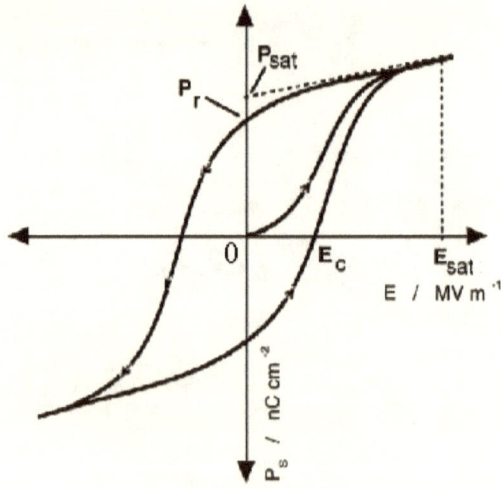

Fig 4.20 P_S versus $\vec{E}$ hysteresis plot

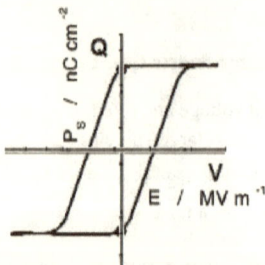

Fig 4.21 P-E Square Hysteresis

This is for one particular temperature, one at which the material is ferroelectric.

What happens if the temperature of the material is raised? The hysteresis loop changes with temperature, becoming sharper and thinner, and eventually disappearing, i.e.,:as in Fig 4.22 for $BaTiO_3$.

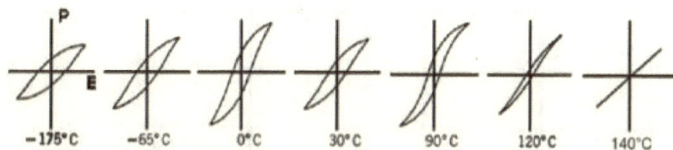

Fig 4.22 P-E Hysteresis change with T for BaTiO$_3$

4.5.3 THREE-TERMINAL DIELECTRIC CELL

McCammon & Work, 1965) designed and fabricated a three-terminal capacitance cell.for the accurate measurement of dielectric constant. Later Fredericks (1971) has also used a 3-T capacitor cell. In Fig 4.23 copying the design features Devanarayanan (unpublished data, 1967) had fabricated a 3-T Capacitor cellat the IISc, Bangalore.

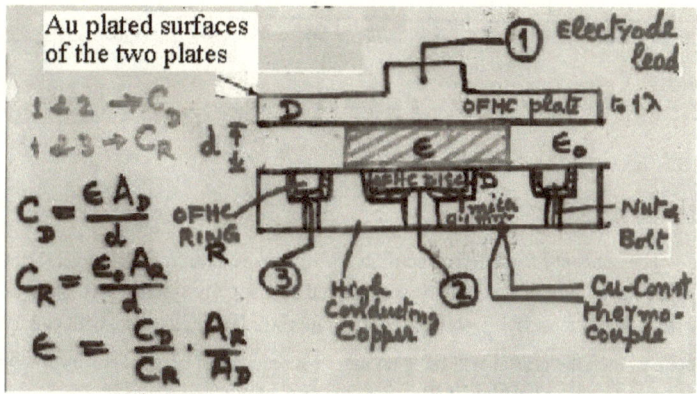

Fig 4.23   3-Terminal Capacitor cell

Measurement of P$_S$

Fig 4.24 is the graph of P$_S$ versus T for BaTiO$_3$,

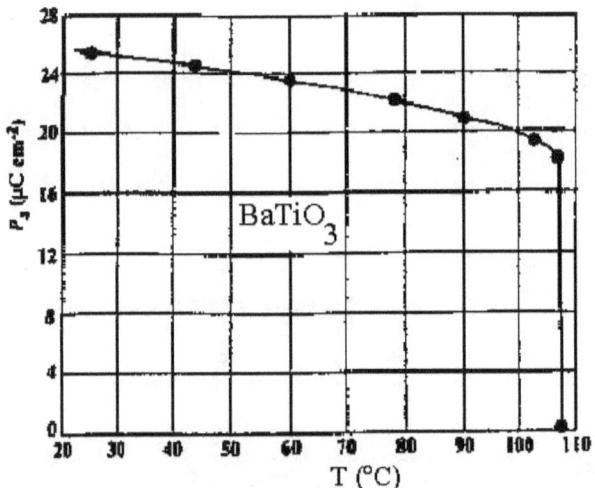

Fig 4.24 P_S *versus* T plot BaTiO$_3$

4.6.   OTHER ASSOCIATED PENOMENA

4.6.1   Anti-ferroelectricity (AFE)

The concept of anti-ferroelectricity (AFE) was first introduced by C. Kittel in 1951. The4 structure of AFE can be represented by a superposition of two or more polarized sub-lattices. If P_{ai} and P_{bi} are two such sub-lattices a and b, with equal and opposite P, then the crystal will have no macro-scopic polarization. Thus a AFE crystal undergoing a phase transition at T_A, as in the case of a FE crystal at T_C,

i)   For $T < T_A$, $(\vec{P}_{ai} + \vec{P}_{bi}) = 0$,

$\vec{P}_{ai} = -\vec{P}_{bi}$.

ii)   For $T > T_A$, $\vec{P}_{ai} = \vec{P}_{bi} = 0$.

If any of the parameters of the polarized sub-lattice is a multiple of high symmetry unit cell parameters, the low temperature anti-ferroelectric phase will exhibit a "super lattice". In such case, the unit cell (super lattice)

= $n.X$ unit cell (Para-electric phase)

= true unit cell of the AFE phase.

The appearance of a super lattice unit cell is, however, not a necessary condition for a AFE transition. If $Z\,(AFE) = Z\,(PE)$, i.e., if anti-parallel P can be brought within one unit cell, then the unit cell is a slightly distorted high temperature unit cell. But no AFE is known to exist in this way, till 1980.

Ammonium dihydrogen phosphate (ADP) $NH_4H_2PO_4$ is the first AFE discovered in 1937. It has $T_A = -125\ ^\circ C$.

PE at RT	$\xleftarrow{\quad}\ \overset{-125\ ^\circ C}{T_A}\ \xrightarrow{\quad}$	AFE
Tetragonal		Orthorhombic

Listed in the Table 4.3 below are AFEs discovered till 1979.

Table 4.3

AFE	T_A
1) Deuterated ADP	-31°C
2) Ammonium Fluoro-beryllate	-97 °C
3) Lead Zirconate	230 °C
4) $NaNbO_3$	354 °C
5) Lead pyro-niobate	15.4 °C
6) Lead Magnesium Tungstate	38 °C
7) $RbNO_3$	219 °C
8) Copper Formate	235 K
9) Silver trihydr-periodate	245 K
10) Tungsten Trioxide	1010 K
11) Cesium Plumbo-chloride	47 °C
12) Rubidium Lithium Sulphate	202 °C
13) $4PbO.GeO_2$	340 °C

The total number of crystals exhibiting AFE has crossed 60.

a) The dielectric constant in the non-ferroelectric phase above T_A does not obey the Curie-Weiss law.
b) There is no hysteresis loop between D and E.
c) But a double hysteresis loop has been observed, as in Fig.

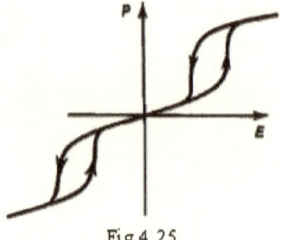

Fig 4.25

4.6.2 FERRIELECTRICITY

In the case of some crystals there are also super structural cells. However, the resulting electric dipole moment of such cells though small is not zero. Crystals exhibiting such a behaviour are called "weak" ferroelectrics, and at a given temperature show FE properties along one axis, and AFE properties along a different axis. Theses crystals are known as ferroelectric materials. The existence of Ferroelectricity and anti-ferroelectricity leads to hysteresis loops seemingly identical with those of ferroelectrics, but are <u>degenerate</u> AFE loops (Fig 4.26). No pure crystal has been found to be ferroelectric. Solid solutions of $NaNbO_3$ and $NaVO_3$, with composition

$Na(Nb_{1-x}, V_x)O_3$ with $x = 0.05$ *to* 0.50 are ferroelectric materials. Another type of such a material is $(Na_{1-x}, Ag_x)(Nb_{0.7}, V_{0.3})O_3$ with $x = 0.05$ *to* 0.15.

4.6.3 Relaxor ferroelectrics

Relaxor ferroelectrics were discovered (Smolenskii) in the 1950s but many of their properties are not understood. In this review, we shall concentrate on materials such as PMN ($PbMg_{1/3}Nb_{2/3}O_3$), which crystallize in the cubic perovskite structure but with the Mg^{2+} ion, and the Nb^{5+} ion, randomly distributed over the B site of the perovskite structure. The peak of the dielectric susceptibility for relaxors is much broader in temperature than that of conventional ferroelectrics, while below the maximum of the susceptibility most relaxors remain cubic and show no electric polarization, unlike that observed for conventional ferroelectrics. Because of the large width of the susceptibility, relaxors are often used as capacitors. Recently, there have been many X-ray and neutron scattering studies of relaxors and the results have enabled a more detailed picture to be obtained. An important conclusion is that relaxors can exist in a random field state, as initially proposed by Westphal, Kleemann and Glinchuk, similar to that which has been studied for diluted antiferromagnets. If a relaxor is cooled from a high temperature, then the Burns temperature is a measure of when slow fluctuations become evident. These fluctuations are connected with the disorder and are known as nano-domains. The Burns temperature is not a well-defined transition temperature. At a lower temperature, there is a well-defined boundary to a so-called random field state when the nano-domains become static but there is no long-range periodic order. This phase may have both history-dependent properties and a skin effect in which the surface of the sample is different from that of the bulk material, as also found in experiments on magnetic systems. Other relaxor systems such as $(PMN)_{1-x}(PT)_x$ for which PMN is mixed with different amounts of the ferroelectric lead titanate (PT), and show that the existence of a random field state enables us also to describe the experimental results for these mixed materials..

Table 4.4

Property	Normal Ferroelectric	Relaxor Ferroelectric
Differences between normal ferroelectrics and Relaxor ferroelectrics.		
1) Dielectric temperature dependence	Sharp 1st or 2nd order	Broad diffused phase transition at Curie maxima
2) Dielectric frequency dependence	Weak Frequency dependence	Strong frequency dependence
3) Dielectric Behavior in paraelectric range ($T > T_c$)	Follows Curie - Weiss law	Follows Curie - Weiss law
4) Remnant polarization (P_R)	Strong P_R	Weak PR
5) Scattering of light	Strong anisotropy	Very weak anisotropy to light
6) Diffraction of X-Rays	Line splitting due to deformation from paraelectric to ferroelectric phase	No X-Ray line splitting

4.6.4   Organic Polymers

Polyvinylidene fluoride (PVDF, $(CH_2-CF_2)_n$) and copolymers of PVDF with trifluoro-ethylene {P(VDF-TrFE)} have found applications as piezoelectric and pyroelectric materials. The piezoelectric and pyroelectric properties of these polymers are due to the remnant polarization obtained by orienting the crystalline phase of the polymer in a strong poling field. Hence the piezoelectric and pyroelectric properties depend on the degree of crystallinity of the polymer and the ferroelectric polarization of the crystalline phase .

The piezo-polymers have some properties which make them better suited for use in medical imaging applications. The density of these polymers is very close to that of water and the human body tissues, hence there is no acoustic impedance mismatch with the body. The piezo-polymers are also flexible and conformable to any shape. However, there are also some problems associated with the piezoelectric polymers including the very low dielectric constant ($K = 5$-10) which could lead to electrical impedance matching problems with the electronics. The dielectric losses at high frequencies are very large for these piezo-polymers. The polymers also have a low Curie point and the degradation of the polymer starts occurring at low temperatures (70-100 °C). The poling efficiency is very low for polymer specimens with large thickness (>1mm).

4.6.4   Ferroelectric Thin Films:

The fabrication of ferroelectric thin films make use of three main methods including physical vapor deposition (PVD), chemical vapor deposition (CVD) and sol-gel processing. In PVD, precursors of the desired film composition are vaporized and deposited on the substrate by one of the sputtering techniques (i.e. rf magnetron sputtering, pulsed vapor deposition (PLD) etc.). This process produces films of very high quality. The CVD technique uses volatile chemical precursors which are vaporized on heated substrates. The advantages of this technique include high deposition rates, pinhole free films and good stoichiometric control. The sol-gel method involves the hydrolysis and condensation of organometallic precursors on the substrate. The deposited sol is annealed at low temperatures to crystallize and densify the film. The main advantages of this process include low temperature processing and low cost. Some other thin film deposition techniques that are used include liquid phase epitaxy (LPE), epitaxial growth by melting (EGM), evaporation methods, molecular beam epitaxy (MBE) and laser ablation.

4.6.5   Ceramic Polymer Composites

The drive for piezoelectric composites stems from the fact that desirable properties could not be obtained from single phase materials such as piezo-ceramics or piezo-polymers. For example, in an electromechanical transducer, the desire is to maximize the piezoelectric sensitivity, minimize the density to obtain good acoustic matching with water, and make the transducer mechanically flexible to conform to a curved surface. Neither a ceramic nor a polymer satisfies these requirements. The requirement can be optimized by combining the most useful properties of the two phases which do not ordinarily appear together.

Piezoelectric composites are made up of an active ceramic phase embedded in a passive polymer.

4.6.6   Ferroelasticity in crystals

The phenomenon of ferroelasticity was first clearly recognized by K. Aizu.(J. Phys. Soc. Japan 27, 1969) who showed that the existence of ferroelasticity could be predicted from a knowledge of the crystal symmetry of the ferroelastic phase and the crystal symmetry of a prototypic paraelastic phase, generally the phase occurring above the Curie temperature. Similarly the existence of ferroelectric properties can be correlated with crystal symmetry. Both ferroelectric and ferroelastic properties can exist simultaneously in the same crystal and in certain instances can be coupled, *i.e.*, the ferroelastic and ferroelectric domains are coextensive (Fig 3.27) and the crystal can be "switched" by either mechanical stress or electrical stress. Moreover, both ferroelectric and ferroelastic properties disappear at a single Curie temperature.

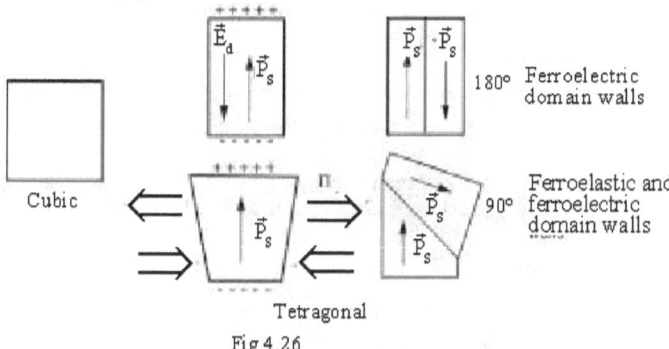

Fig 4.26

. The transition is accompanied by critical behaviour of the elastic compliance (the effective susceptibility for the strain). Ferroelastic materials are defined by having switchable domains, or twins, which may be switched on application of an external field: stress. Such domain microstructures often result from phase transitions.

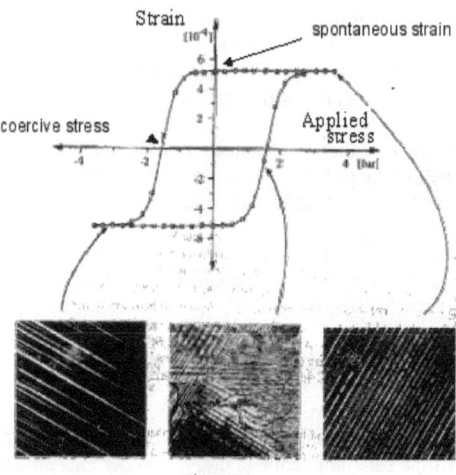

Fig 4.27

Ferroelasticity is the most common nonlinear effect in natural materials and plays a major role in the mineralogical behavior of the Earth's crust and mantle. It produces interfacial twin walls that act as sinks and sources for defects and that show localized effects such as superconducting twin boundaries and ferroelectricity, even when such effects do not exist in the bulk. The movement of twin walls under elastic forcing is creep-like, with some superimposed jerks due to pinning and unpinning by defects and jamming by other twin boundaries.

L. Brixner (1984) discovered that compositions of the formula $Pb_{3-x}Sr_x(PO_4)_2$, wherein x = 0.01 to 0.8 can be made by heating the stoichiometric amounts of ammonium phosphate, lead carbonate and strontium carbonate. The compositions can be grown in the form of single crystals which have ferroelastic properties (Aizu species 3mF2/m and which can be used in mechanically actuated optical switches and the like.

4.6.7 Antiferromagnetic Ferroelectrics

Ferroelectrics are commonly antiferromagnetic, rarely ferromagnetic. Alexandrite, or Cr-chrysoberyl (Cr_2BeO_4), is an example of a magneto-ferroelectric mineral (Newnham *et al.*, 1978).

4.7 FERROELECTRIC CERAMICS

Preparation of ceramics is like this:

i) The raw materials (metal oxides or metal carbonates) are first weighed according to the stoichiometric formula of the ferroelectric ceramic desired. The raw materials should be of high purity. The particle size of the powders must be in the submicron range for the solid phase reactions to occur by atomic diffusion.

ii) The powders are then mixed either mechanically or chemically. Mechanical mixing is usually done by either ball milling or attrition milling for a short time. Chemical mixing, on the other hand, is more homogeneous as it is done by precipitating the precursors in the same container.

iii) During the calcination step the solid phase reaction takes place between the constituents giving the ferroelectric phase.

For example for the calcination of PZT, the starting raw materials PbO, TiO_2 and ZrO_2 are mixed in the molar ratio of 2:1:1, pressed into lumps and then calcined in ambient air at 800 °C to obtain the perovskite phase. The calcining temperature is important as it influences the density and hence the electromechanical properties of the final product. The higher the calcining temperature, the higher the homogeneity and density of the final ceramic product. However, calcining PZT at T > 800 °C could lead to lead loss, resulting in a detrimental effect on the electrical properties. So proper calcination at the right temperature is necessary to obtain the best electrical and mechanical properties.

iv) After calcining, the lumps are ground by milling. The green bodies should have a certain minimum density before they can be sintered. The desired shape and a minimum green density can be provided by various techniques including powder compaction, slip-casting, and extrusion. The choice of the method depends on the type of powder used, particle size distribution, state of agglomeration, desired shape, and thickness of the part.

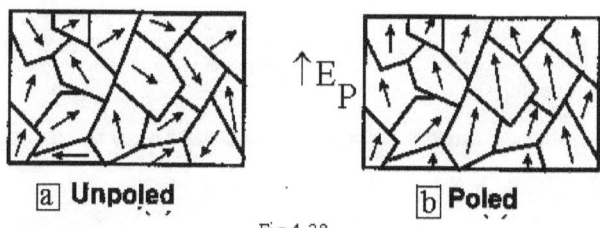

Fig 4.28

Fig 4.29 is the schematic of the poling process in piezoelectric ceramics: (a) In the absence of electric field the domains have random orientation of polarization; (b) the polarization within the domains aligns in the direction of the applied field

4.8. Ferroelectrics *versus* Ferromagnetics

Table 4.5

Basic similarities and differences between FE and FM	
Ferroelectrics (FE)	Ferroemagnetics (FM)
1) The electric dipoles are embedded in an environment whose α Polarizability is T dependent.	Magetic dipoles due to spin of atomic electrons are not environment dependent.
2) P_S cannot be observed at static conditions due to twinning or free surface twinning or free surface charges.	M_S can be measured directly.
3) $\vec{P}_S$ is due to the electric dipole moment $\iiint \rho(r)\, r\, d\tau$ of nuclear and electronic charge distribution $\rho(r)$ of dielectic crystals	M_S is due to exchange forces and spins of electrons os 3d shell, quantum principles and dynamics.
4) Reversibility of $\vec{P}_S$ is a consequence of the fact that the polar structure, of a ferroelectric, is a slightly distorted one of the non-polar structure.	
5) $\vec{P}_S$ reveral by external $\vec{E}$ is not always possible.	$\vec{M}_S$ is always reversible with by applying $\vec{H}$.
6) $\vec{P}_S$ decreases with increase in T.	$\vec{M}_S$ is T independent. decreases with increase in T.
7) $\vec{P}_S = \dfrac{10^{-29}}{3}\,Cm^2 = 1D$	$\vec{M}_S$ in Bohr magneton, $\mu_B = 9.27\,x10^{-24}\,Am^2$.
8) Domain structure ? Yes.	Domain structure ? Yes.
9) Domain walls are few A°thick.	Domain walls are 10^2 - 10^4 A° thick.
10) D-E hysteresis at $T < T_C$.	E-H hysteresis at $T < T_C$.
11) E_C is T and ω dependent.	H_C is T and ω independent.
12) FE wall between adjacent domains, $\theta_{FE} = 90°$. Wall thickness $\approx$ 1 lattice constant.	Bloch wall between adjacent domains. $\theta_{Bloch} \neq 90°$. In the case of FM iron, wall thickness $\approx$ 300 lattice constants.
13) See Fig below	

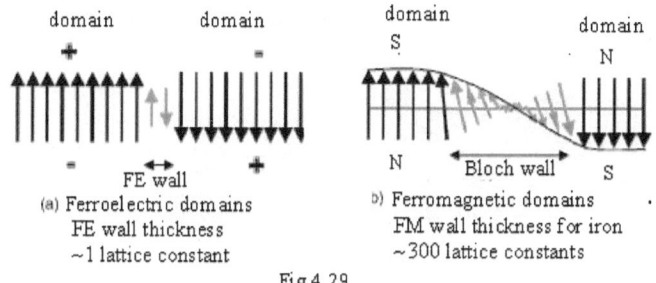

(a) Ferroelectric domains
FE wall thickness
~1 lattice constant

b) Ferromagnetic domains
FM wall thickness for iron
~300 lattice constants

Fig 4.29

4.9 APPLICATIONS OF FERROELECTRIC SUBSTANCES.

Ferroelectric crystals have the following properties due to different polarizing states after the transition at the T_C. Some ions shift spontaneously to new equilibrium positions.

i) An external electric field can shift these ions. So
    a) Polarizability of ions can have very high value in a certain direction near the T_C,
    b) Dielectric constant can reach large value near the T_C,
    c) Refractive index can be very large.

ii) Both the spontaneous deformation and the spontaneous polarization are shown by a ferroelectric crystal at the T_C. These two are mutually dependent directly. This coupling between deformation and polarization is an electro-mechanical one. Ferroelectric crystals exhibit piezoelectricity. Only those piezo-electric coefficients which are unusually large at $T > T_C$, corresponding to e.m coupling below T_C.

iii) Electro-optic and elasto-optic effects can be very large near the T_C due to the coupling between polarization and spontaneous deformation in the crystal, and because the high polarizability can be strongly influence by E or mechanical stress.

iv) Blow the T_C, a ferroelectric is a strong pyroelectric.

These unusually strong characteristics are desirable properties for the development of devices using ferroelectric materials. Given below are a number of devices making use of FE properties:

1) Ferroelctric capacitors for electronic circuits.
    $(Ba_{1-x}, Sr_{x-y}, Ca_y)(Ti_{i-m}, Nb_{m-n}, Sn_n)O_3$ ceramic compounds are found to be popular, as a selective combination of the elements is known to reduce the $T_C = 120°C$ for $BaTiO_3$ to get lowered by ~ 20°C. A multi-layer capacitor (MLC) is

schematically as in Fig 4.30. Each dielectric layer is as thin as $< 20\mu$. Relaxors find more suitable.

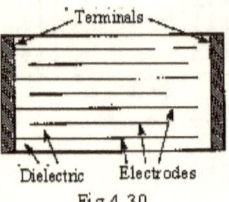

Fig 4.30

2) Phonograph pick-up heads in using Rochelle salt during 1920s and after.
3) Biomorphs and multimorphs (Displacement transducers)
4) Ultrasonic transducers,
5) Vibrations transducers,
6) Resonant transformers,
7) Data storage devices
8) Non-linear optical devices
9) Second harmonic generator (SHG),
10) Pyroelectric bolometers, (Pyroelectric Detectors) : Single crystals of triglycine sulphate (TGS), $LiTaO_3$, and $(Sr,Ba)Nb_2O_6$ are widely used for heat sensing applications.
11) Pyroelectric vidicon,
12) Optical memories (page composers, for example)
13) Ferroelectric display, Thin Film Optical Memory Displays: The material requirements for thin film optical memory and displays include large electro-optic coefficients and/or strong photo-sensitivities for the film.
14) Ferroelectric transistor,
15) Ferroelectric thin films : Some of the most important electronic applications of ferroelectric thin films include nonvolatile memories, thin films capacitors, pyroelectric sensors, and surface acoustic wave (SAW) substrates. The electro-optic devices being studied include optical waveguides and optical memories and displays.
16) Ferroelectric Memories: The ability of ferroelectric materials to switch robustly from one polarization state to another forms the basis of a new thin film technology for storing data. Non-volatile FRAM (Ferroelectric RAM). They are almost replacing semiconductor RAMs.
17) Ferroelectric Thin Film Waveguides. A great deal of work has been done on making ferroelectric thin film waveguides from $LiNbO_3$ and $Li(Nb,Ta)O_3$ using LPE, EGM, and MBE methods. PZT and PLZT thin films are even better candidates for optical waveguide applications because of their large electro-optic coefficients.
18) Gas Ignitors
19) Accelerometers
20) Impact Printer Head: Dot matrix impact printers driven by multilayer piezoelectric ceramic actuators have been successfully produced on a large commercial scale.

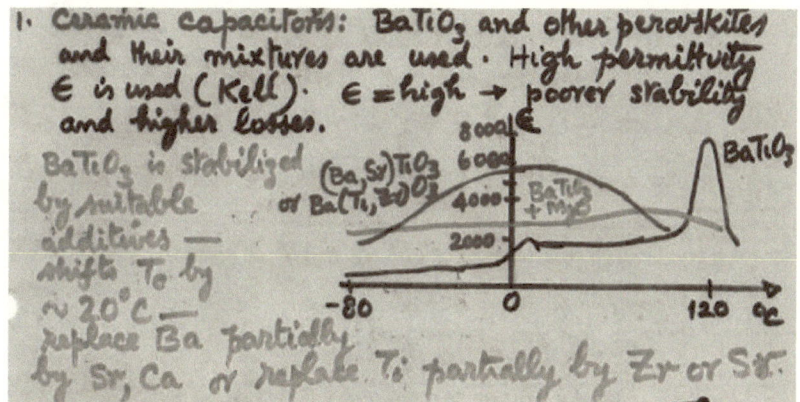

1. Ceramic capacitors: $BaTiO_3$ and other perovskites and their mixtures are used. High permittivity ε is used (Kell). ε = high → poorer stability and higher losses.

$BaTiO_3$ is stabilized by suitable additives — shifts T_0 by ~ 20°C — replace Ba partially by Sr, Ca or replace Ti partially by Zr or Sn.

REVIEW QUESTIONS

R.Q 4.1   Given a pyroelectric transducer containing a tourmaline plate with $\lambda = 1.33$ CGSE and permittivity $\varepsilon_3 = 7.5 \perp$ to (0001), 1 mm thick, and cut parallel to the (0001) plane, is connected to a milli-voltmeter. It has a sensitivity of $10^{-3} V/\text{div}$. Determine the minimum temperature increment that can be measured? (Ans: $\Delta T \approx 10^{-5} K$).

&&*&*&*&*&*&*&*&*&

Chapter 5

THEORIES OF FERROELECTRICITY
IN CRYSTALS

Chapter 5

THEORIES OF FERROELECTRICITY

IN CRYSTALS

A common sense interpretation of the facts suggests that a superintellect has monkeyed

with physics, as well as with chemistry and biology, and that there are no blind forces

worth speaking about in nature. — Fred Hoyle

5.1 FERROELECTRIC PHASE TRANSITIONS: A THEORETICAL TREATMENT

To explain the behaviour of ferroelectric crystals and to answer the phenomenon of ferroelectricity in crystals various attempts have been made to formulate a theory. From literature, one may classify three broad theories to explain satisfactorily ferroelectric phase transformations in various crystals. They are

a) Macroscopic Theory,
b) Model Theories, and
c) Microscopic Theory.

Of these I shall deal with the Macroscopic theory in detail , only mention the Model theory , and the microscopic theory in detail.

5.2 MACROSCOPIC (OR PHENOMENOLOGICAL OR THERMODYNAMIC) THEORY

Devonshire in 1949 was the first to develop the thermodynamic theory, and he revised it in 1954. The macroscopic (*i.e.*, equilibrium) properties of a crystal are such as P_S , χ , E, and their temperature dependent properties. The thermodynamic theory attempts to relate these properties tone another. The discussion is based on the so-called fundamental relation of Gibbs, which can be the general Free energy (*viz.*, Gibbs Free energy, G(F)) for stress-free ferroelectric crystals with the applied field, G(F) .applied parallel to the polar axis.

Landau's expansion is

$$G(P,T) = G_0 + \frac{1}{2}AP^2 + \frac{1}{4}\xi P^4 + \frac{1}{6}\varsigma P^6$$

with higher order terms neglected. The coefficients are at fixed stress and temperature dependent. In the Free energy expansion all odd powers of P are zero, since +P and –P are equivalent. Thus the stability properties of the crystal are from the quadratic terms in G.

The electric field

$$E(P) = \frac{dG}{dP}\Big|_{\text{at } P = P_s}$$

The curve representing G versus P depends strongly on the number of terms in the expansion of G(P), the signs and magnitudes of the three coefficients A, ξ, ς. This is schematically shown in Fig.

Thermodynamically, phase transitions in solids can be classified under two categories, *viz.*, First- and Second- order transformations. It is desirable to deal with the two one by one.

5.3.1   Second-Order Phase Transitions:

In FE crystals $T > T_C$ it is in PE phase whereas $T < T_C$ the crystal is in the FE phase.

a)   <u>PE phase</u>: If $A > 0$, the curve has a minimum at $P = 0$. The crystal is stable over the entire range of P, for $T > T_C$.

$$\chi^{-1} = \frac{d^2G}{dP^2}\Big|_{E=0} = A.$$

Since $\varepsilon \propto \chi$,

$$\varepsilon^{-1} = A$$

b)   <u>FE Phase</u>:   If $A < 0$, then G(P,T) has more than one solution as will be seen below.

$$E = \frac{dG}{dP_S} = 0 \text{, in the FE phase.}$$

$$= AP_S + \xi P_S^3 + \varsigma P_S^5 \neq 0$$

The coefficient $\xi > 0$ does not add anything to G by the term coefficient ς. Therefore giving $\varsigma = 0$, $(A + \xi P_S^2)P_S = 0$. This means either $P_S = 0$, or $P_S^2 = \frac{-A}{\xi}$.

Letting the temperature dependence of A is expressed as

$$A = k(T - T_0),$$

where T_0 is a specific temperature at which $A = 0$.

For $T \geq T_0$, the only real root of the equation $(A + \xi P_S^2)P_S = 0$ is at $P_S = 0$, because A and ξ are both positive.

This means $\underline{T_C = T_0}$.

For $T < T_0$, $P_S^2 = \frac{-A}{\xi}$,

This solution is physically significant only if A and ξ are opposite in signs. As ξ is assumed to be positive, A < 0, for temperature T< T_0. Hence a double minimum potential curve. (Fig. 5.1).

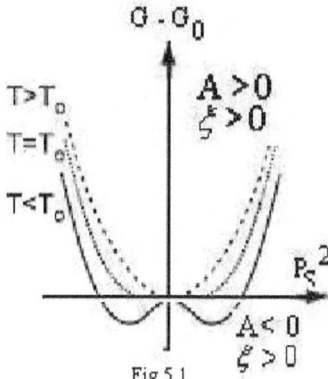

Fig 5.1

$$P_S^2 = -\frac{k}{\xi}(T - T_0) = K(T_0 - T),$$

where K is a constant.

This means $|P_S| = K_0(1 - \frac{T}{T_0})^{1/2}$, and

At $T = T_0 = T_C$, $P_S = 0$.

This relation between P_S and T is shown in Fig.5.2.

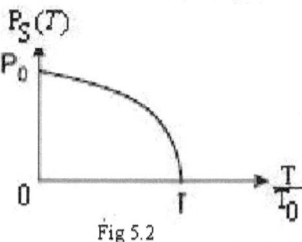

Fig 5.2

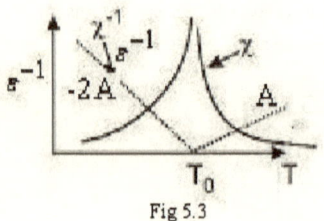

Fig 5.3

$$\varepsilon^{-1} = \frac{dG}{dP}\bigg|_{E=0} = \frac{d^2G}{dP^2}\bigg|_{E=0}.$$

For $T > T_C$, $P_S = 0$;

$\therefore \quad \varepsilon^{-1} = A$, reciprocal susceptibility.

For $T < T_0$, $P_S^2 = \frac{-A}{\xi}$

$$\varepsilon^{-1} = \frac{d}{dP_S}(A P_S + \xi P_S^3) = (A + 3\xi P_S^2)$$

$$= (A + 3\xi \frac{-A}{\xi}) = -2A$$

The variation of ε^{-1} versus T is reproduced in Fig.5.3

This kind of variation in P_S is characteristic of II order phase transitions. Examples of ferroelectric crystals have second- order phase change are

1) Triglycine sulphate, ($T_0 = T_C = 49°C$).
2) KDP ($T_0 < T_C = 123$ K),
3) Rubidium bisulphate ($T_0 = T_C = -15°C$).

5.3.2   First-order Phase Transformation:

It will be seen below that as $\xi < 0$, the term congaing ς should not be neglected, as otherwise $G_0 \to \infty$.

$$\therefore \quad G(P,T) = G_0 + \frac{1}{2} AP^2 + \frac{1}{4} \xi P^4 + \frac{1}{6} \varsigma P^6$$

i) PE phase: If $A > 0$, the curve has a minimum at $P = 0$. The crystal is stable over the entire range of P, for $T > T_c$. The reciprocal susceptibility

$$\chi^{-1} = \frac{d^2G}{dP^2}\Big|_{E=0} = 0, \text{ for } T > T_c.$$

$$A = k(T - T_0)$$

But $\chi = (\varepsilon - 1) k_0$

$$\therefore \varepsilon = \frac{C}{T=T_0}, \text{ which is the } \underline{\text{Curie-Weiss Law}}.$$

Thus the Curie-Weiss Law is obeyed in the PE phase. This is also the case with the second-order phase changes. Therefore, the Curie-Weiss Law is obeyed a ferroelectric crystal in the PE phase, irrespective of the order of phase change to the FE phase.

ii) <u>FE phase</u>:

$$E = \frac{dG}{dP_S} = AP_S + \xi P_S^3 + \varsigma P_S^5 = 0.$$

This means either $P_S = 0$, or $(A + \xi P_S^2 + \varsigma P_S^4) = 0$.

At $T = T_C$, $P_S = 0$.

$$G(PE) = G_0 = G(FE)$$

i.e., $(\frac{1}{2}AP^2 + \frac{1}{4}\xi P^4 + \frac{1}{6}\varsigma P^6) = 0$, at T_C.

Or, $P_S^2 = (-\frac{3}{4}\frac{\xi}{\varsigma})$, at T_C.

$$\therefore (A + \xi(-\frac{3}{4}\frac{\xi}{\varsigma}) + \varsigma(-\frac{3}{4}\frac{\xi}{\varsigma})^2) = 0$$

i.e., $\underline{A\varsigma = \frac{3}{16}\xi^2}$, at T_C.

This relation shows that $\xi < 0$.

$$\therefore P_S^2 = (-\frac{3}{4}\frac{\xi}{\varsigma})\Big|_{T=T_C}$$

$$\therefore |P_S| = 2[\frac{k}{\varsigma}(T_C - T_0)]^{1/2}$$

Therefore a sudden drop in the value of P_S to $P_S = 0$ occurs at T_C, and $\underline{T_C > T_0}$.

It can be shown that

$$\varepsilon = k'(T_C - T_0), \text{ at } T_C.$$

ε^{-1} versus T curve is as shown in Fig.5.5.

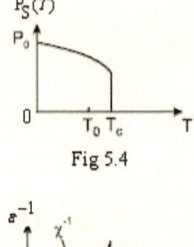

Fig 5.4

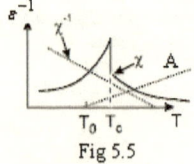

Fig 5.5

Specific heat with temperature for I order phase change is as shown in Fig.5.6.

$T \rightarrow$

Fig 5.6 $KH_2(PO_4)$

Second-order phase change in specific heat is displayed in Fig 5.7.

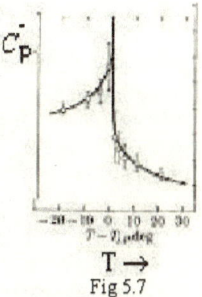

$T \rightarrow$

Fig 5.7

Table 5.1

P_S $(\mu C.cm^{-2})$		ε	C $(T - T_C)$	C(expt) $(T > T_C)$
1) TGS	2.3	43	4450	3250
2) $BaTiO_3$	26.0	160	2.03×10^5	1.73×10^5
3) $LiTaO_3$	71	46	1.2×10^5	1.6×10^5
4) $Sr_{0.5}Ba_{0.5}Nb_2O_6$	27	400	1.72×10^5	3.5×10^5

It is evident from the treatment above that the basic assumption that A is temperature dependent, having the form $A = k(T - T_0)$, though the theory is silent on this aspect. Otherwise the phenomenological theory explains satisfactorily of ferroelectric properties both above and below the Curie point. So the theory is particularly useful in situations when certain quantities cannot be measured directly, example the change in Entropy and specific heat at the T_C, a knowledge which is indispensable to know, the nature of the transition. For Lithium selenite, one of such case, the values are given below:

	P_S $(\mu C.cm^{-2})$	ε	C(expt) $(T > T_C)$ K	T_C K	T_0 K
1) TGS	2.3	43	3250	40.4	40.4
2) $BaTiO_3$	26.0	160	1.73×10^5	393	370
3) $KNbO_3$			2.7×10^5	683	623
4) $PbTiO_3$			1.1×10^5	763	693
5) $LiTaO_3$	71	46	1.6×10^5		
6) $Sr_{0.5}Ba_{0.5}Nb_2O_6$	27	400	3.5×10^5		
7) KDP			0.3×10^4	123	123
8) SbSI			2.95×10^5	292.2	289
9) $(NH_4)HSO_4$			461	260	260

5.3.2   MODEL THEORIES:

In the Model theory, the existence of spontaneous polarization P_S requires, in general, a physical model the dipole moments of the different unit cells are oriented along a common direction in the crystal. A rigorous treatment of the realistic theoretical model is extremely difficult and models that permit a theoretical approach are very far from reality. Slater, Mason, Matthias and others have developed models for KDP, $BaTiO_3$, Rochelle salt, *etc.* Further this requires a model for each crystal.

5.3.3   MICROSCOPIC THEORY

A seemingly fundamental theory of ferroelectricity in crystals was proposed by Cochran based on the dynamical theory of crystalline lattices. Theoretically thermo dynamical properties of a substance are derived from the Hamiltonian, which contains the detailed laws of interaction between its various microscopic degrees of freedom; lattice phonons contribute the thermodynamic properties of all solids. I have for the understanding of the present section, given an account of the preliminaries on the lattice dynamics in Chapter 2B.

A crystal can be stable against small deformations only if all the normal modes have real frequencies in its lattice (Born & Huang, 1969). In 1960, Cochran and Anderson independently demonstrated that in some crystals (which are ionic or partly ionic)a long wavelength transverse optic (TO) phonon branch of the lattice may become imaginary in the harmonic approximation, resulting in an instability of the lattice with respect to this normal mode, and causing a ferroelectric phase transition. This normal mode is referred to as "SOFT MODE" or 'Cochran mode'. The soft mode concept was first suggested by Dr. CV Raman (1940), in his work to explain the $\alpha \leftrightarrow \beta$ quartz transition. The theory behind this argument is that in ionic crystals, these excitations are accompanied by polarization oscillations which create local fields interacting with the ions through long-range Coulombic forces. For a given normal mode, these long-range forces have the same magnitude, but of opposite sign than the short-range forces, the crystal becoming unstable against this mode. Above the T_C anharmonic interactions provide stability to the system, making the observable quasi-harmonic frequency real and positive, but temperature dependent, *i.e.,* "soft". The dynamical matrix of the crystal (derivable from the Hamiltonian) is meaningful only if its characteristic determinants vanish. The characteristic determinants form the detailed dispersion relations $\omega = \omega(\vec{q})$.

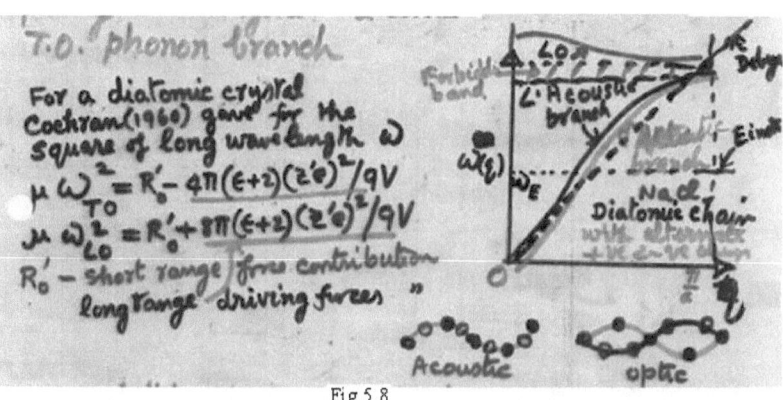

Fig 5.8

The relation between the static dielectric constant $\varepsilon(0)$ and the dynamical properties of the dielectric system is given by the "Kramer's-Kronig" dispersion relation

$$\varepsilon(0) - \varepsilon(\infty) = \frac{2}{\pi} \int_0^\infty \frac{\varepsilon''(\omega, \vec{q}, T)}{\varepsilon'} \, d\omega'$$

where ε' and ε'' are real and imaginary parts of $\varepsilon(\omega, \vec{q}, T)$.

This expression leads to the "Lyddane-Sachs-Teller" (LST) relation

$$\frac{\varepsilon(0)}{\varepsilon(\infty)} = \prod_{i=1}^{N-1} \frac{\omega_i^2 \big|_{LO}}{\omega_i^2 \big|_{TO}}$$

LO and TO stand for longitudinal optic and transverse optic branches. The product is for all normal modes of the optic branch. Being harmonic in origin, ω_{LO} and $\varepsilon(0)$ are temperature independent. Anharmonicity in phonons near T_C leads ω_i to be "anomalous" dependence of ω_{TO}.

$$\omega(\vec{q})_{TO}^2 \big|_{anom} = \gamma(T - T_0)$$

where T_0 has the usual significance. Cochran showed that the ferroelectric transition is accompanied by 'softening' of the anomalous TO mode at the Brillouin zone centre. i.e., at $\vec{q} = 0$. Thus the LST relation leads to

$$\varepsilon(0) = \varepsilon(\infty) \left. \frac{\omega^2 \big|_{LO}}{\gamma(T-T_0)} \right|_{\bar{q}=0} = \frac{C}{(T-T_0)}$$

which is similar to the Curie-Weiss Law. More than 90% of the $\varepsilon(0)$ is accounted by the soft mode. In the case of anti-ferroelectric transition, the softening of the anomalous mode occurs at the Brillouin zone (BZ) boundary, $viz.$, $\bar{q} = \frac{2\pi}{a}$. This means the LST relation is not valid, and there is no dielectric anomaly is usually observed. The mode may become soft somewhere.

$0 < \bar{q} < \frac{2\pi}{a}$, holds good for some ferroelectric transitions, and such materials are called "improper ferroelectrics".

In this case one may consider two distortive structural changes: 'Ferro-distortive' and 'antiferro-distortive' transitions.

$\omega_{TO}\big|_{anom}$ can be measured by

    1)   Raman / Brillouin scattering of light.
    2)   Inelastic neutron scattering, where optical phonons take part in the scattering.

Example

PbZrO$_3$ ceramic sample with $\ell = 51.04$ mm and breadth 6.20 mm, at room temperature, has thickness 2.02 mm and $\rho = 7.7$ $gm.cm^{-3}$. $f_s = 29.10$ kHz, $f_p = 30.13$ kHz

$v = 2 f_s \ell = 298.800$ $cm.s^{-1}$.

$\rho s_{11} = 1/v^2$; gives $s_{11} = 14.5x10^{-12} m^2 N^{-1}$.

$k_{31}^2 = \frac{\pi}{4} \frac{\Delta f}{f_s} (= \frac{d_{31}^2}{s_{11}\varepsilon_{33}})$; $\kappa = 1400 = \frac{\varepsilon'}{\varepsilon_o}$;

$k_{31} = \frac{\pi}{2} \sqrt{2.54x10^{-2}} \approx 0.301$;

$d_{31} = 0.301 / \sqrt{(14.5x10^{-12}) \frac{1400}{36\pi 10^9}} = 130x10^{-22} MKS$.

5.3.4   Ferro-distortive Transitions.

In $this$ $case$ $Z(PE) = Z(FE)$. This transition is connected with the condensation of the Cochran mode (which may be either progressive or diffusive) at the BZ centre. BaTiO$_3$, KDP,

$NaNO_2$, *etc.*, belong to this type of ferroelectric, known as 'proper ferroelectrics'. The order parameter here is the P_S. AFE of this type is yet to be found.

5.3.5   Anti-ferroelectric Distortive Transitions:

$Z(FE) = n. \ Z(PE)$ is for this case. AFE distorted transition is relate to the condensation of a low frequency TO mode, with a non-zero wave vector, which may be either propagating or diffusive at the BZ boundary. Improper FE like Gadolinium Molybdate is a typical example of this type of transition. The order parameter in 'frozen -in' Zone boundary mode. All known AFE s like ADP, $PbZrO_3$ belong to this class. The order parameter is the sub-lattice polarization.

A universal theory of ferroelectrics is yet to be formed.

5.3.6   Curie Principle and Ferroelectric Transitions:

Owing to the complexity of Landau.s treatment, seen above, one can consider the symmetry group of the crystal. Most of the ferroelectrics lower their crystal point group symmetry as a result of the phase change at the T_C as temperature is lowered to the FE phase. $\vec{P}_S$ aligns / appears in the unit cell. The $\vec{P}_S$ may have any direction. If any of the two parameters, *viz.*, PE symmetry, FE symmetry and the number of orientations of $\vec{P}_S$, is known the third can be determined by applying the Curie Principle. I.S. Zheludev *et al.* have studied the FE states of several crystals by assuming the orientations of $\vec{P}_S$.

According to Neumann's Principle (1885), 'Symmetry elements of any physical property must include the symmetry elements of the point group of the crystal'.

Curie's principle in itself may be formulated in the physics of crystals as follows: the symmetry group of a crystal under an external influence ($\hat{K}$) is given by the greatest common subgroup of the symmetry group of the crystal without the influence ($\hat{K}_0$) and of the symmetry group of the external influence ($\hat{G}$) considering also the mutual position of the symmetry elements of these groups:

$$\boxed{\hat{K} = \hat{K}_0 \cap \hat{G}}$$

Curie's principle expressed in other words: *a crystal under an external influence will exhibit only those symmetry elements that are common to the crystal without the influence and the influence without the crystal*

Worked out Problem 5.1

$\hat{G}$ = Symmetry group of the space group of the crystal in its disordered state, G, in the FE ordered group, $\hat{H}$.

R = # orientations of Ps, N= # of symmetry elements in the Group,

$$\boxed{\hat{G} \cap \hat{H}},$$

$$m3m \cap \uparrow \vec{E} \xrightarrow[\infty m \parallel [001]]{} 4mm, \ r = \frac{N_G}{N_H} = \frac{48}{8} = 6$$

The lost symmetry elements evidently appear as the number of Ps orientations.

$$m3m \cap \uparrow \vec{E} \xrightarrow[\infty m \parallel [1\overline{1}0]]{} mm, \ r = \frac{N_G}{N_H} = \frac{48}{4} = 12,$$

$$Amm2 \cap \uparrow \vec{E} \xleftarrow[\infty m \parallel [111]]{-90} R3m$$

$$m3m \cap \uparrow \vec{E} \xrightarrow[\infty m \parallel [hk0]]{} m, \ r = \frac{N_G}{N_H} = \frac{48}{6} = 8.$$

For TGS: $G=2/m$, $H=m$, $r = \dfrac{N_G}{N_H} = \dfrac{4}{2} = 2$.

5.4   Uni-axial Ferroelectrics:

   If the $\vec{P}_S$ is parallel to the unique non-polar axis of the crystal then the material is uni-axial ferroelectric. KDP is an example to this type.

5.4.1   Multi-axial Ferroelectrics:

   If the $\vec{P}_S$ js parallel to n equivalent unique non-polar axes giving 2n possible directions, as in the case of $BaTiO_3$.

&^&^%%^^&&^%&

Chapter 6

BEHAVIOUR PHYSICAL PROPERTIES DURING PHASE TRANSITIONS PN FERROELECTRIC CRYSTALS

Chapter 6

BEHAVIOUR OF PHYSICAL PROPERTIES

DURING PHASE TRANSIONS IN

FERROELECTRIC CRYSTALS

"The true purpose of education is to train the mind to think. For that

reason it is priceless".   Albert Einstein

6.1.   Introduction

"A ferroelectric is defined as a material that shows spontaneous and reversible dielectric polarization in a temperature range".

Changes in physical properties are known to accompany symmetry change when a crystal undergoes a phase transition. Such physical properties are: dielectric susceptibility, dielectric constant, piezoelectric coefficients, elastic modulii, lattice constants, coefficients of thermal expansion, specific heats, birefringence, electro-optic coefficients, isomer shift, quadrupole coupling coefficients, soft mode frequency, *etc.* A number of theoretical work has been available in the literature on this subject (for *e.g.*, Aizu, 1964, 1965; Devonshire, 1964).

6.2.1 Ferroelectric Domains:

Ferroelectric materials consist of domains that are spontaneously polarized in one of the symmetry-permitted directions of the material. Domains are regions of uniform polarization in a crystal, *i.e.*, region of uniform alignment of electric dipoles in a ferroelectric crystal, is a ferroelectric domain. Domains were predicted in a ferroelectric crystal before they were actually observed, because

(i)   The coercive field E_C is small in magnitude,

(ii)   Presence of some super structural lines in X-ray diffraction patterns of ferroelectrics,

(iii)   Presence of some re-polarization jumps.

In the presence of any external stress, a ferroelectric crystal breaks up into domains of different directions of spontaneous polarization. Thus instead of a single domain crystal one gets a multi-domain crystal. This process is also referred to as 'twinning', and one gets a twinned crystal. The Free energy of a twinned crystal is lower than the corresponding single domain crystal. Free energy of a crystal has to be minimum for it is to be stable. The boundary between adjacent domains is called "domain wall".

6.2.2   Geometry of Domains:

The directions of $\vec{P}_S$ vector in neighbouring domains make up definite angles, *i.e.*, geometry of the domains. In a crystal plate this depends upon the number of preferred direction that the crystal posses in its FE phase, and which are equivalent in the PE phase. While forming a twinned state, the crystal has to observe among other conditions, the condition of the electrical boundary neutrality, so that free energy is minimized. This means the projection of the $\vec{P}_S$ vector on the boundary from the side of one domain is

equal and opposite in sign to that of the other adjacent domain. In other words, the "head-to-tail" arrangement has to be observed. But in a real crystal, the 'head-to-head' and "tail-to-tail" arrangement had also been observed.- The charge on the wall had been compensated by conduction, and the wall was "locked" in place. The other important condition for a stable domain boundary is that a continuity of and matching of the crystal lattice at the wall. (Fig 6.1).

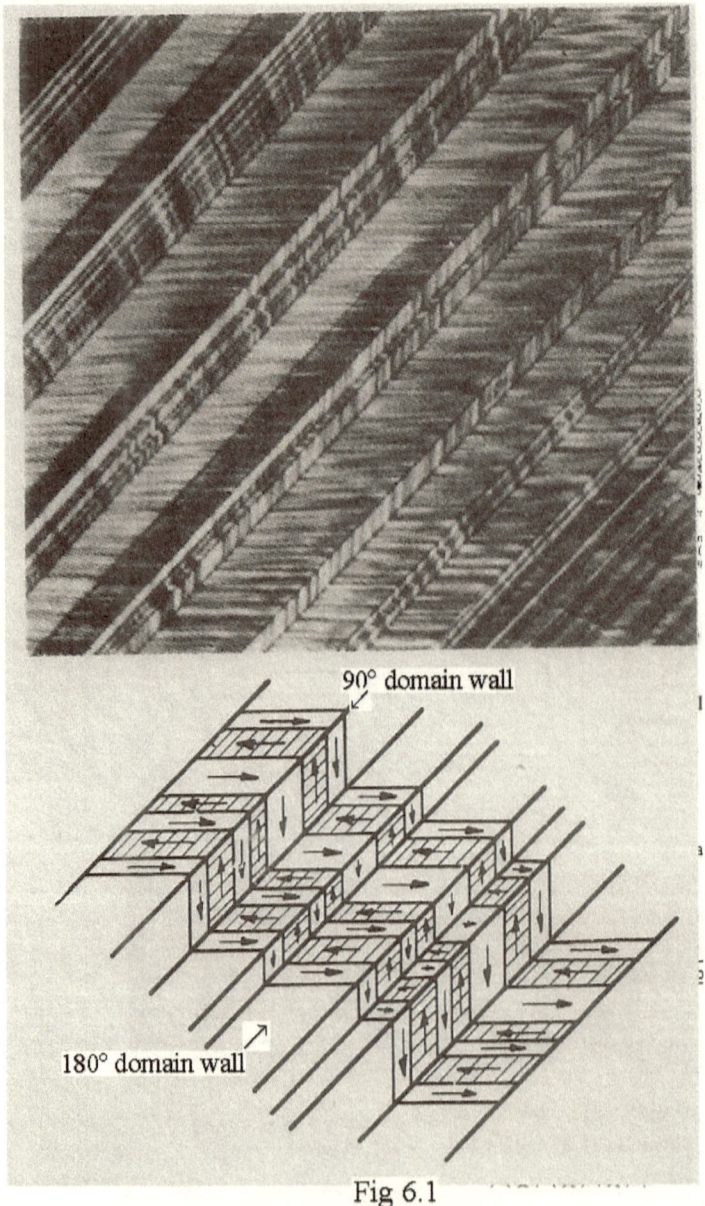

Fig 6.1

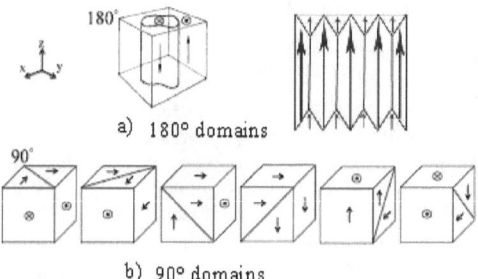

Fig 6.2   90° and 180° domains in baTiO$_3$

The directions of the $\vec{P}_S$ vector on either side of the domain wall make up definite angles, and, therefore, the domain walls can be distinguished as follows:

The Domain wall that separate two adjacent domains with $\vec{P}_S$ vector anti-parallel to each other is called '180°-wall'. If the neighbouring $\vec{P}_S$ vectors are mutually at right angles then it is the '90°-wall' which is separating them.(Fig 6.2). 60° (*i.e.*, 120°) wall, *etc.*, have also been formed in multi-axial ferroelectrics. The thickness of a 90°-wall is greater than that of the 180°-wall. In 180°-wall thicknesses the $\vec{P}_S$ can reverse within one unit cell dimensions.

6.2.2.1 Domain wall thickness, N, in BaTiO$_3$ is (Merz, 1954)

$$N \cong \sqrt{\left(\frac{2 \, x10^{-14}}{C_{33} Z_z^2 \, a^3} \right)} \approx 1 \text{ lattice constant}$$

where C_{33} =elastic constant, Z_z =spontaneous strain at room temperature, and

a = lattice constant.

6.2.2.2 Domain size

Equilibrium size of 90° domain, (Artl and Pertsev, 1991)

$$d = \sqrt{\left(\frac{128\pi g \sigma}{C_{11} S_s^2 \, a^3} \right)} \approx 0.8 \mu m$$

S_S =spontaneous strain during cubic-tetragonal phase change, $S_s = (1 - a/c)$, a and c are lattice constants, C_{11} =average longitudinal elastic constant, g =crystal size, and

σ = domain wall energy.

Each two adjacent domains are separated by compositionally homogeneous interface having a typical width of $1-10$ nm. These interfaces usually meet the condition of electrostatic compatibility, i.e. the normal component of the spontaneous polarization is continuous across the interface. They carry no net bound charge and are hence called neutral domain walls (NDW). Alternatively, charged domain walls (CDW) may exist, which carry bound charge due to a jump of this polarization component at the wall, violating the condition of electrostatic compatibility

6.3   Experimental work

Investigations on ferroelectric and anti-ferroelectric domains was used to solve, in general, three problems.

a) Study of domain geometry,
b) Study of changes in domains, and
c) Study of domain dynamics.

A number different technique has been developed for revealing domain structure. The are

(1) Using an optical polarizing microscope (Forsbergh in $BaTiO_3$, 1949),

(2) Chemical methods (Hooton & Merz in $BaTiO_3$, 1955), (Chemicals means). Rub & dip technique

Anti-parallel domains were made visible. Etch rate in $BaTiO_3$ with hydrochloric acid depends on the positive end of the dipoles etch rapidly whereas the negative end slowly.

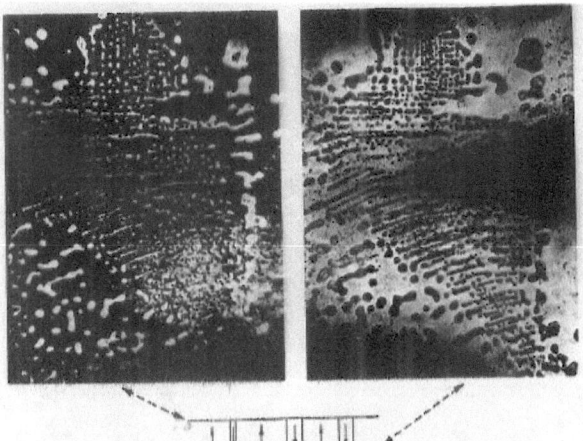

Fig 6.3 Etched top and bottom surfaces of a BaTiO$_3$ crystal, showing 180° domains.(Hooton &Merz)

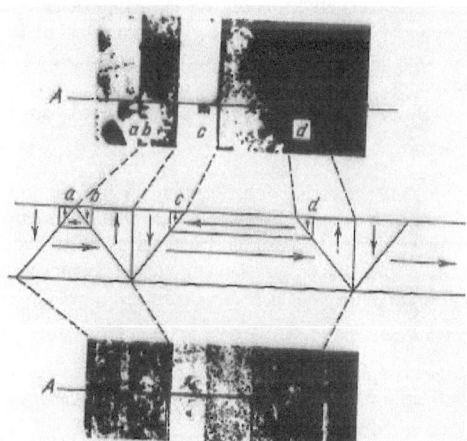

Fig 6.4 Upper and lower etched faces of a BaTiO$_3$ plate and orientation of domains in the interior

Water is found preferentially etch TGS when rubbed on a cloth of wet and dry portions.

(3) Colloidal suspensions in insulating organic liquids. (Pearson & Feldman, 1959) Yellow sulphur in hexane deposits on negative dipole ends.. Pb$_3$O$_4$ (lead oxide) deposits on positive ends. The FE materials studied initially were GASH, TGS, Rochelle salt.

(4) X-ray powder pattern technique (1959),

(5) Electron-mirror method (Zheludev *et al.*,1958),

(6) Transmission electron Microscopy (TEM) (1963),

Tanala *et al.* used hot ortho-phosphoric acid to thin $BaTiO_3$.

(7) Selective crystallization of anthraquinone on TGS (Kobzareva *et al.* in 1970.) Needle shaped crystals crystallize on different manners on domains of positive or negative ends. 30 ml of glacial acetic acid plus 20 drops of concentrated HCl and 15 drops of H_2O on TGS surface for 10-15 sec used for sublimation of anthraquinone at $100\,°C$.

Then optical microscopic examination is made.

(8) Solid dew method (Toshev, 1963), The FE crystal plate with $\vec{P}_S \perp$ to plate, polish with silk cloth, moistened with saturated mother solution, dry with filter paper, and immerse in liquid nitrogen and examine with a microscope.

(9) X-ray Topography (Caslovasky & Polcarova, 1964), Long's technique of transmission X-ray diffraction is used for method of anomalous dispersion of X-rays. Studied $180°$ domains in $BaTiO_3$..

(10)   Condensation of vapour of chemically inactive liquid (Fousek *et al.*, 1966), in TGS. Isobutyl alcohol vapour – existence strong EFG at the domain boundaries condensation of vapour occurs.

(11)   Electro-luminescent method (Zheludev *et al.*, 1961), TGS, Rochelle salt, GASH. Liquid paste of silicone oil and ZnS (electro-luminophor) plus glass plate of film containing stannic oxide conductor pressed to the surface, and the other surface of the crystal is given an electrode

(12)   Optical Topography (Bhide, 1961), using phase contrast microscopy in $BaTiO_3$.

(13)   Method using S.H. light ,

The polarizing microscope method is the only direct viewing method for domain studies, yielding complete information, with detailed results about the domain structure. The principle of this method is the property of optical birefringence of the crystal, between crossed Nichols. One can either study the domain structure under spontaneous birefringence or by induced strain in the crystal. Study of domain states has given the following results.

a) $\vec{\nabla}.\vec{P} = 0$, in the bulk of the crystal, $\vec{\nabla}.\vec{P} \neq 0$ only at the surface,

b) A 'fatigue' or 'decay' effect is shown. Repeated domain reversal cause decreased magnitude of $\vec{P}_S$, but increased value of E_c.

c) Domain dynamics study, known as switching of domains give

i)   What happens during the $\vec{P}_S$ reversal?

ii)   The speed at which switching takes place,

iii)   Dependence on domain switching, on temperature, electric field, crystal size, etc.

iv)   How domains form,

v)   At what speed the domains are propagated in the crystal, and

vi)   Barkhausen pulses in ferroelectrics connected with different types of switching

vii)   (Example, 90° -switching).

6,4   Study on Domains using Polarized light Microscope

6.4.1 Studies on BaTiO$_3$

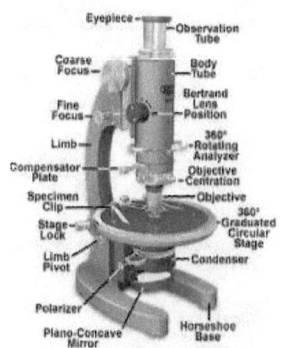

Monocular Polarized Light Microscope

Fig 6.5

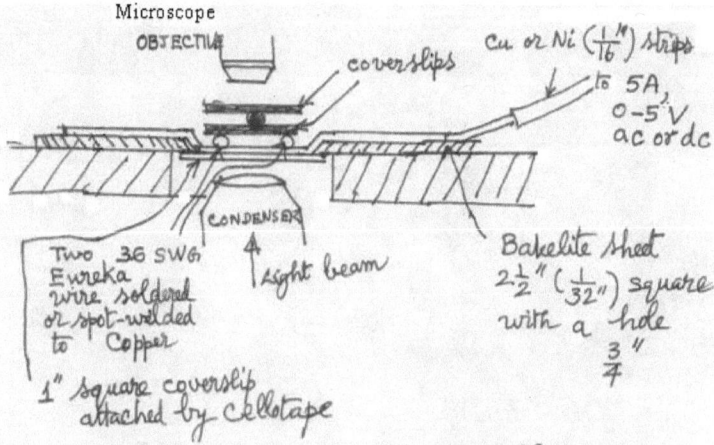

Fig 6.6 HOT STAGE FOR POLARIZING MICROSCOPE

Ferroelectric domains are observed in transparent single crystal thin sample plates, in $BaTiO_3$ and other crystals. A number of studies have been carried out. The author has seen the photographs of domain in $BaTiO_3$ studied by Prof. Zheludev (in 1963).

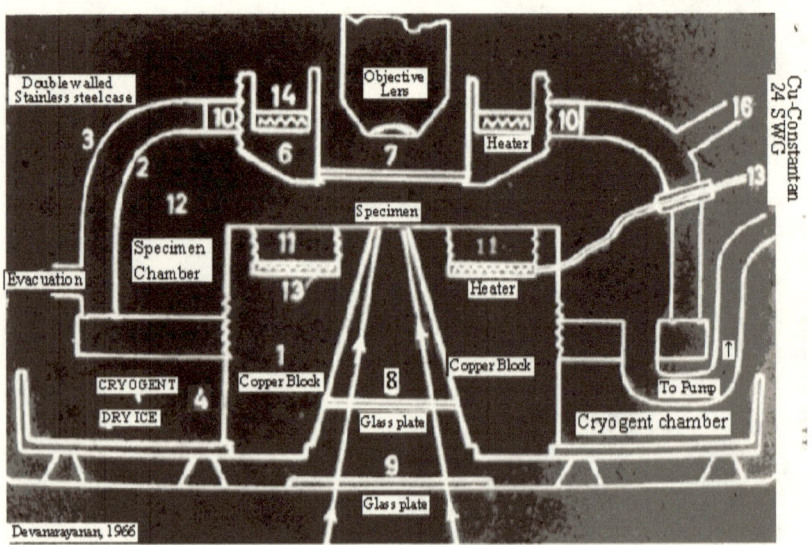

Fig 6.8 Cryostat (using Dry Ice) stage for Polarizing Microscope
(S. Devanarayanan, 1966

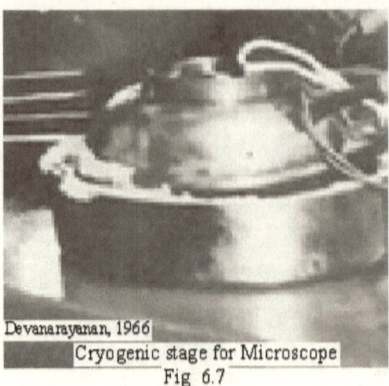

Cryogenic stage for Microscope
Fig 6.7

He had also showed (in a series of lectures on Ferroelectricity, at Physics Dept., Indian Institute of Science, Bangalore) a cinematic photo showing the domains when the crystal was undergoing phase transition from tetragonal to cubic phase at the Curie temperature.

The relation between the reversal of ferroelectric domain and the change of phase by nucleation and growth of other system is pointed out by Kay (1969).The relative simplicity of the ferroelectric system together with the large number of observation technique available make it suitable for testing nucleation theories and kinetics of domains in single crystals.

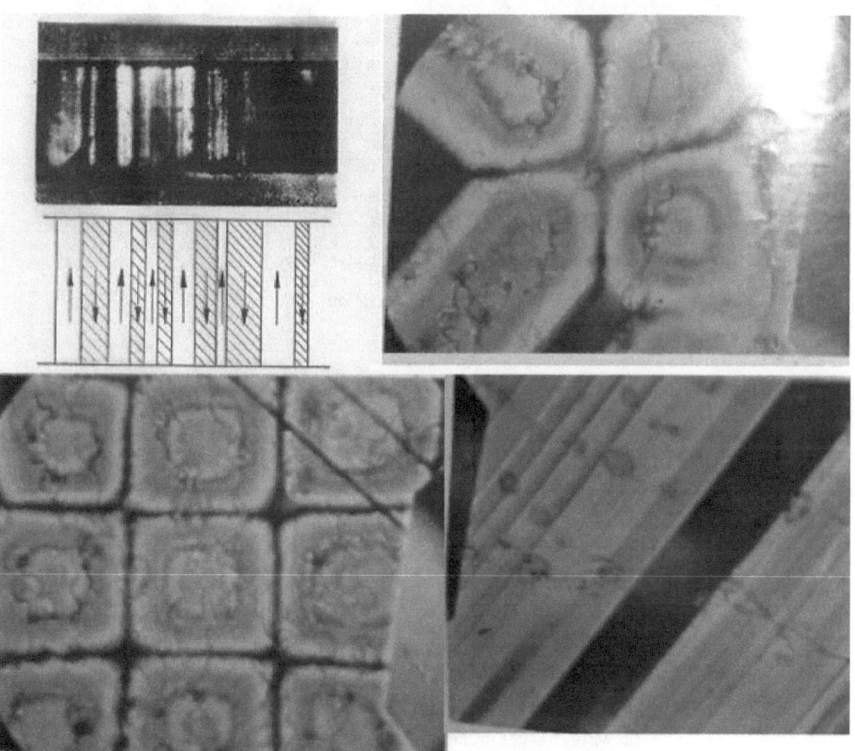

Fig 6.9 Edges of a domain crystal of BaTiO$_3$ showing antiparallel domains [Hooton & Merz]

The present author had designed and fabricated (in 1965) a hot-stage as well as a cryo-stage for the polarized light microscope (Carl-Zeiss) in the Department of Physics of the Indian Institute of Science. BaTiO$_3$ single crystalline samples, gifted by Zheludev were

182

used for examining domain structure. The domains were viewed, between crossed Nicols, while the sample was cooled from 130 °C through the $T_C = 120$ °C to the Tetragonal phase and at $T_C = 5$ °C to the orthorhombic phase (Fig 6.9 and Fig 6.10).

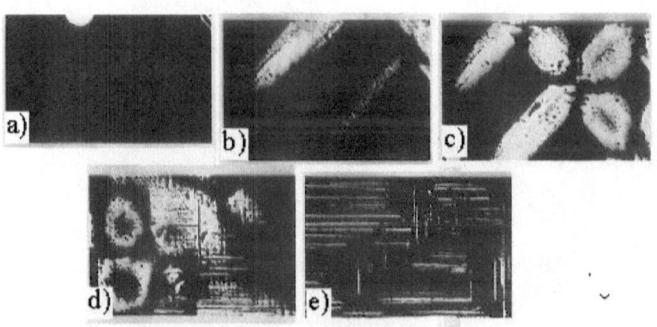

Fig 6.10 Domains in crystal of BaTiO$_3$
showing changes as the crystal passes thorugh
the transition temp. $T_C = 120$ °C
The crossed Nicols have ∥ and ⊥ polarizations

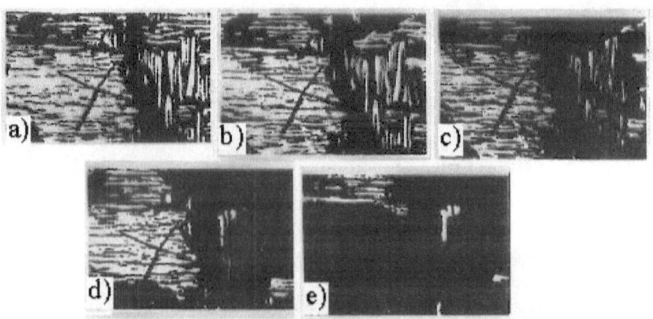

Fig 6.11 Domains in crystal of BaTiO$_3$
showing changes as the crystal passes thorugh
the transition temp. $T = 5$°C
The crossed Nicols have ∥ and ⊥ polarizations

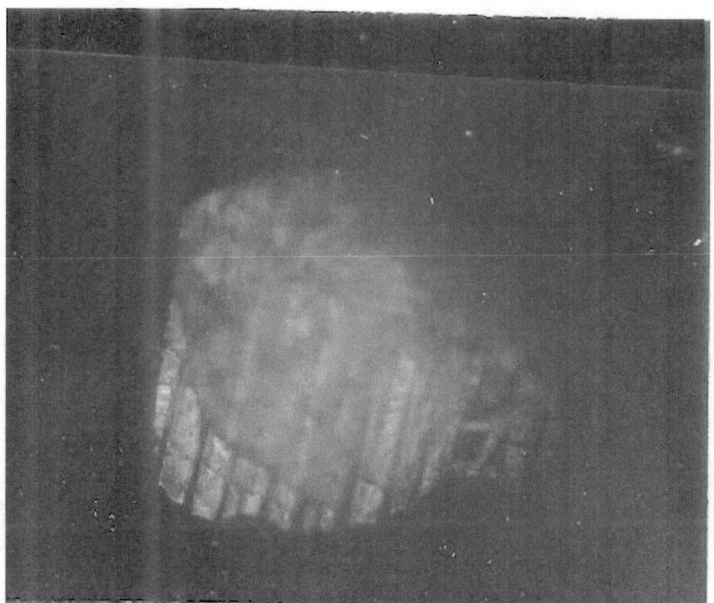

Fig 6.13 BaTiO₃ domains, Highspeed panchromatic, 24 Din,
ultraspeed 200 ASA 123 12Mar1964 (by the author)

The study of domains while the crystal undergoes phase transition at the T_C were
viewed with hot stage fabricated by the author (Devanarayanan, 1964, unpublished data)
with design of Steward (1952))with very slow rate of heating and while cooling at very
low rate. It was seen that there was an 'intermediate phase'(designated as Devan phase)
between the paraelectric and ferroelectric states. This intermediate phase in the crystal is
the possible domain wall motion as and when nucleated and grown, while cooling the
crystal, appeared as a fluid medium where a wave is moving (the wave appeared just
similar to the advancing sea wave at the time of spreading on sea coast).

6.4.2   Investigations on $KNbO_3$

$KNbO_3$ single crystalline flakes presented by IS Zheludev were used for
observing domains with a polarized light microscope (Devanarayanan, unpublished data,
1964)

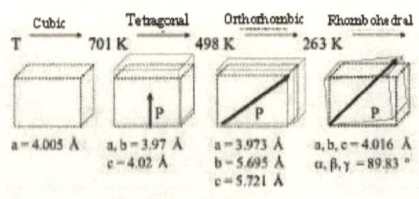

Fig 6.14 KNbO₃

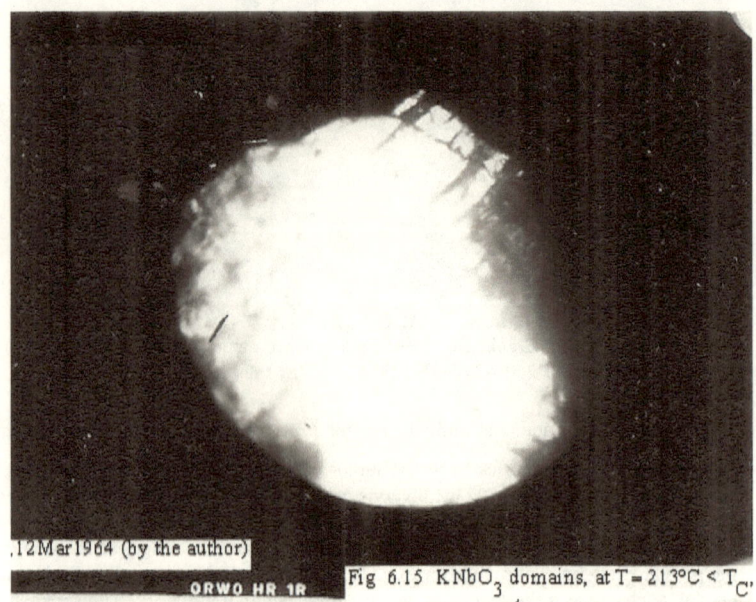

.12Mar1964 (by the author)

ORWO HR 1R

Fig 6.15 KNbO₃ domains, at T = 213°C < T_C'

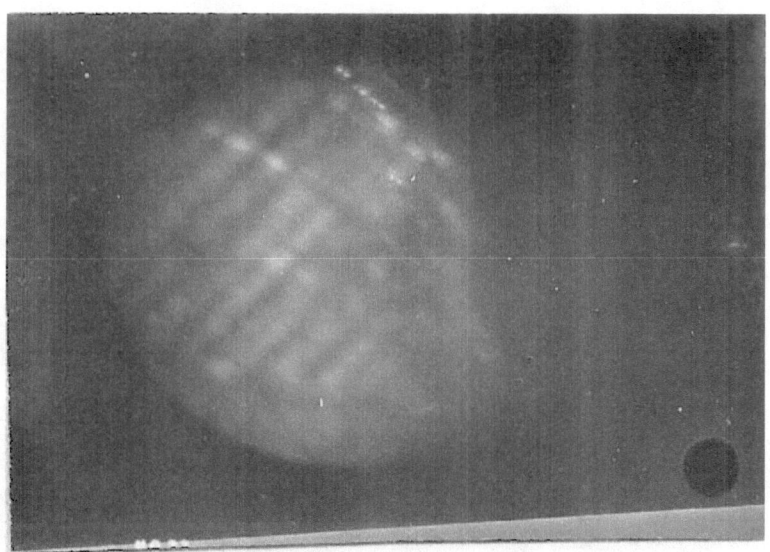

Fig 6.16 KNbO$_3$ domains

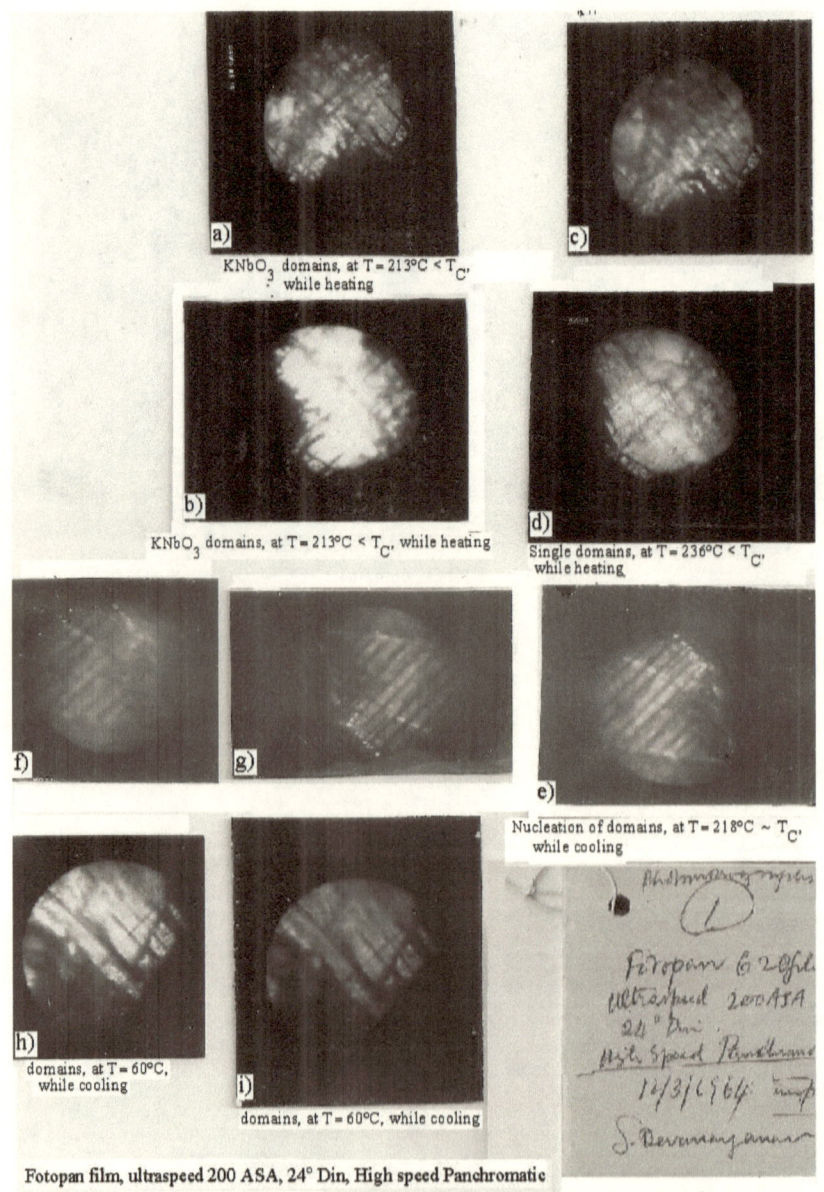

Fotopan film, ultraspeed 200 ASA, 24° Din, High speed Panchromatic

Fig 6.17 KNbO$_3$ domains, 12Mar1964 (by the author)

6.4.3 The results of domain dynamics study

Studies by Little, Cameron, Peacock, Miller and others on $BaTiO_3$, TGS, *etc.*, will be give in a nutshell as follows:

6.4.3.1 Domain switching involves the following steps:

1) Nucleation of domains: When an electric field is applied, domains are first formed mainly at the surface (Fig 6.18a) in the form of needles. (For more details, Kay, 1969).

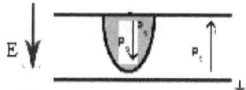

Fig 6.18a Nucleation

2) Forward growth of the domains through the wall thickness of the crystal, (Fig 6.18.b).
3) Domains which grow forward expands mid-ways while formation of new domains continues to take place, (Fig 6.18c).

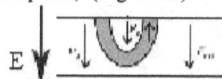

Fig 6.18c Growth

4) Coalescence of domains: when the domains have expanded far enough, sideways, they begin to join together until all the unswitched region is completely 'over-run' by them. (Fig 6.18d).

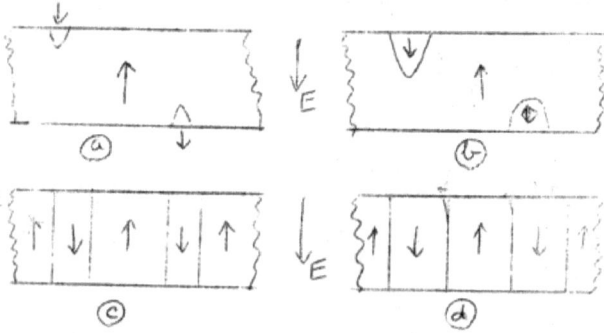

Fig 6.18 Domain Switching

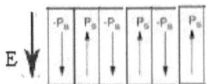

Fig 6.18d Coalescence

6.4.3.2 Kinematics of domains are studied by

    i)   Electrical method
    ii) Miller-Savage method and chemical means,
    iii) Stroboscopic method, and
    iv) Ultrasonic method.

        The electrical method has been most popular. The circuit is as shown below (Fig 6. 19), with detailed setup in Chapter 4.

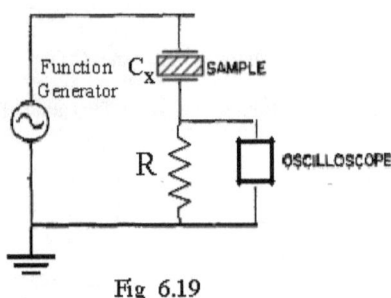

Fig 6.19

For the $\vec{P}_S$ reversal to take place the circuit condition is

$$RC_X \ll t_S,$$

where t_S is the switching time.

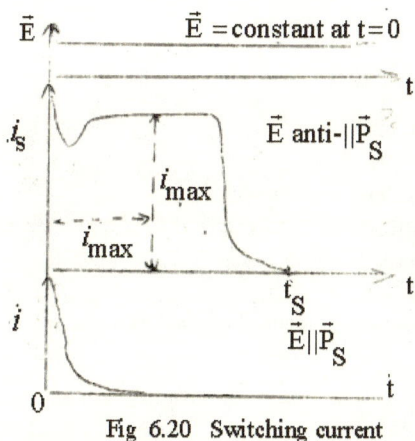

Fig 6.20 Switching current

If i_s = switching electric current, which flows through the crystal Capacitor C_x, during the switching, one obtains the curve of Fig 6.20. This is depicted in the case of $BaTiO_3$ as in Fig 6.20 above.

$$i_{max} \propto t_s^{-1},$$

and $\qquad$ $\ln(i_{max})$ *versus* E^{-1} plot

is seen to be a straight line.

One has to consider the switching with different magnitudes of the electric field E.

a) <u>At Low Electric fields;</u>

The experimental data can be empirically represented as the nucleation rate, defined as

$$\text{nucleation rate} = (1/t_n),$$

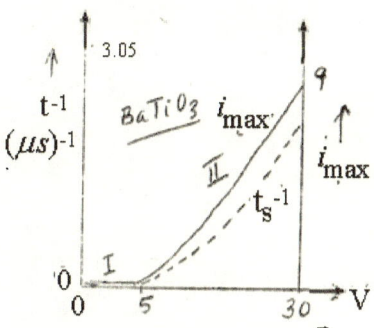

Fig 6.21 Switching at low $\vec{E}$

where t_n = nucleation time.

$$t_n^{-1} = t_s^{-1} = t_\infty^{-1} e^{-\alpha E}$$

$$t_n = t_s = t_\infty \, e^{+\alpha E}$$

where t_∞ and α are constants independent of E.

$\alpha = \alpha(T)$ = activation field, depends on temperature. The results are presented in Fig.6.21 for $BaTiO_3$. This lack of a 'threshold field' is encountered in all ferroelectrics.

For $E < E_C$, domain switching takes quite long time. This is depicted as Region I in (Fig 6.21).

b) <u>At higher values of E</u>

$$t_n^{-1} = t_s^{-1} \cong t_d^{-1} = \frac{v}{d} = \frac{\mu(E-E')}{d}$$

v = velocity, μ = mobility, E = applied field, E' = a constant field and d = a dimension. (This is represented in Region II of Fig 6. 21).

Thus for $E < E_c$ domain switching occurs in a fairly long time. This lack of a definite 'threshold field' is encountered in all ferroelectric.

c) At very high fields E :

Stadler showed that

$$t_s = k \, E^{-n}$$

K= a constant, $n = 1.5$. At very high fields, this is represented in Fig 6. 21, as region III.

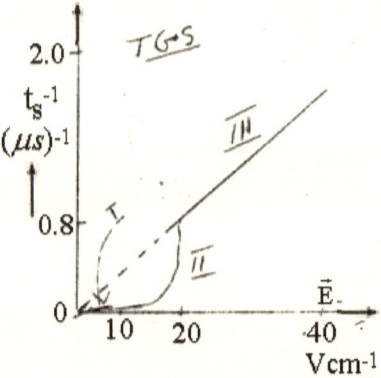

Fig 6.22 Switching at high $\vec{E}$

The domains are found to move at supersonic speeds, in TGS. (Fig 6. 23)

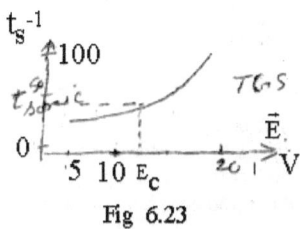

Fig 6.23

Both electric and magnetic fields to study switching of ferroelectric domains in $Pb(Ti,Zr)O_3$-$Pb(Fe,Ta)O_3$ single crystal lamellae are reported recently by Evans *et al.*, (2009).

6.4.3.4 Interaction of Defect and domain structure

In TGS crystals in FE and PE phases are reported (Distler, *et al.*, 1968) showed many important properties in this class of crystals.

6.4.3.5 Piezo-response Force Microscopy (PFM)

This is a local probe based method for non-destructive high resolution ferroelectric domain imaging. This has proved to be a powerful tool form characterization of ferroelectric thin films, ceramics and single crystals. Recent advances in application of this PFM for studying a mechanism of polarization reversal at the nano-scale, domain dynamics, degradation effects, and size-dependent phenomena. (Example, Review by Kalinin, *et al.*, 2006)..

6.5 STUDY OF PHASE TRANSITIONS IN FERROELECTRIC AND RELATED CRYSTALS.

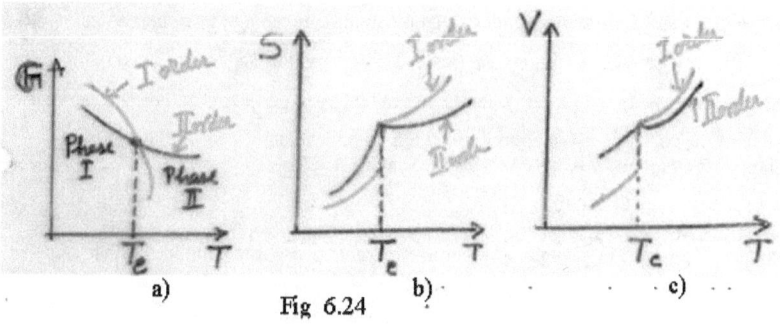

Fig 6.24

There are several studies reported in the literature. A few selected investigations are included in this Chapter.

The three external parameters, *viz.*, pressure, temperature and electric field, can alter the crystallographic directions of ferroelectric domains.

6.5.1 Pressure:

If a unidirectional pressure is applied along the *y* axis of a well-formed crystal, the [001] and [100] twins grow in volume at the expense of the [010] twins by a migration of the twin boundaries in a direction perpendicular to their own plane. The boundaries become progressively less numerous until the whole crystal is a composite of [001] and [100] twins only.

A pressure (compressive stress) may only rotate the polar direction by $90°$, and it results in a large strain. Domain walls grow as spikes (Merz, 1954) or wedges (Mathias & Hippel, 1948).

If a spike is taken to be a lamina with thickness *w* and length *L*, the minimum stress, $\acute{o}0$, needed to induce 90° domain spikes under compressive stress (Loge & Suo, 1996) is

$$\sigma^o = \frac{2\Gamma_{90}}{w\gamma_s}$$

where Γ_{90} = the 90° domain wall energy and γ_s = spontaneous strain. The forward velocity, *V*, induced by the stress field, σ, is

$$V = M\left(\gamma_s\sigma - \frac{2\Gamma_{90}}{w}\right).$$

with the domain wall mobility, $M = 4.8 \times 10^{-4} m^3 s^{-1} N$. As the temperature of the crystal is increased, it becomes more susceptible to applied pressures. Above the Curie temperature, T_C, when the crystal is normally isotropic (cubic), small pressures will make it anisotropic (tetragonal). Therefore, it should be possible to produce single crystals from twinned crystals by simultaneous pressure along the *x* and *y* directions, especially if the crystals are at the same time cooled slowly through the Curie temperature. T_C (Kay,1948). Samara(1966) has shown that the inverse proportionality between T_C and pressure, *p*, suggests that the unidirectional pressure should decrease the T_C value of a single $BaTiO_3$ crystal (instead of raising it as Kay alleged originally). The inverse relationship of T_C and *p* has been verified by experiments and by the negative value of the cubic-tetragonal phase transition volume change, ΔV, in the Clausius-Clapeyron equation (Shirane *et al,*, 1955; Samara, 1966)

$$\frac{\partial T_C}{\partial p} = \frac{\Delta V}{\Delta S} = \frac{T_C \Delta V}{Q}$$

with ΔS = the entropy change, Q = cubic-tetragonal phase transition latent heat, and ΔV = $-0.062 A^{o3}$ / unit cell. Consequently, pressure favours the smaller volume, a stabilization of the cubic phase, or a decrease in T_C.

6.5.2 Electric Fields.

If electric fields are applied along the y direction between opposite cube faces of a well-formed crystal, [001] twins grow in volume at the expense of the [100] and [010] twins by a migration of the twin boundaries in a direction perpendicular to their own plane(Kay, 1948). Unlike the effect of a unidirectional pressure, an electric field may rotate the polar direction by either 90° or 180°. A 180° polar rotation does not result in any strain (Loge & Suo, 1996). The minimum electric fields, E_{90}^0 and E_{180}^0, needed to drive 90° and 180° domain spikes are

$$E_{90}^0 = \frac{2\Gamma_{90}}{P_s w'}$$

$$E_{180}^0 = \frac{\Gamma_{180}}{P_s w}$$

with Γ_{90} = 90° domain wall energy, Γ_{180} = 180° domain wall energy, P_s = spontaneous polarization, and w = domain thickness. The forward velocities, V_{90} and V_{180}, of 90° and 180° domain spikes induced by the electric fields, E_{90}^0 and E_{180}^0, are

$$V_{180} = M\left(2P_s E_{180}^0 - \frac{2\Gamma_{180}}{w}\right)$$

$$V_{90} = M\left(P_s E_{90}^0 - \frac{2\Gamma_{90}}{w}\right)$$

with M = the domain wall mobility $M = 4.8\,x10^{-4}m^3 s^{-1}N$.

The stronger the field, the more effective the aligning process is. A complete saturation may be obtained in the case of a twinned crystal. This may result in a single domain crystal whose c axis becomes completely oriented in a direction parallel to the applied field, and all the twin boundaries disappear. The voltage required for saturation depends on the complexity of the specimen and the time of application of the field. A typical value is around 15 $kVcm^{-1}$ When the field is completely reduced to zero, the crystal may remain perfectly single if complete saturation was previously achieved, but more often partial relaxation of the orientation occurs, although the crystal never returns completely to its original twin configuration.(Kay,1948). Slow relaxation of the crystal can be achieved by keeping it in a closed electrical circuit. When crystals are at the same time cooled slowly through the Curie temperature, the boundaries are seen to be in a state of violent agitation. The stress involved with high voltages often crack the crystal in an irregular way, which does not obviously conform to single crystallographic directions. The application of a strong field, E, raises the Curie temperature, T_C, in accordance with a linear relationship (Shirane et al., 1955)

$$\frac{\partial T_C}{\partial E} = -\frac{\Delta P}{\Delta S} = \frac{T_C P_{SC}}{Q}$$

with ΔS = the entropy change, Q = cubic-tetragonal phase transition latent heat, and ΔP = the discontinuous jump of the spontaneous polarization, P_S, at T_C upon heating

$(0 - P_{SC} = -P_{SC})$..

Consequently, Merz (1954) has observed that at temperatures higher than T_C ($> T_2$), the cubic crystal is paraelectric (non-ferroelectric) and the *P-E* plot gives a straight line, where $T_2 = T_C + 11.7°C$ (Fatuzzo & Merz, 1967). At lower temperatures, ($T_C > T > T_2$), the material can be driven to the ferroelectric state when a strong enough field, E_C, is applied, which gives a double-hysteresis *P-E* loop. At temperatures near T_C, however, these two loops overlap. The temperature dependence of the *P-E* plot Using thermodynamics, the phenomenological representation of the *P-E* plot as a function of temperature can be expressed as (Merz, 1954).

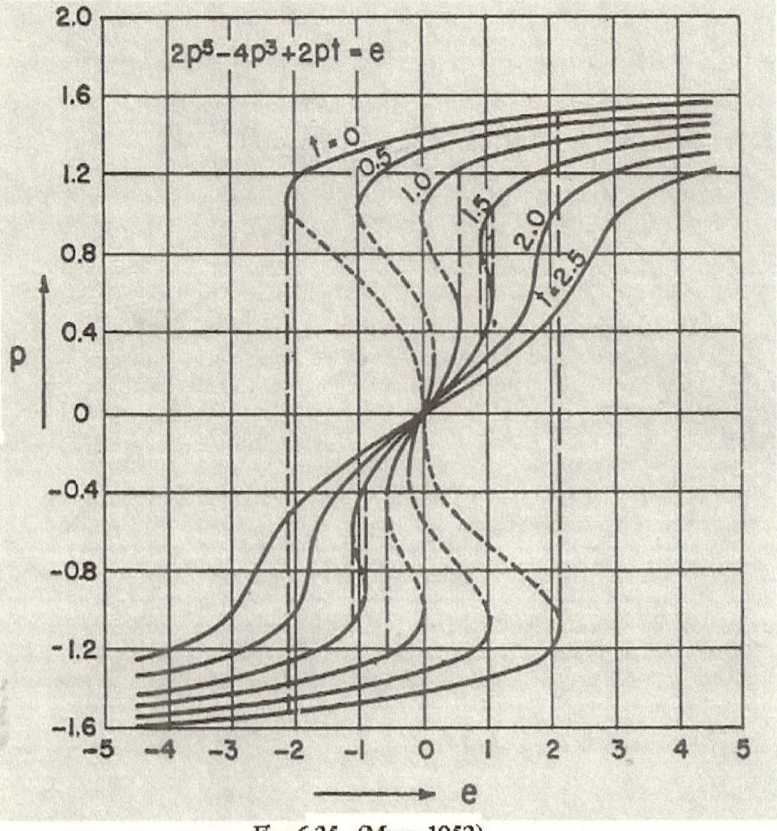

Fig 6.25 (Merz, 1953)

$$e = 2p^5 + 4p^3 + 2pt$$

where

$$p = \sqrt{(3C/B)}P$$

$$e = \sqrt{(27C^3/B^5)}E$$

$$t = (3\beta C/B^2)(T - T_0)$$

with P = electrical polarization; β, B, and C = constants, and T_0 = extrapolated temperature of the reciprocal dielectric constant plot of $1/\varepsilon(T)$, where $T_0 \cong T_C - 7.7°C$ (Fatuzzo & Merz, 1967). The plot of eq $e = 2p^5 + 4p^3 + 2pt$, is shown in Figure 6.21 (Merz, 1953; Kanzig, 1957). In general, P_S can be obtained by extrapolation of the P-E curve back to the E_C ordinate.(Shirane et al., 1955). Since the P-E plot is a straight line at temperatures $> T_2$, $P_S = 0$ regardless of the strength of the applied field. At lower temperatures, $T_C > T > T_2$, P_S decreases as E_C increases. At

$T_C > T > T_2$ P_{SC} and $E_C = 0$. Figure 6.21 shows that, once below T_C, P_S increases (with $E_C =$) 0) as T decreases.

6.5.3. Temperature.

Kay (1948) has found that as the $BaTiO_3$ crystal plate is slowly heated near the T_C, the twin boundaries often become slightly curved. They no longer conform exactly to simple crystallographic directions. The small amounts of [001] tetragonal twins have the a-cell dimensions parallel to the thinnest dimension of the plate and are able to expand as T_C is approached. However, the large [100] twins have both a and b cell dimensions in the plane of the crystal plate, and the necessity for their expansion opposes the tetragonal-cubic transition. This will result in stress restricting the transition of neighboring portions of the [100] tetragonal component. Some of the boundaries begin to migrate irregularly through the crystal. This explains the temperature range of the tetragonal-cubic transition from 108 to 117.5 °C. This also results in the reorientation from [100] to [010] and [001] directions. These patches with "dislocations" or "faults" become smaller and fainter and eventually disappear, leaving the crystal completely isotropic (cubic) at T_C. When it is cooled, the crystal returns to a composite of tetragonal twins whose arrangement usually varies after each cycle. The more rapid the cooling is, the greater the complexity will be. Slow cooling alone, however, has rarely produced truly single tetragonal crystals. This is probably due to strain centers resulting from local lattice imperfections characteristic of each individual crystal. Since the decrease in temperature and the strain effects tend to cooperate and produce the anisotropic (tetragonal) phase, the temperature range of the cubic-tetragonal transition should be less than the case of raising temperature. It is found to be from 114.8 to 109.8 °C. The value of c/a, as determined by the X-ray diffraction, decreases as the temperature increases due to the contraction of c and the expansion of a ($b = a$) near the tetragonal-cubic phase transition. At the transition, T_C, the tetragonal parameters are

$a = 4.0051 \pm 0.0008$ Å and $c = 4.0206 \pm 0.0007$ Å and the cubic parameter is
$a = 4.0096 \pm 0.0002$ Å $(a = b = c)$..

6.5.4. Lattice parameters *versus* Temperature in $BaTiO_3$

As the temperature is varied, ionic displacements in the lattice take place, resulting changes in lattice dimensions. Above T_C, the $BaTiO_3$ crystal has a simple cubic array of corner-sharing TiO_6 octahedra with barium ions filling the holes between. At the cubic tetragonal transition near T_C, there is an expansion in the direction of the Ti^{4+} displacements and a contraction in perpendicular directions. (Kay & Vousden, 1949), and is displayed in Fig 6.26. The cubic parameter is $a = b = c = 4.0096 \pm 0.0002$ Å, and the tetragonal parameters are $a = b = 4.0051 \pm 0.0008$ Å and $c = 4.0206 \pm 0.0007$ Å (Kay, 1948). The atomic displacements are oscillations about a nonpolar site; after a displacive transition.

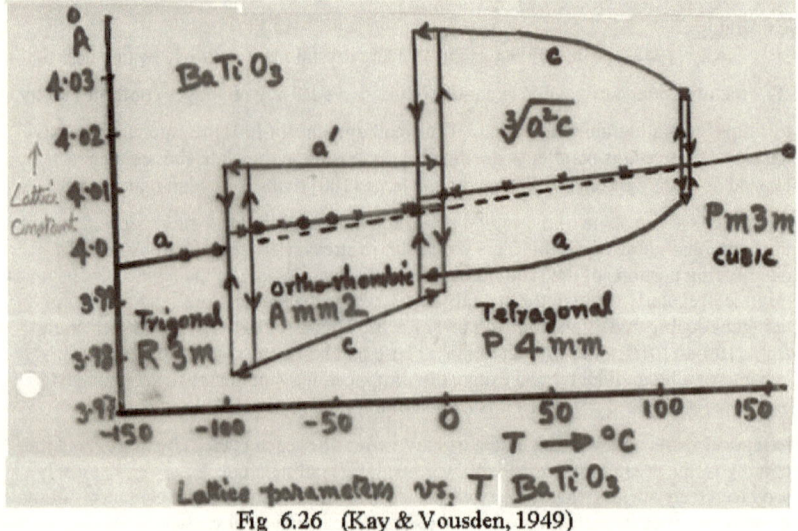

Fig 6.26 (Kay & Vousden, 1949)

6.5.5. Effect of temperature on dielectric constant

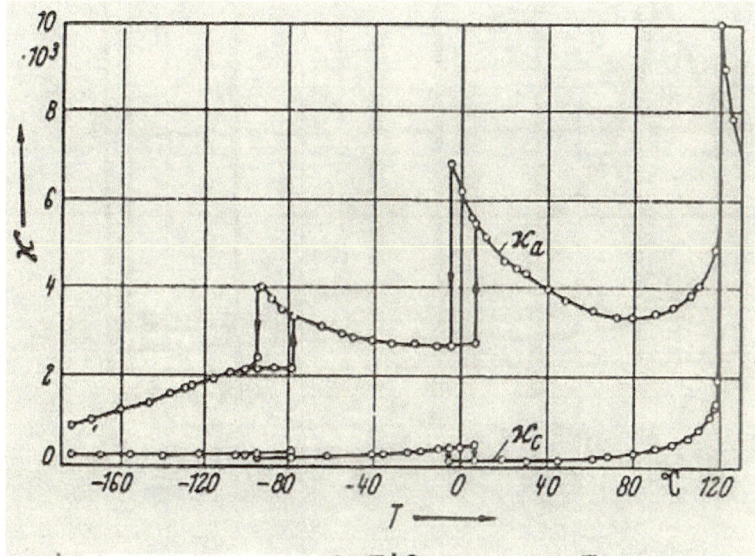

Fig 6.27 (Merz, 1953) BaTiO$_3$, к vs T

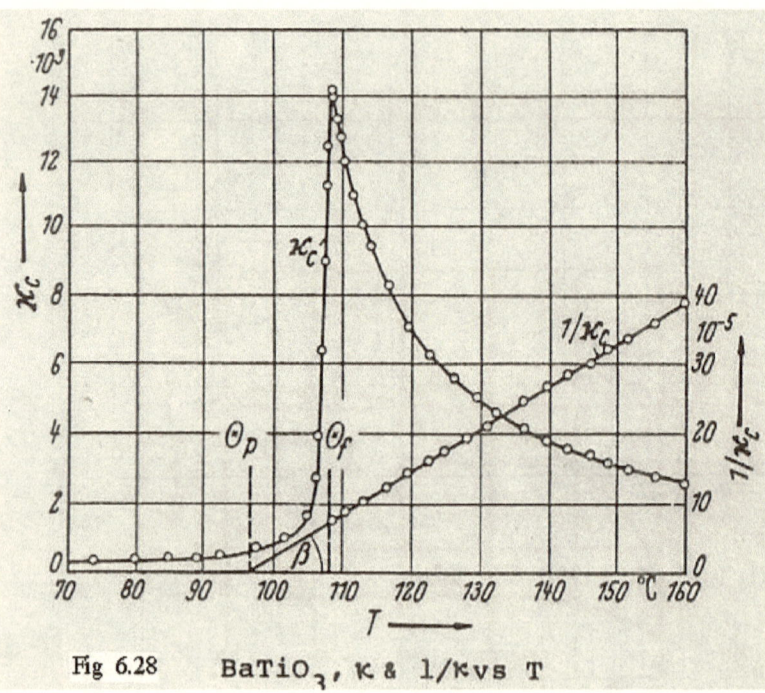

Fig 6.28   BaTiO$_3$, κ & 1/κ vs T

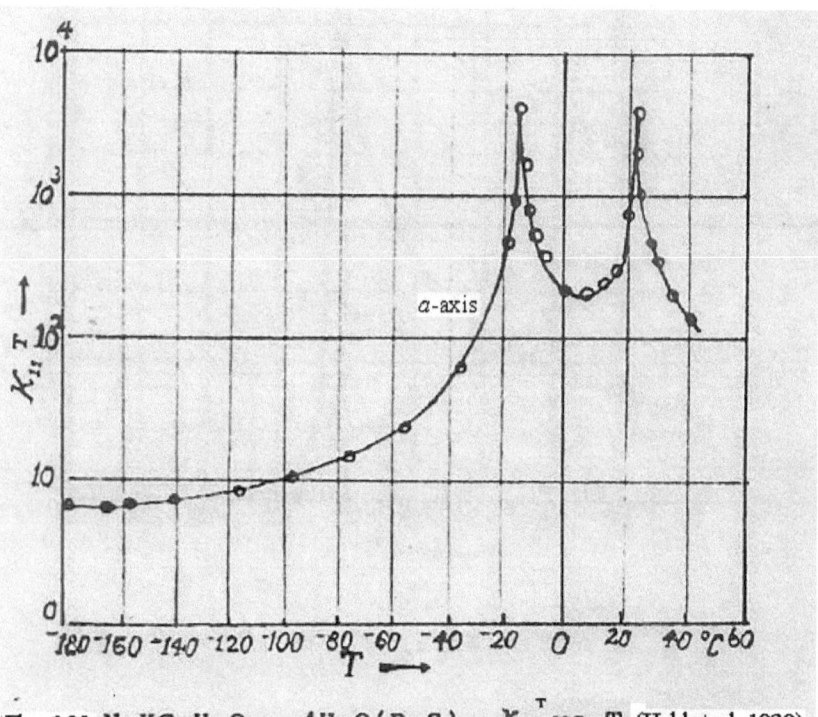

Fig 6.29 $NaKC_4H_4O_6 \cdot 4H_2O(R.S)$, K_{11}^T vs T (Hablutzel, 1939)

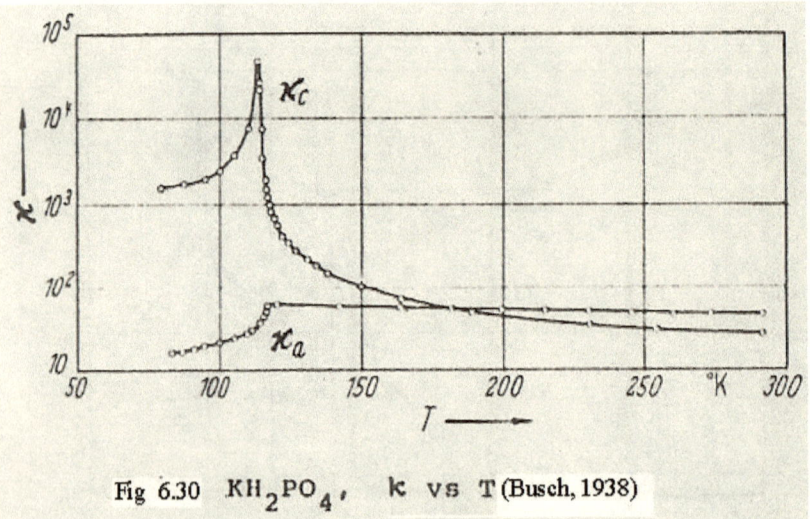

Fig 6.30 KH_2PO_4,   k vs T (Busch, 1938)

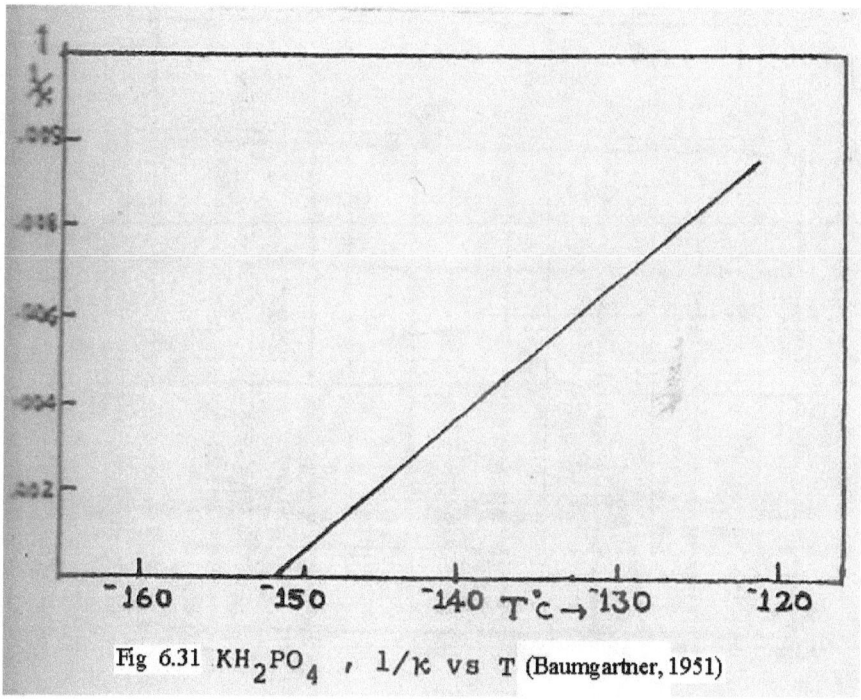

Fig 6.31 KH_2PO_4 , $1/K$ vs T (Baumgartner, 1951)

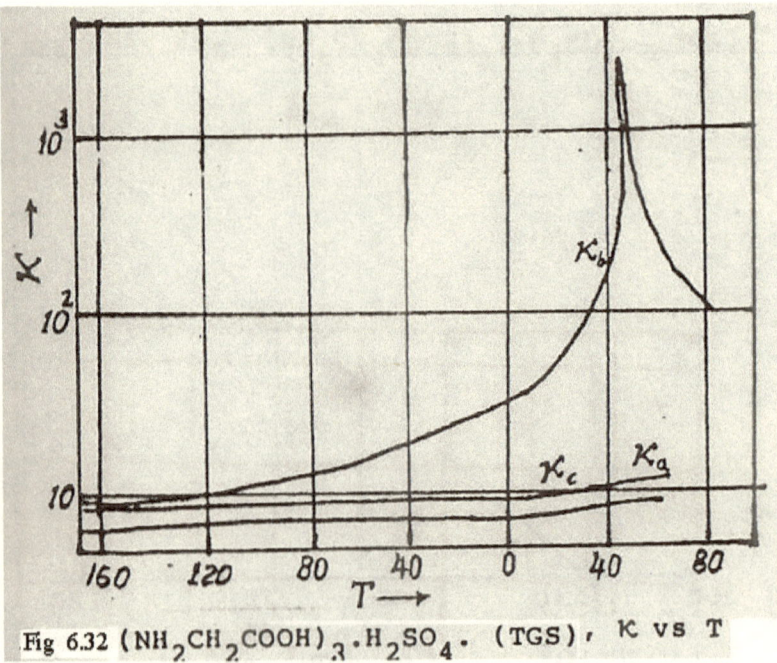

Fig 6.32 $(NH_2CH_2COOH)_3 \cdot H_2SO_4$. (TGS), K vs T

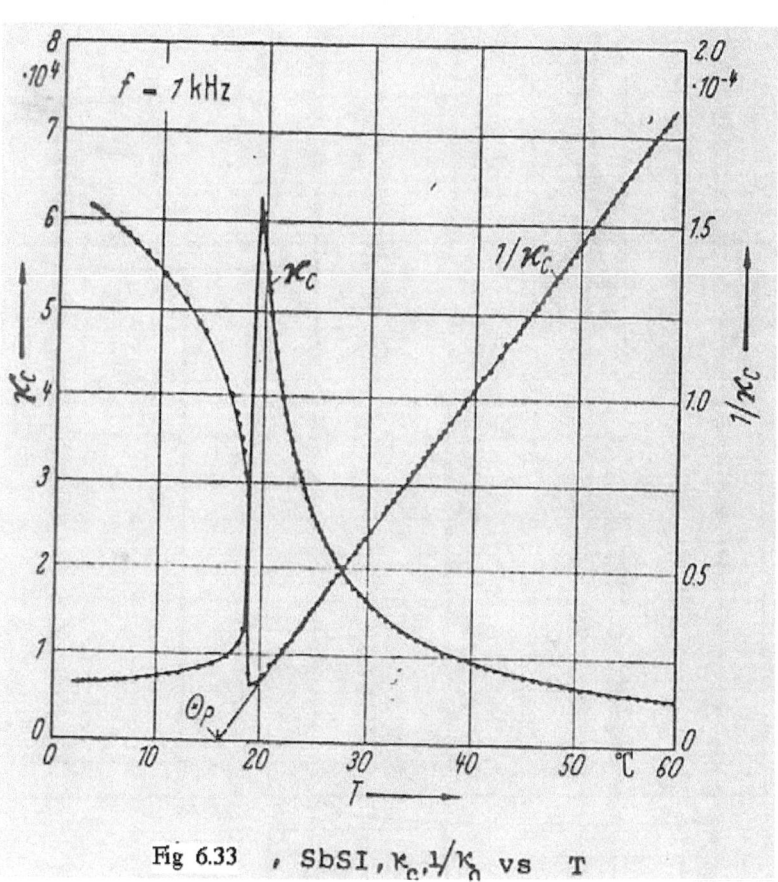

Fig 6.33 , SbSI, κ_c, $1/\kappa_c$ vs T

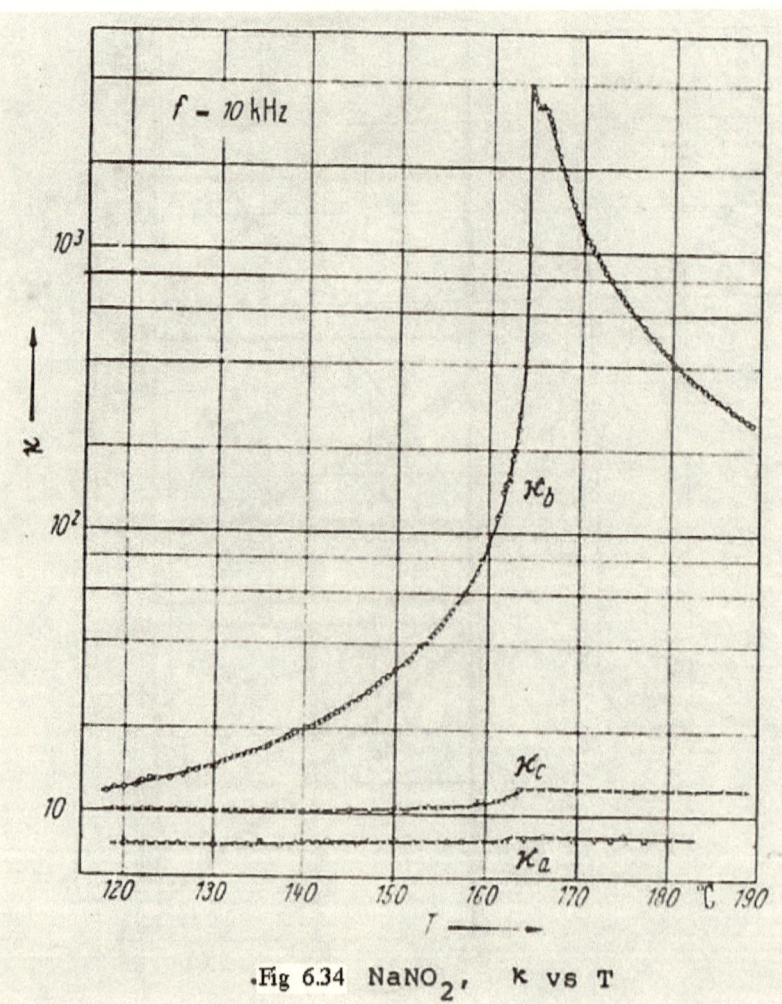

.Fig 6.34 NaNO$_2$, κ vs T

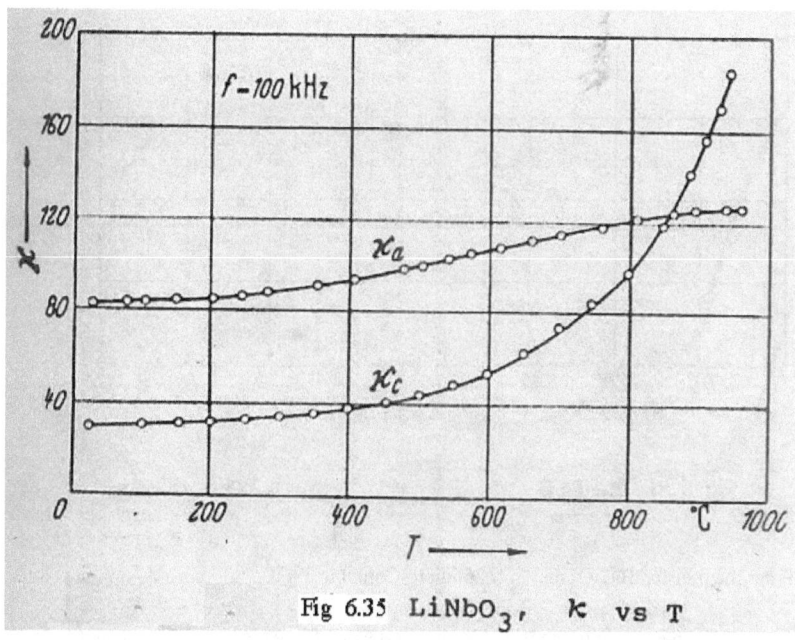

Fig 6.35 LiNbO$_3$, k vs T

6.5.6 Spontaneous Polarization and Temperature

The P_S versus T plots for some of the representative crystals like BaTiO$_3$, TGS, KDP, RS, DRS, SbSI, PbTiO$_3$, KNbO$_3$, (NH$_2$)$_2$SO$_4$, MASD, and NaNO$_2$, are reproduced in Fig.6.36 to Fig.6.42.

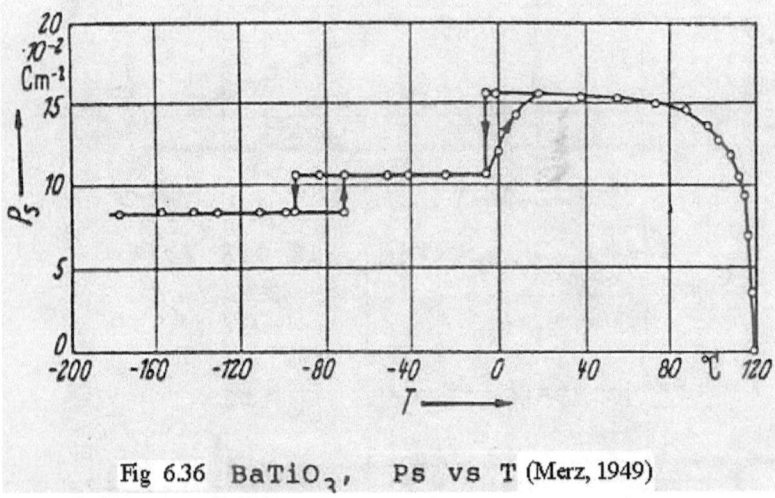

Fig 6.36 BaTiO₃, Ps vs T (Merz, 1949)

From the plot for $BaTiO_3$ the $P_S = 26$ micro-Coulomb cm^{-2}.

It is interesting to see that this observed value for P_S is in quite good agreement with the calculated value of 15.6 $\mu C cm^{-2}$, above.

First order Transition

In a first order transition the polarization varies continuously, until the Curie temperature at which there is a discontinuity.(Fig 6.42 a).

Second order Transition

In a second order transition, the order parameter itself is a continuous function of temperature, but there is a discontinuity in its first derivative at T_c .(Fig 6.42b).

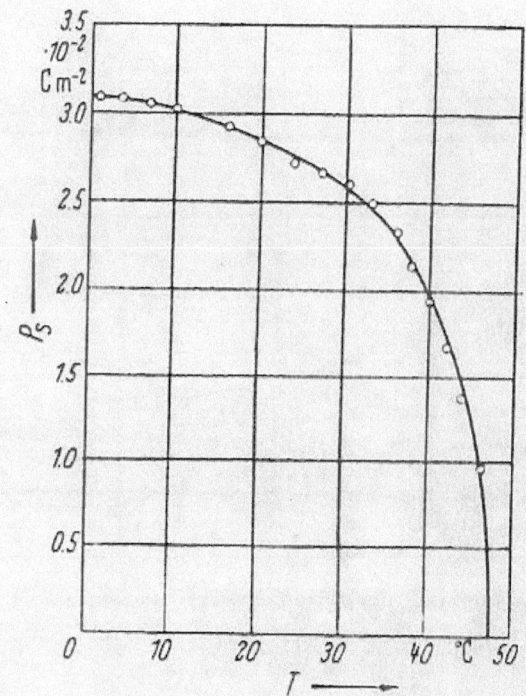

Fig 6.37. $(NH_2CH_2COOH)_3H_2SO_4$ (TGS), P_S vs T

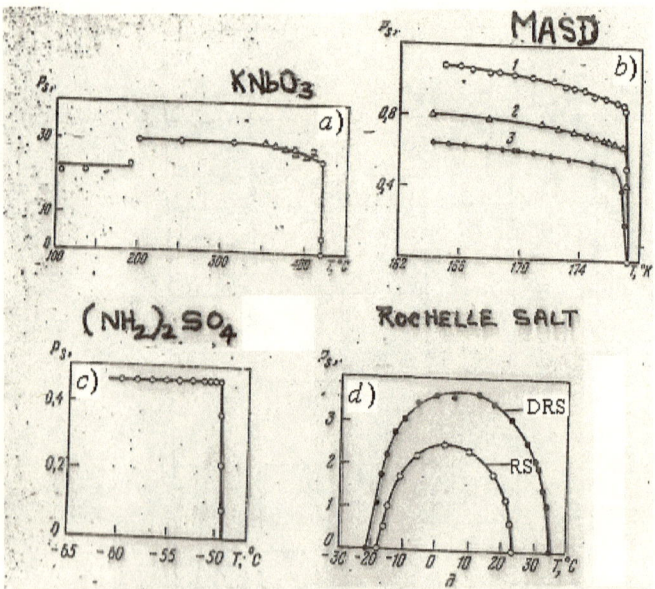

Fig 6.38   Spontaneous Polarization *versus* Temperature

a) $KNbO_3$ b) $(CH_3NH_3)Al(SO_4)_2.12H_2O$

c) $(NH_4)_2(SO_4)$

d) Rochelle Salt and Deuterated RS

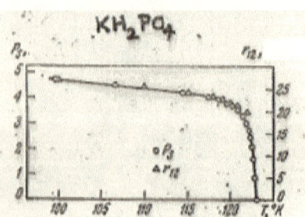

Fig 6.39 Spontaneous Polarization vs. Temperature

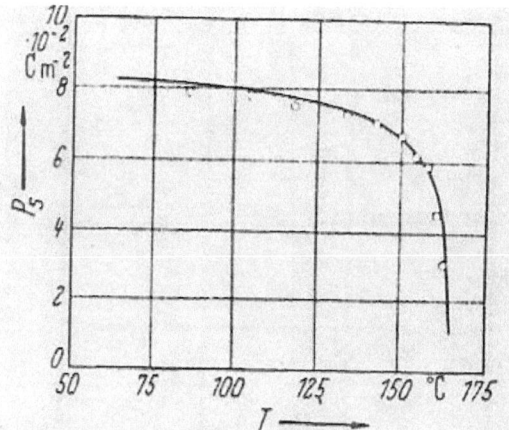

Fig 6.40 , NaNO$_2$, Ps vs T

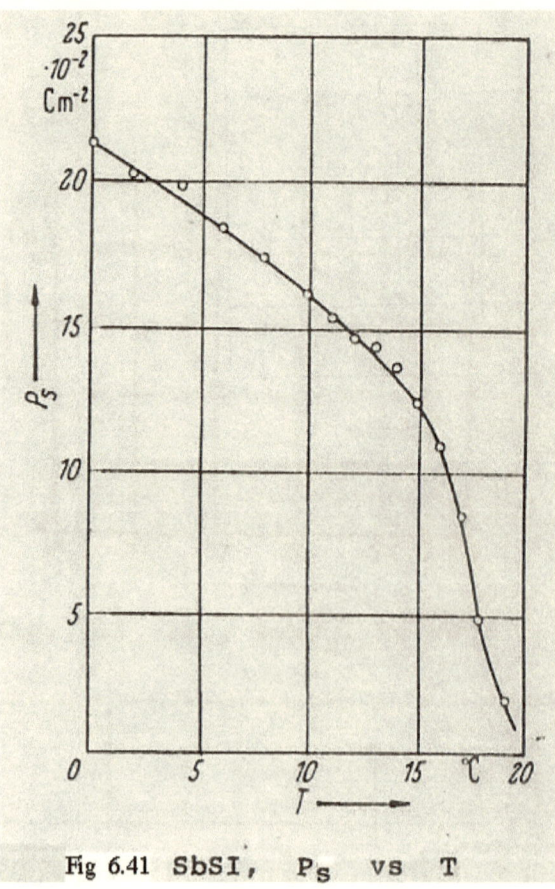

Fig 6.41 SbSI, P_s vs T

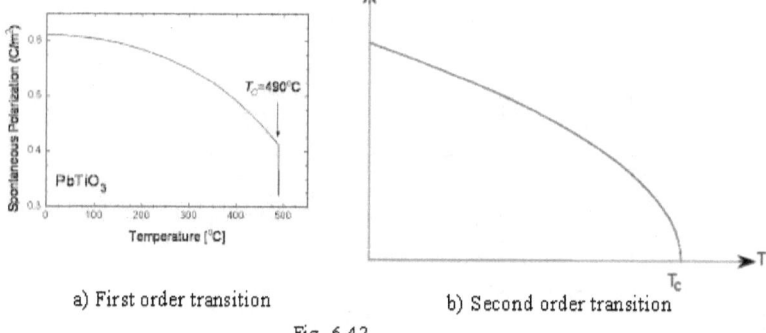

a) First order transition   b) Second order transition

Fig. 6.42

.6.5.7   <u>Specific heats versus Temperature</u>

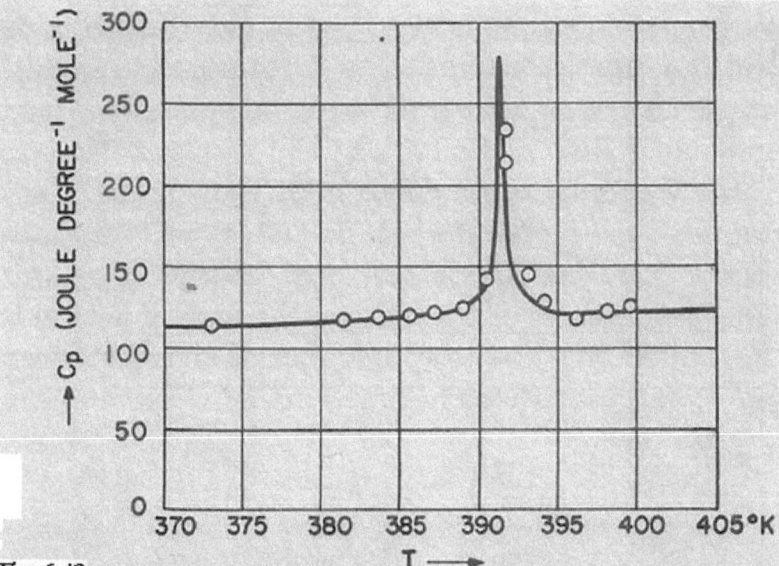

Fig 6.43

Specific heat anomaly of polycrystalline BaTiO₃ at the tetragonal to cubic phase transition (after Volger [1])

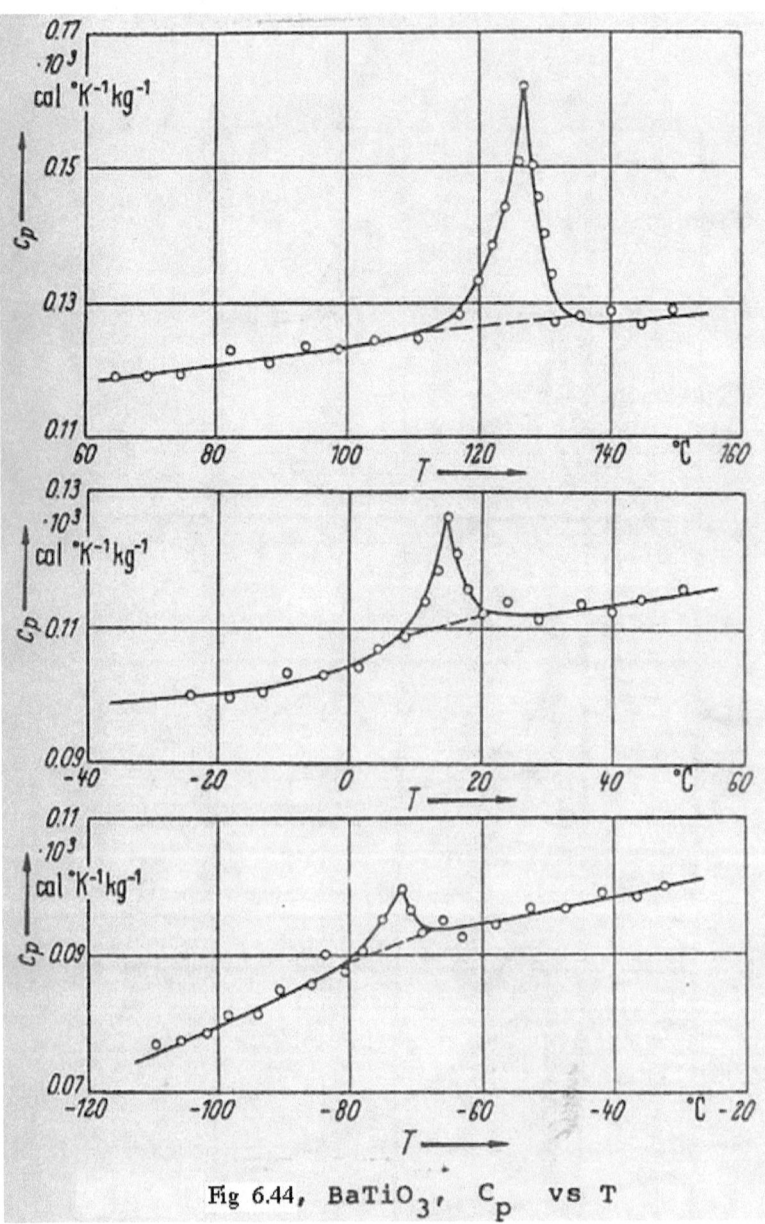

Fig 6.44, BaTiO$_3$, C$_p$ vs T

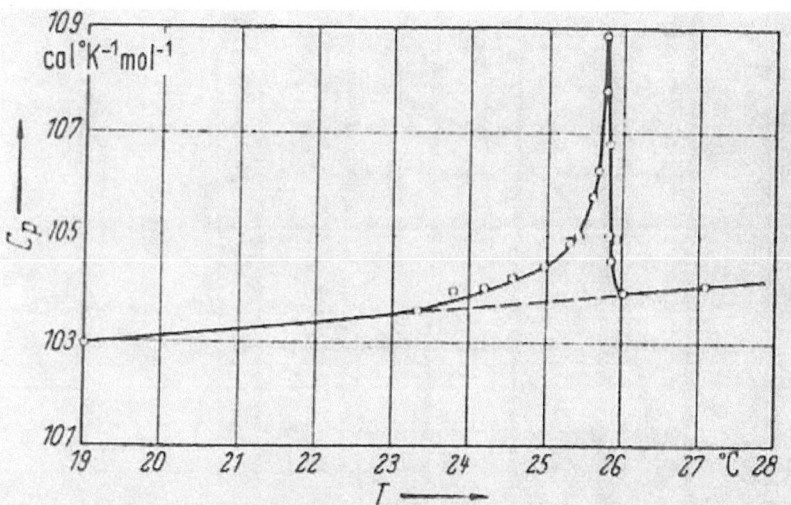

Fig 6.45 NaKC$_4$H$_4$O$_6$·4H$_2$O(R.S.), Cp vs T

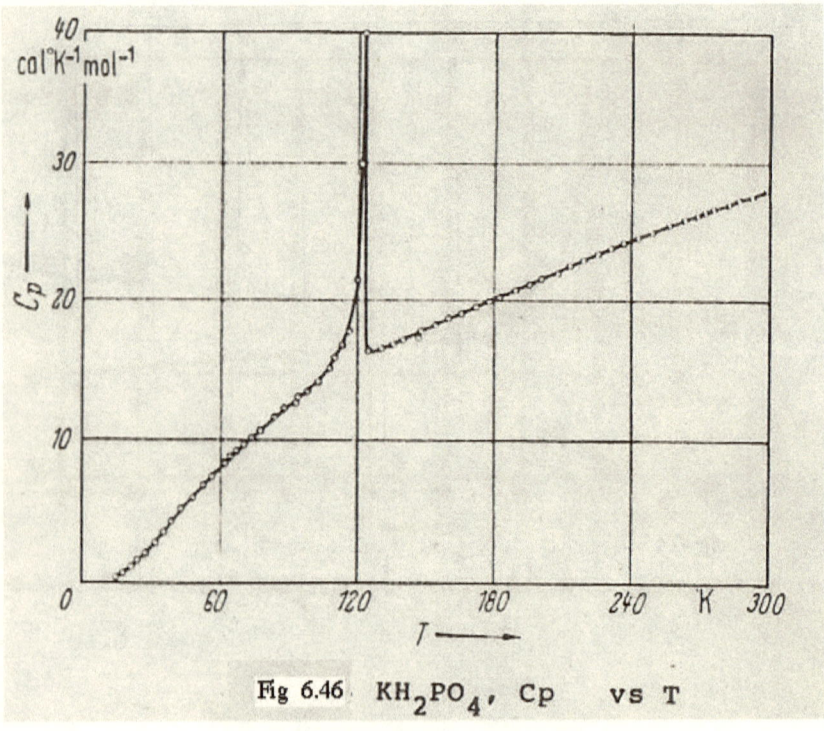

Fig 6.46 KH_2PO_4, Cp vs T

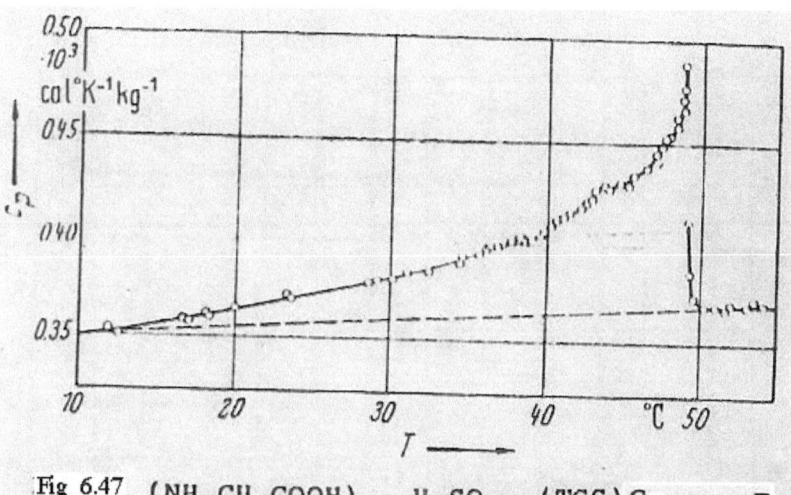

Fig 6.47 $(NH_2CH_2COOH)_3 \cdot H_2SO_4 \cdot$ (TGS) C_p vs T ,

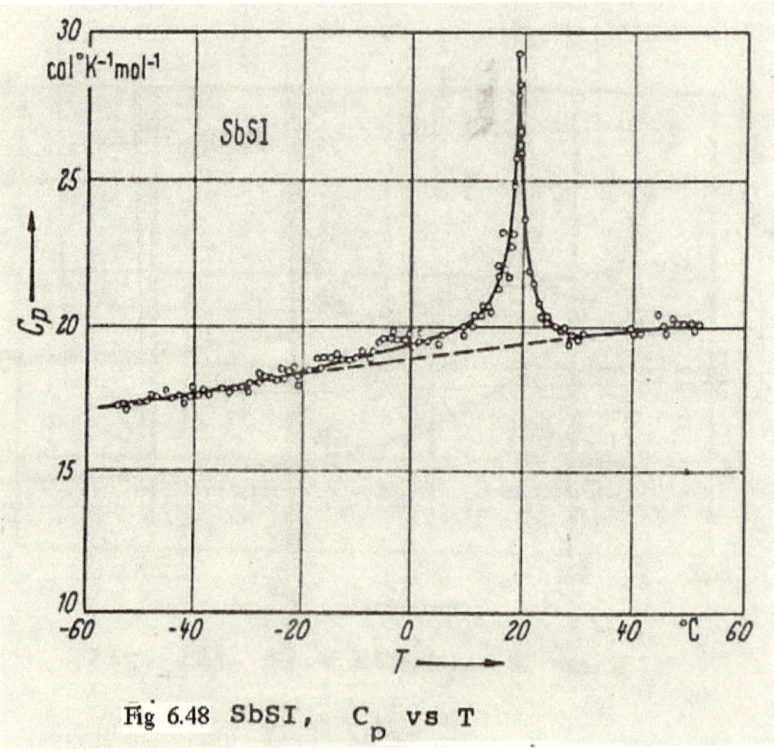

Fig 6.48 SbSI, C$_p$ vs T

Volger , 1952

In Gd$_2$(MoO$_4$)$_3$ Ann Fouskava (1969) reported the measurements on C$_p$ in the range 100-200°C, by Perkin-Elmer differential calorimeter.. The surplus entropy of transition $= \Delta S = 0.22 \pm 0.08 \ Cal.mole^{-1}.°C^{-1}$.

6.5.8 Thermal Expansion *versus* Temperature:

Ferroelectric NaH$_3$(SeO$_3$)$_2$ is a typical most complicated crystal to study by means of a Fizeau optical interferometric dilatometer (Devanarayanan, 1968, 1969). It undergoes phase transition as detailed below.

$$NaH_3(SeO_3)_2 : \begin{array}{|c|c|c|} \hline \text{Monoclinic} & T_C = -79°C & \text{Triclinic} \\ \hline \text{Space Group P2}_1/a & P_S\|\ a\text{--}axis,[310] & \text{Space Group } C1 \\ \hline \end{array}$$

Devanarayanan (1969) has made thermal expansion measurements along the following four directions in the crystal, as shown in Fig. 6.49.

(i)   Along the b-axis, *i.e.*, crystalline [010] direction,
(ii)   Along the Z-axis, *i.e.*, c-axis of the crystal,
(iii) Along the X-axis, *viz.*, normal to the c-axis and in the (010) crystal plane, and
(iv)   Along W-axis which is inclined to c-axis by 151°43' and lying in the (010) plane. This direction is normal to a natural external plane surface of the crystal.

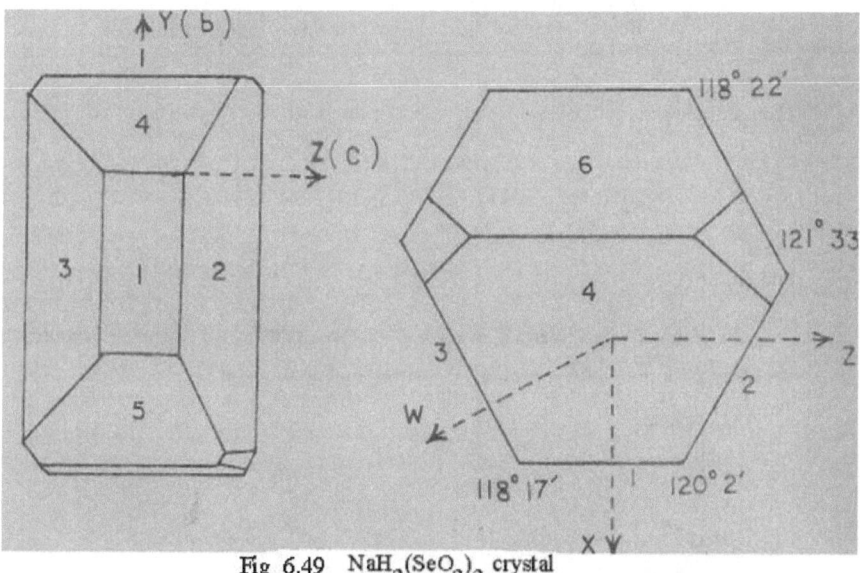

Fig 6.49 $NaH_3(SeO_3)_2$ crystal

The Fizeau's optical interferometer was used. The interference fringes were formed with three pyramids of the crystal placed in an isosceles triangle in between two fused quartz discs with highly polished surfaces from the Thermal Syndicate, England. The crystal were cut to the desired size by using 22 teeth/cm fretsaw blade of length 5.3". For grinding and polishing rough crystals, Linde's Al_2O_3 (0.3 mi), and for water soluble

crystals, ethylene glycol was used. The measurements on height of the pyramids, precision gauge with micrometer scale was used. Temperatures were detected through proper thermocouples and Precision vernier Potentiometer (Leeds and Northrup Co.) and a deflection mirror galvanometer.

From Section 2.10.5 using $\theta_1 = 90°$, $\theta_2 = 0°$ and $\theta_3 = 151°43'$.

$$[\theta] = \begin{bmatrix} 1 & 0 & 0 \\ 0 & 0 & 1 \\ 0\text{-}2245 & -0.8345 & 0.7754 \end{bmatrix}$$

$$[T] = \begin{bmatrix} 1.0000 & -0.0004 & -0.00014 \\ 0.2688 & 0.9255 & -1.11700 \\ -0.0004 & 0.9982 & 0.00115 \end{bmatrix}$$

Thus at $-60°C$, the thermal expansion quadric is calculated as

$$\begin{bmatrix} 7.993 & 0 & 12.772 \\ 0 & 72.0 & 0 \\ 12.772 & 0 & 16.281 \end{bmatrix} x\ 10^{-6}\ °C^{-1}$$

From this the principal values of α at $-60°C$ for the $NaH_3(SeO_3)_2$ crystal are (Fig 6.50)

$$[\alpha_i] = \begin{bmatrix} -1.3 & 0 & 0 \\ 0 & 72.0 & 0 \\ 0 & 0 & 25.6 \end{bmatrix} x\ 10^{-6}\ °C^{-1}$$

and $\varphi = 36°1'$.

Defining the total thermal expansion, $\epsilon = \frac{\Delta L}{L}$, of the crystal, when temperature was changed from $-80°C$ to $-30°C$, the corresponding values are

$$\begin{bmatrix} 59.804 & 0 & 73.110 \\ 0 & 369.2 & 0 \\ 73.110 & 0 & 82.706 \end{bmatrix} x\ 10^{-5}$$

$\epsilon_1 = 2.8x\ 10^{-5}$, $\epsilon_2 = \epsilon_b = 369.2x\ 10^{-5}$, $\epsilon_3 = 145.3x\ 10^{-5}$ and $\varphi = 40°33'$.

The mean values of the coefficients, for $-80°C$ to $-30°C$, are

$\bar{\alpha}_1 = -0.6x\ 10^{-6}\ °C^{-1}$, $\bar{\alpha}_2 = 74.0x\ 10^{-6}\ °C^{-1}$, and $\bar{\alpha}_3 = 29.0x\ 10^{-6}\ °C^{-1}$.

The polar diagram of the thermal expansion of the crystal is shown in Fig. 6. 51.

This when superposed on the projection of the crystal structure is as shown in Fig.6.52 and Fig 6.53.

The volume coefficient

$$\beta = (\alpha_1 + \alpha_2 + \alpha_3) = 96.3x\ 10^{-6}\ °C^{-1}$$

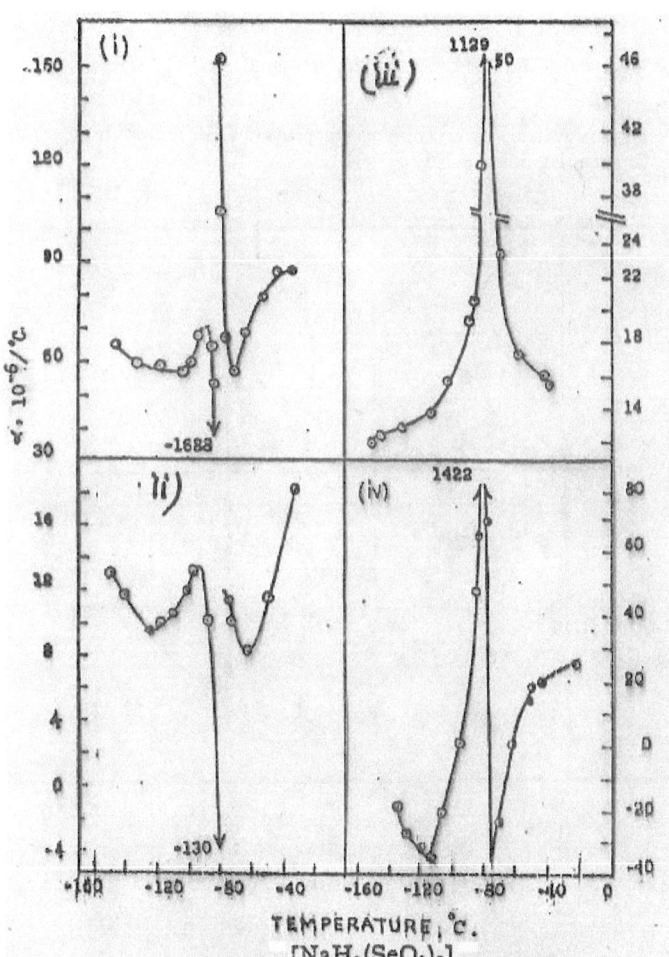

Fig 6.50 Observed variation of mean coefficient of thermal pansion (α) with temperature along different directions [Directions: (i) b; (ii) X; (iii) Z; and (iv) W]

(Devanarayanan, S. & PS. Narayanan, 1968a, 1969)

At T= -60 °C, for $NaH_3(SeO_3)_2$, α-tensor is $\begin{bmatrix} 7.993 & 0 & 12.774 \\ 0 & 72.0 & 0 \\ 12.774 & 0 & 16.281 \end{bmatrix} 10^{-6}/°C$

(Devanarayanan, 1969). This tensor appears in a 2-D projection appear as polar diagram, shown in Fig.6.51.

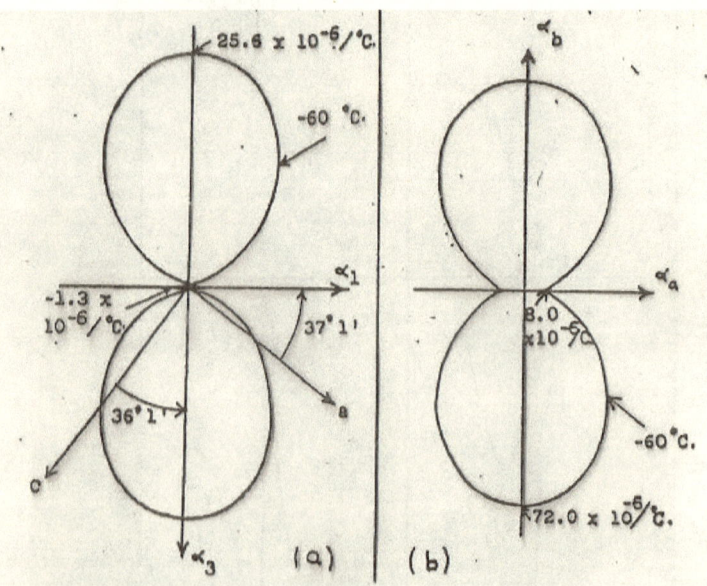

[$NaH_2(SeO_3)_2$]

Fig 6.51— Variation of α with orientation in the (010) and (001) planes [(a) (010) plane; and (b) (001) plane] (Devanarayanan, S. & PS. Narayanan, 1968a, 1969)

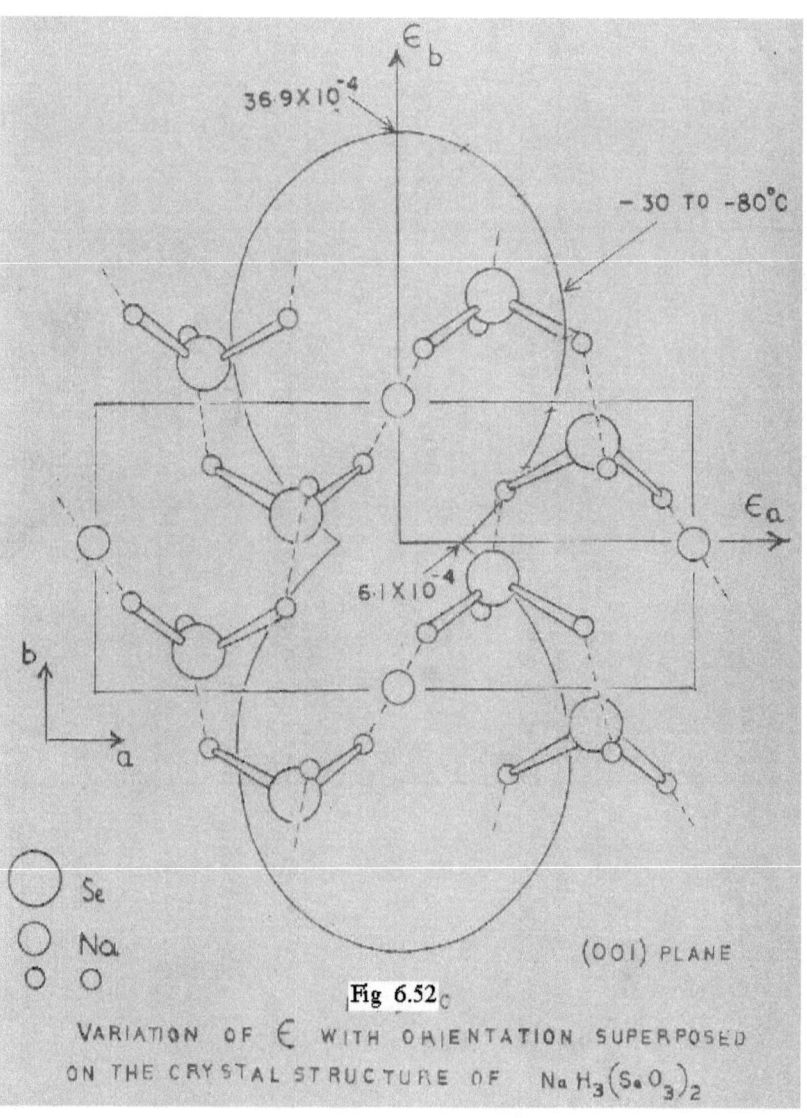

36.9×10⁻⁴

- 30 TO -80°C

ϵ_b

ϵ_a

6.1×10⁻⁴

b

a

Se

Na

(OOI) PLANE

Fig 6.52c

VARIATION OF ϵ WITH ORIENTATION SUPERPOSED
ON THE CRYSTAL STRUCTURE OF $NaH_3(SeO_3)_2$

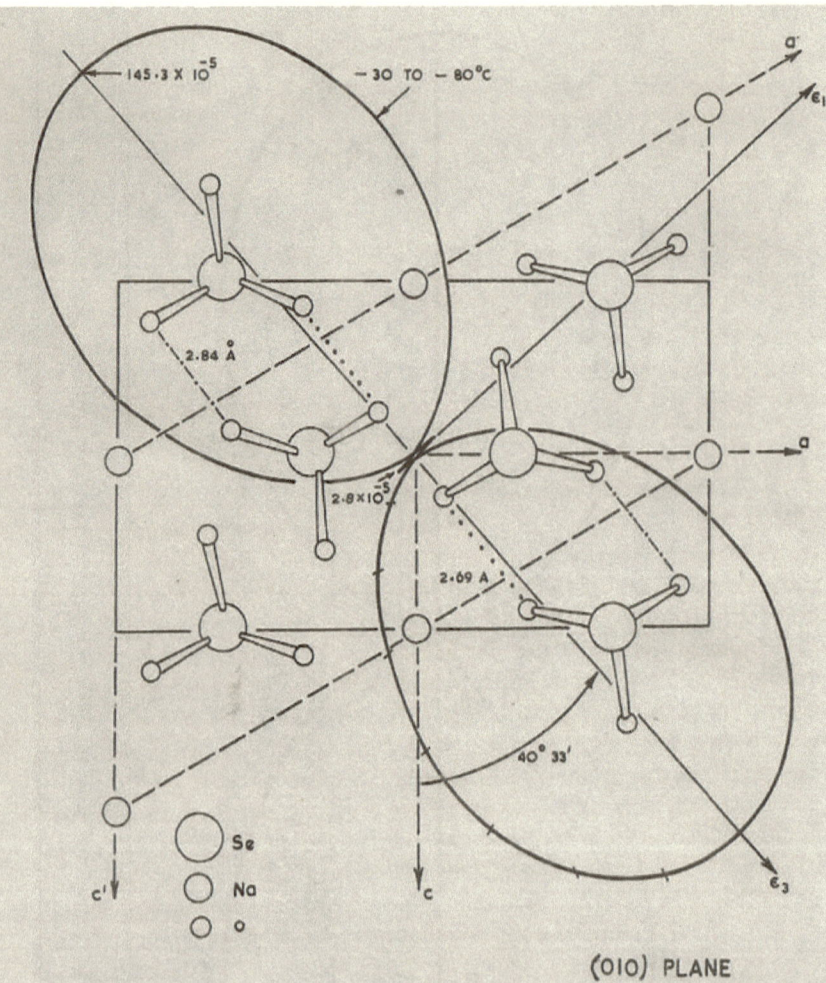

Fig 6.54. Variation of Thermal Dilatation ε with Orientation Superposed on the Crystal Structure of $NaH_3(SeO_3)_2$.

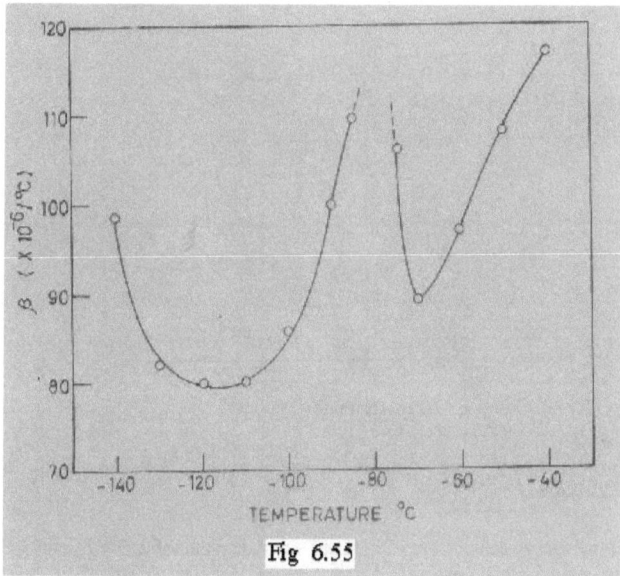

Fig 6.55

In the case of TGS (Ganesan, 1962) has observed that there is a correlation between the direction of maximum thermal expansion and the orientation of the H-bond in the crystal.

On the other hand, the study on $NaH_3(SeO_3)_2$ concludes that ihe direction of least expansion (negative thermal expansion anomaly), ϵ_1, coincides almost with the direction of maximum dielectric anomaly, *i.e.*, [310] or [3$\bar{1}$0] , along the ferroelectric axis in the crystal (Blinc *et al.*, 1965).

Thermal expansion of TGS

TGS:

Monoclinic	$T_C = 49°C$	Monoclinic
Space Group $P2_1/a$	$P_S \parallel b-axis$	Space Group 2

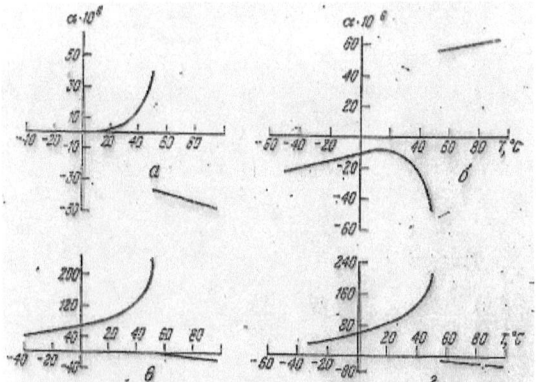

Fig 6.56 Thermal expansion coefficients versus T in TGS (Ganesan, 1962).

Thermal Expansion of LiTaO$_3$

The principal thermal expansion coefficients of single crystals of LiTaO$_3$ and LiNbO$_3$, both belonging to 3m were measured by X-ray diffraction method, in the range 25 °C to 500 °C. The c- or z-axis expansion value showed a definite maximum at ≈ 250 °C for LiTaO$_3$.

T_R =reference temperature. The expansion coefficients are fitted as a polynomial

$$\frac{d}{dT_R} = 1 + a(T - T_R) + b(T - T_R)^2 + c(T - T_R)^3$$

Table 6.1

$$\frac{d}{dT_R} = 1 + a(T - T_R) + b(T - T_R)^2 + c(T - T_R)^3, \quad \text{range } 0°C - 500°C$$

	T_C	T_R	cut	a	b	c
LiTaO$_3$	601°C	25°C	X	$1.61x10^{-5}$	$7.5x10^{-9}$	-
			Y	$1.54x10^{-5}$	$7.0x10^{-9}$	-
			Z	$0.22x10^{-5}$	$-5.9x10^{-9}$	-
LiNbO$_3$	1165°C	25°C	X	$1.44x10^{-5}$	$7.1x10^{-9}$	-
			Y	$1.59x10^{-5}$	$4.9x10^{-9}$	-
			Z	$0.75x10^{-5}$	$-7.7x10^{-9}$	-
Quartz	NA	25°C	X	$0.98x10^{-5}$	$20.4x10^{-9}$	-
			Z	$0.56x10^{-5}$	$9.4x10^{-9}$	-
			Z	$0.22x10^{-5}$	$-5.9x10^{-9}$	-

Thermal expansion of KDP

Thermal expansion of KDP single crystals were studied by X-ray diffraction at 110 – 250 K by Boiko & Dhat (1969). $\alpha_{[100]}$, $\alpha_{[110]}$ and $\alpha_{[001]}$ become negative at ~ 1.5K above T_C. During the phase transition the unit cell volume is found to change continuously.

Thermal expansion of Ferroelectric LiN$_2$H$_5$SO$_4$

Thermal expansion of Ferroelectric LiN$_2$H$_5$SO$_4$ was studied (Devanarayanan, 1966, 1969) from -160°C to 220 °C.

$$\text{LiN}_2\text{H}_5\text{SO}_4 : \xrightarrow{\text{Ferroelectric, - 40°C to 120°C}}_{\text{Orthorhombic, } C_{2v}^9 - Pbn2_1, Z = 4, \text{ FE } \| c\text{-axis}}$$

At room temperature, the principal value along the a-, b-, and c-axes are:

$$[\alpha_{ii}] = \begin{bmatrix} 18.7 & 0 & 0 \\ 0 & 13.5 & 0 \\ 0 & 0 & 43.0 \end{bmatrix} x \, 10^{-6} \, °C^{-1}.$$

226

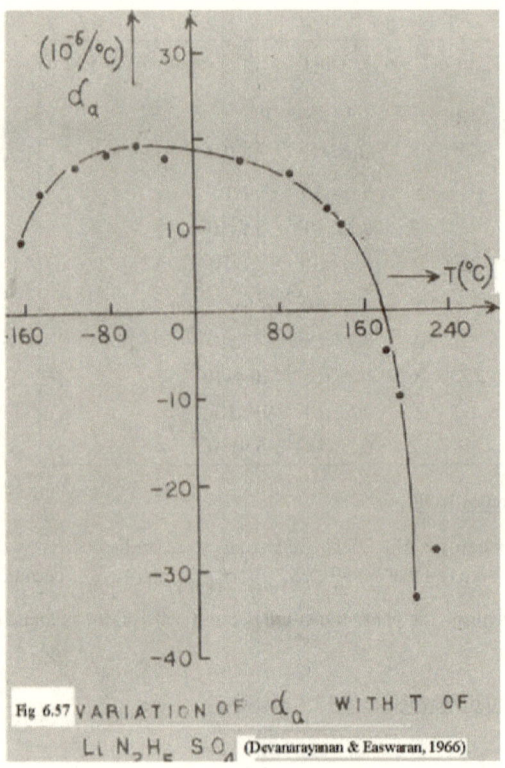

Fig 6.57 VARIATION OF α_a WITH T OF

Li N$_2$H$_5$ SO$_4$ (Devanarayanan & Easwaran, 1966)

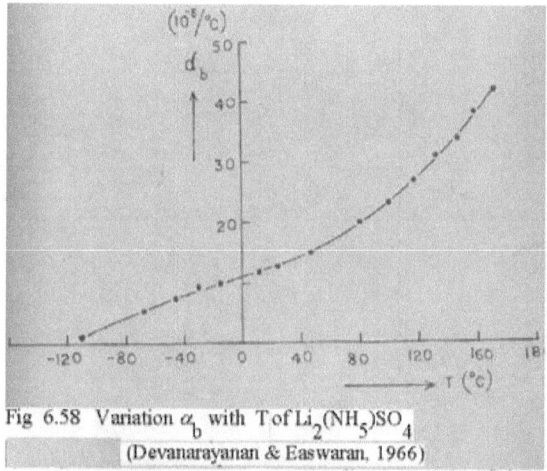

Fig 6.58 Variation α_b with T of $Li_2(NH_5)SO_4$
(Devanarayanan & Easwaran, 1966)

α_c variation with T is seen to be anomalous between -160°C and -60°C, and also a kink around +128°C (Fig 6.57, Fig 6.58, Fig 6.59). The anomalies are interpreted as due to 'homomorphous transitions' in the crystal (Zwikker, 1954), in which there is a re-orientation of the molecule hydrazinium ion in which the $-NH_3^+$ group about the N-N axis. α is found to be highly anisotropic in the (010) plane of the crystal (Fig.6.61), and in the (001) plane superposed on to the crystal structure (Fig 6.62). It is concluded that the ferroelectric phase is confined in the range -40°C to +120°C. The presence of kink at +128°C' The homomorphous transition -175°C to -40°C is referred to as $\alpha \rightarrow \beta$ type, as seen in both the α versus T plot as well as in the β versus T plot. Similarly, at +128°C both the plots indicate kinkspointing out a high temperature polymorph $\beta \rightarrow \gamma$ in the crystal. This is caused by the onset of hindered rotation of the $-NH_2$ group about the N-N axis in the crystal. Both the studies of hysteresis and the thermal expansion conclude that ferroelectricity in the crystal is confined to the β – polymorph.

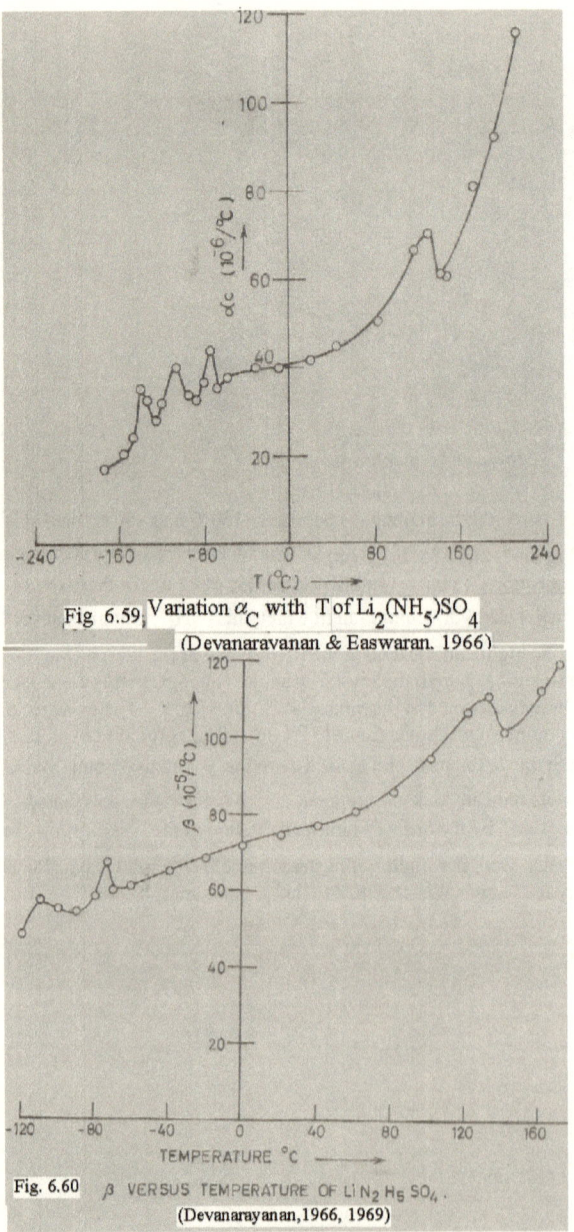

Fig 6.59 Variation α_C with T of $Li_2(NH_5)SO_4$
(Devanaravanan & Easwaran. 1966)

Fig. 6.60 β VERSUS TEMPERATURE OF $LiN_2H_5SO_4$.
(Devanarayanan,1966, 1969)

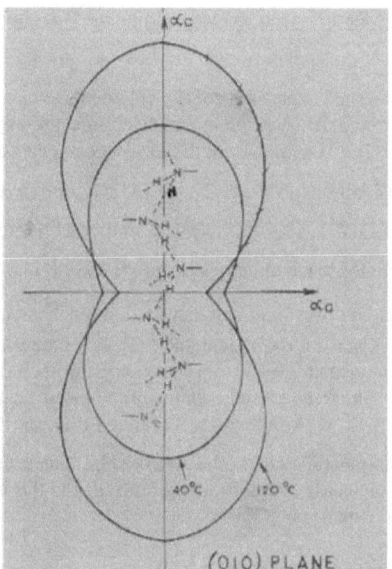

Fig. 6.61. Polar Diagram in (010) plane of α in Li(N$_2$H$_5$)SO$_4$
(Devanarayanan,1966, 1969)

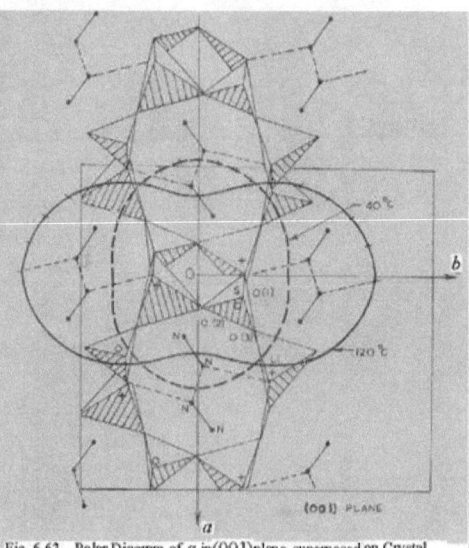

Fig. 6.62. Polar Diagram of α in (001)plane superposed on Crystal
structure of Li(N$_x$H$_x$)SO$_4$ (Devanarayanan,1966, 1969)

6.5.9 Thermal Expansion coefficient Measurements in Determining the Direction of FE axis in a FE crystal

Measurement of linear thermal expansion coefficient and its temperature variation in a ferroelectric (FE) crystal through its Curie point can determine not only the nature of the phase transition but also the T_C. A summary of the experimental data available on representative FE crystals (no isomorphs), viz., $BaTiO_3$, $LiTaO_3$, $PbTa_2O_6$, KDP (KH_2PO_4), TGS ($(NH_2CH_2COOH)_3.H_2SO_4$), $NaNO_2$ and $NaH_3(SeO_3)_2$, have been listed in a Table. Also included are symmetries in the PE and FE phases, T_C, direction of $\vec{P}_S$ (FE axis), direction of predominant dielectric anomaly and the direction of maximum negative thermal expansion coefficient anomaly. It is noteworthy from a perusal of the Table 6.2 (Devanarayanan, 1969). that in all crystals the FE-axis has the same direction as the one along which the crystal exhibits pronounced 'negative expansion' anomaly. In $NaH_3(SeO_3)_2$, however, the Triclinic FE crystal structure being a pseudo-symmetric of the PE monoclinic phase, the observed FE axis is the "a'-axis" $<313>$ direction along which dielectric anomaly is pronounced, and this coincides with the principal axis of the thermal expansion ellipsoid

Table. 6.2 (Devanarayanan, 1969)

Material	Symmetry of the Phase		Curie Temp. T_C °C	Ferroelectric axis	Direction of maximum dielectric anomaly C	Direction of Principal axis having negative anomaly	Remarks	References
	Paraelectric	Ferroelectric						
1. Barium Titanate Ba TiO₃ (Perovskite)	Cubic m3m	Tetragonal 4mm	120	°tetra (i.e. °out)	°tetra & a tetra	°tetra	At room temperature $c_t \neq t_a$ yet near T_c, $c_t = t_a$	1-7, 10, 11, 12 and 15.
2. Lithium Tantalate Li Ta O₃	Trigonal 3m		600	c	c	c		13, 14 and 15.
3. Lead Metaniobate Pb Nb₂ O₆ (Alkali tungsten Bronze type)	Tetragonal 4/mmm	Orthorhombic	570	b ortho	b ortho and a ortho	b ortho	$t_a = t_c$ in the region of the T (Two-dimensional ferroelectric).	1, 15, 16, 17 and 18.
4. Lead meta-tantalate Pb Ta₂O₆	Ortho-rhombic mm2	Orthorhombic mm2	265	⊥ to [001]	⊥ [001] ≅ b axis	b axis (⊥ to [001])	"Unidirectional" ferroelectric. $t_c \neq t_a$	19, 20
5. Potassium dihydrogen phosphate KH₂PO₄ (KDP)	Tetragonal 42m	Orthorhombic mm2	~150	°tetra	°tetra	°tetra		1, 15, 21
6. Triglycine sulphate (NH₂CH₂COOH)₃.H₂SO₄ (TGS)	Monoclinic 2	Monoclinic 2/m	49	b mono	b mono	b mono		4, 15, 22 & 23.
7. Sodium Nitrite NaNO₂	Ortho-rhombic mmm	Orthorhombic mm2	160	b	b	c		15, 21, 25, 26, 27 and 28
8. Sodium selenite NaH₃(SeO₃)₂	Mono-clinic 2/m	Triclinic 1	~79	[313]	a'	a'	The projection of [313] on (010) plane in ∥ a'-axis	Chapter V of this thesis

along which negative maximum is reported. The conclusion is the first ever reported (Devanarayanan, 1969) observation in literature that behaviour of the thermal expansion property can uniquely determine the direction of the Polar axis in a FE crystal.

6.5.10 Thermal conductivity K_c of ferroelectric

K_c of Rochelle salt single crystal was measured in the range -110 $^\circ C$ to +33 $^\circ C$, (Dj. Kristic and R. Blinc (1969) Phys. Lett., 30A, 387-88.). Significant anomalies in K_c ($Cal.cm^{-1}.s^{-1}.^\circ C^{-1}$) were found at both the Curie temperatures. At the lower T_C there is a dip, with $K_c \approx 15$, whereas at the upper T_C there is a peak $K_c \approx 20..$

6.5.11 Spontaneous Birefringence of Barium titanate

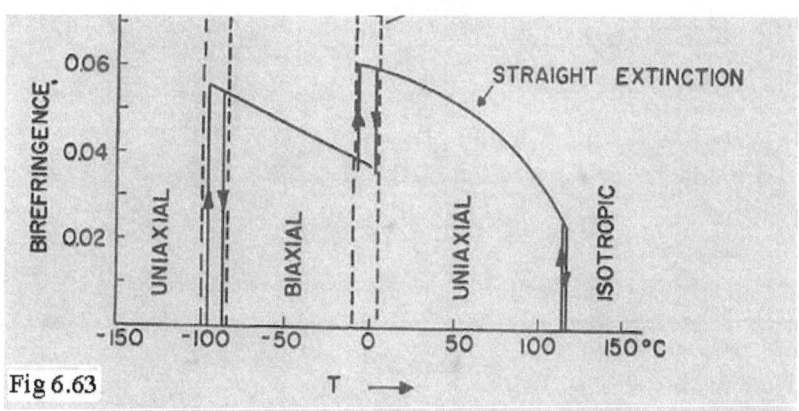

Fig 6.63

Spontaneous birefringence of BaTiO$_3$ (after Kay and Vousden).

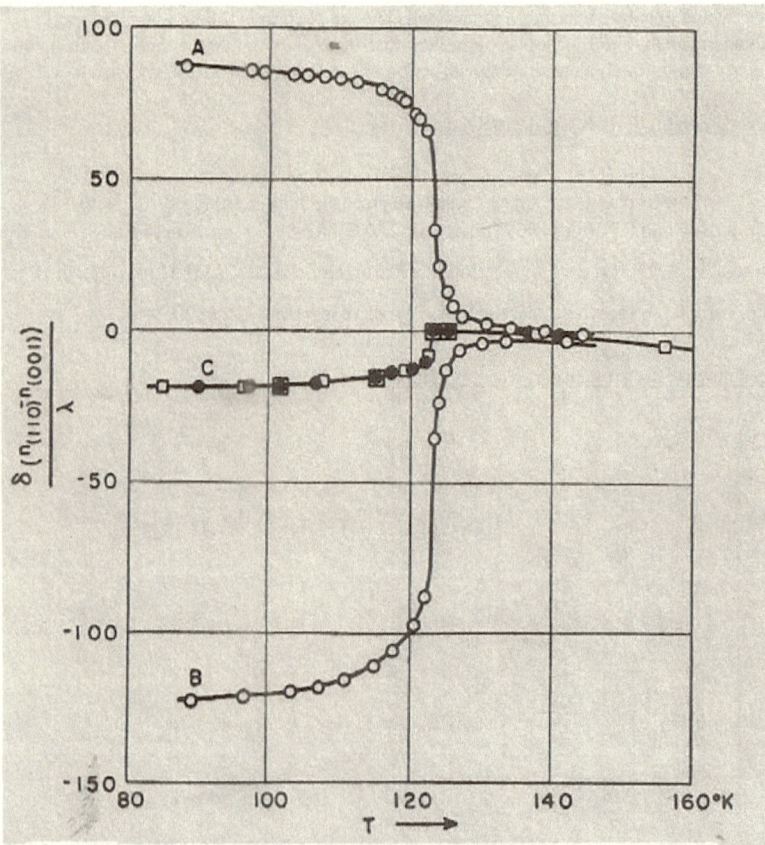

Fig 6.64. Spontaneous change of birefringence $(n_{(110)} - n_{(001)})/\lambda$ of KDP crystal. (A) Single domain with "up" polarization achieved by a bias of $+3kV/$ cm. (B) Single domain with "down" polarization achieved by a bias of $-3kV/$ cm. (C) Mean value between curves A and B, identical with curve for insulated crystal (after Zwicker and Scherrer).

6.5.12 Electro-optic coefficients

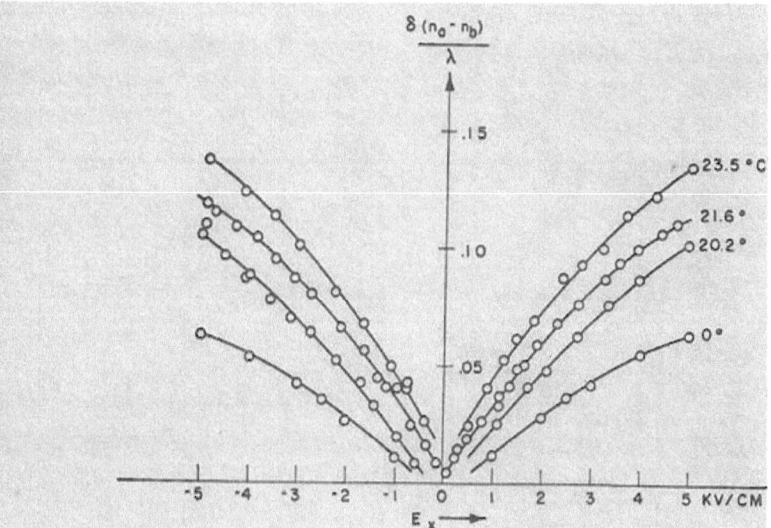

Fig 6.65 Linear electro-optic effect in Rochelle salt "induced" by the Spontaneous polarization. E along ferroelectric axis [100], light traveling along [001]. Near the origin the figure would have to be completed with quadratic hysteresis loops (after Mueller).

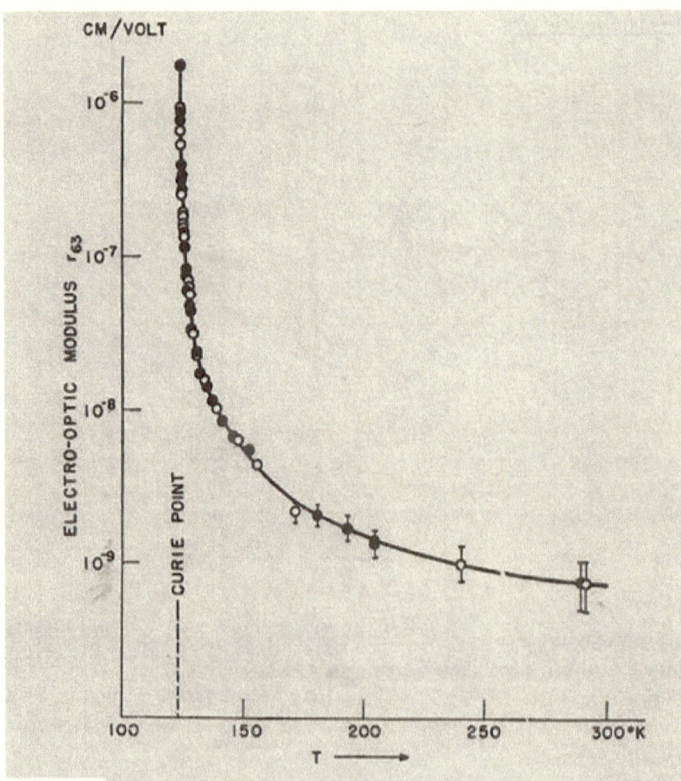

Fig 6.66 Curie-Weiss law of the electro-optical modulus r_{63}' of KH_2PO_4
(Zwicker & SCherrer, 1944)

6.5.13 Elastic and Piezolectric modulii

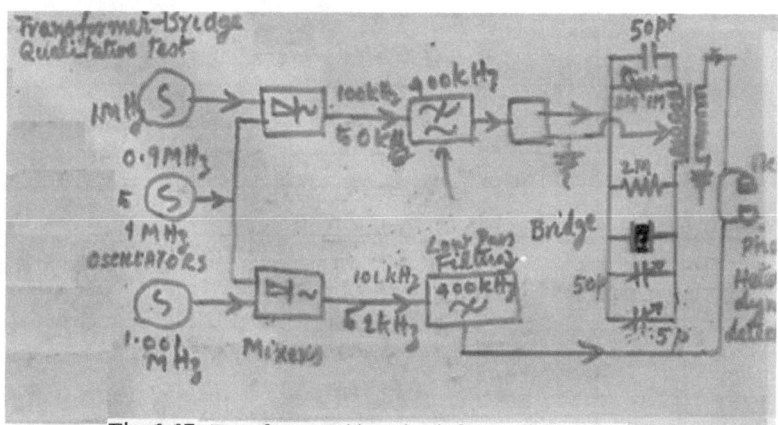

Fig 6.67 Transformer Bridge Circuit for Qualitative Test

Consider the case of KDP crystal.

$$KDP : \quad \frac{\text{Tetragonal}}{\text{Point Group, } D_{2d} - I\bar{4}2d} \quad \xleftarrow{T_C = 123K} \quad \frac{\text{Orthorhombic}}{\text{Point Group } Fdd2}$$

The paraelectric phase above T_C is not centro-symmetric, means the crystal is piezoelectric. The matrix for the piezo-electric strain coefficient is

$$
\begin{bmatrix}
& & & & X_Y \downarrow & \\
0 & 0 & 0 & d_{14} & 0 & 0 \\
0 & 0 & 0 & 0 & d_{14} & 0 \\
0 & 0 & 0 & 0 & 0 & d_{36}
\end{bmatrix}
$$

Single stress X_Y is associated with the polarization along the Ferroelectric axis. Therefore,

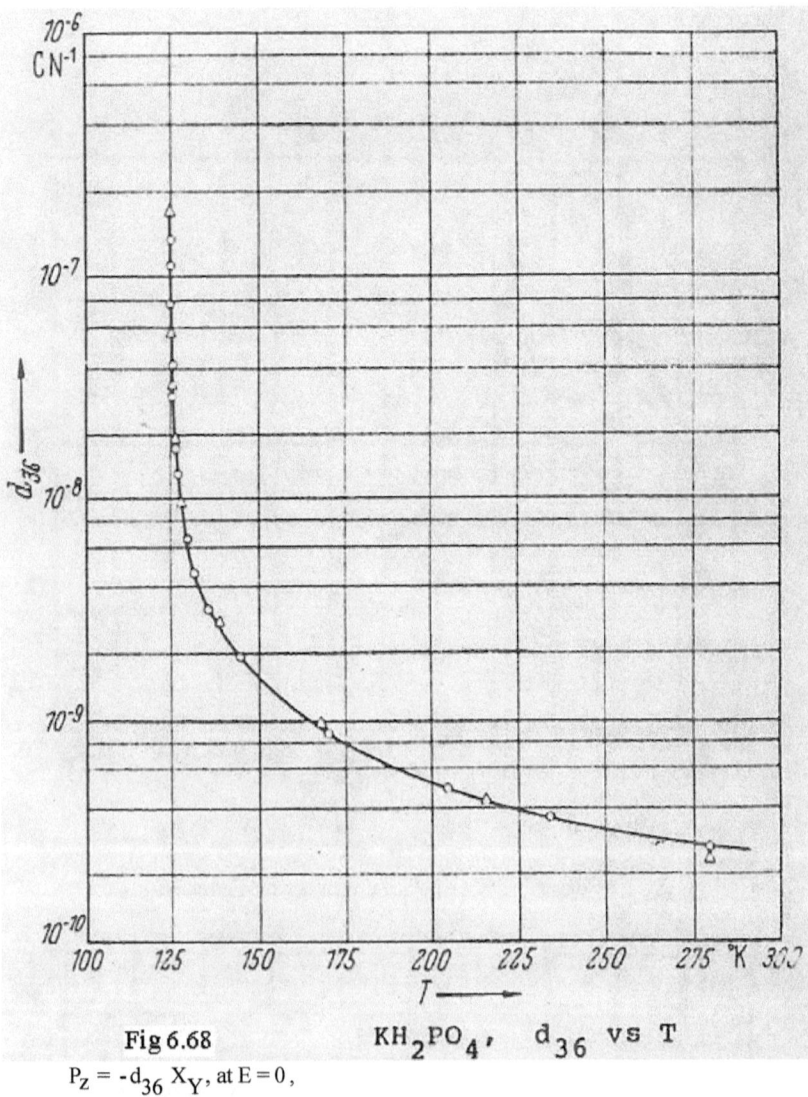

Fig 6.68 KH_2PO_4, d_{36} vs T

$P_Z = -d_{36} X_Y$, at $E = 0$,

In the paraelectric phase. The temperature behaviour of the piezoelectric coefficient d_{36} .is shown in Fig.6.68   (Hellwege & Hellwege, 1969).

In the paraelectric phase it follows the Curie-Weiss Law,

$$d_{36} = d_{36}^0 + \frac{B}{T-T_0},$$

where $B = 1.26 \times 10^{-4}$ cgs units .(Kanzig, 1957), and $T_0 = T_C$. $d_{36} \approx 8 \times 10^{-8}$ cgs esu is temperature dependent. At T_C, $d_{36} \approx 69.6 \times 10^{-8}$ cgs esu

d_{14} also shows anomalous behaviour near T_C. (Kanzig, 1957)(Fig 6.).

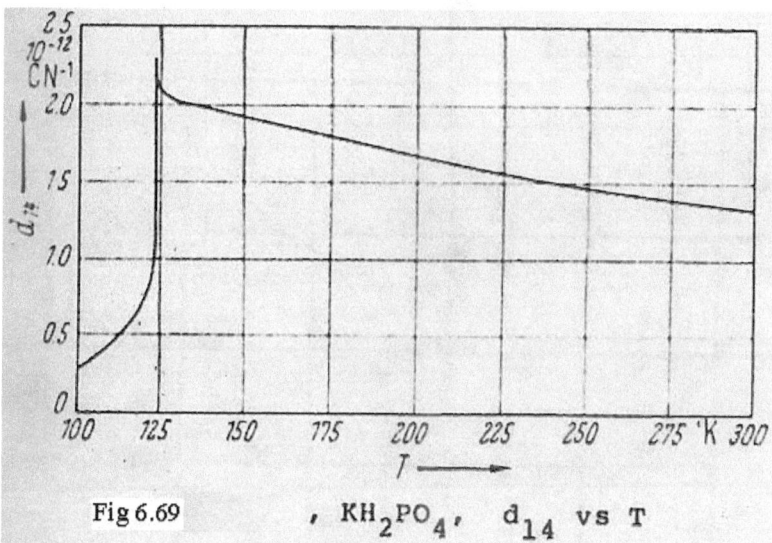

Fig 6.69   , KH_2PO_4, d_{14} vs T

In the ferroelectric phase the tensor changes to

$$\begin{bmatrix} 0 & 0 & 0 & 0 & d_{15} & 0 \\ 0 & 0 & 0 & d_{24} & 0 & 0 \\ d_{31} & d_{31} & d_{33} & 0 & 0 & 0 \end{bmatrix}.$$

$d_{31} = d_{36}$. The value of d_{36} vanishes below T_0. Hence this is the true piezoelectric constant, and exhibits no anomaly.

Elastic and Piezoelectric constants in $BaNaNb_5O_{15}$ are measured (Warner, AW, Coquin,, GA &JL Fink (1969), J. Appl. Phys., 40, 4353-56) at carious crystallographic orientation of plate-shaped samples

The elastic modulii of KDP crystal vary with temperature (Hellwege & Hellwege, III/2, 1969) in the paraelectric phase is represented by its tensor,(Fig 6.70.).

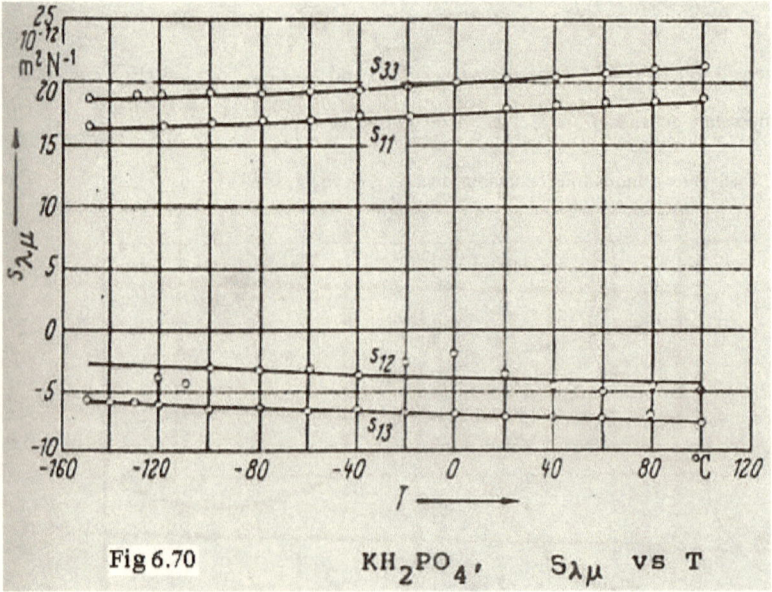

Fig 6.70 KH_2PO_4, $S_{\lambda\mu}$ vs T

For elastic compliance constants s_{ij}

$$\begin{bmatrix} s_{11} & s_{12} & s_{13} & 0 & 0 & 0 \\ s_{12} & s_{11} & s_{13} & 0 & 0 & 0 \\ s_{13} & s_{13} & s_{33} & 0 & 0 & 0 \\ 0 & 0 & 0 & s_{44} & 0 & 0 \\ 0 & 0 & 0 & 0 & s_{44} & 0 \\ 0 & 0 & 0 & 0 & 0 & s_{66} \end{bmatrix}$$

Similarly for elastic stiffness constant c_{ij}

$$\begin{bmatrix} c_{11} & c_{12} & c_{13} & 0 & 0 & 0 \\ c_{12} & c_{11} & c_{13} & 0 & 0 & 0 \\ c_{13} & c_{13} & c_{33} & 0 & 0 & 0 \\ 0 & 0 & 0 & c_{44} & 0 & 0 \\ 0 & 0 & 0 & 0 & c_{44} & 0 \\ 0 & 0 & 0 & 0 & 0 & c_{66} \end{bmatrix}$$

Both s_{66} and c_{66} show anomalous temperature dependence.

In the ferroelectric phase,

$$\begin{bmatrix} s_{11} & s_{12} & s_{13} & 0 & 0 & 0 \\ 0 & s_{22} & s_{23} & 0 & 0 & 0 \\ 0 & 0 & s_{33} & 0 & 0 & 0 \\ 0 & 0 & 0 & s_{44} & 0 & 0 \\ 0 & 0 & 0 & 0 & s_{55} & 0 \\ 0 & 0 & 0 & 0 & 0 & s_{66} \end{bmatrix}.$$

c_{66} of the paraelectric phase drops to zero at the T_C. (Fig 6.). (Hellwege & Hellwege, 1969; Zheludev, 1971).

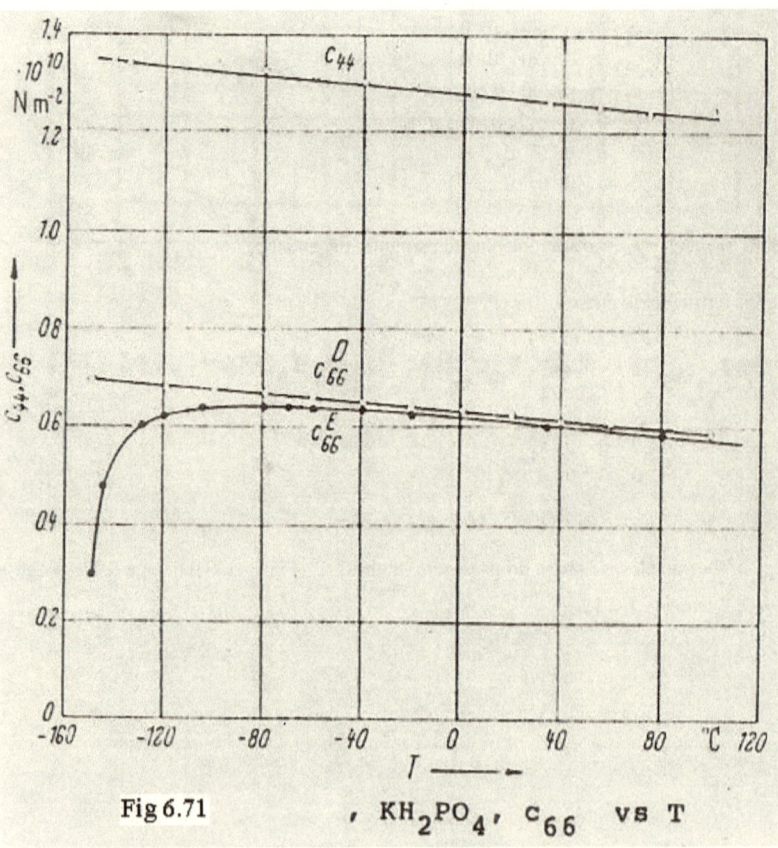

Fig 6.71　　　　　　　　, KH_2PO_4, c_{66}　vs T

In the case of Rochelle salt,($Na\ K\ C_4H_4O_6.4H_2O$)

RS:	Orthorhombic Space Group $P2_12_12_1$	$\xleftarrow{\quad T_C = 297K \quad}_{P_S \parallel a-axis}$	Monoclinic Space Group $P2_1$	$\xleftarrow{\quad T_C = 255K \quad}$	Orthorhombic Space Group $P2_12_12$

In the paraelectric phase, the crystal is non-centrosymmetric and show piezoelectric behaviour.

$$\begin{bmatrix} 0 & 0 & 0 & d_{14} & 0 & 0 \\ 0 & 0 & 0 & 0 & d_{25} & 0 \\ 0 & 0 & 0 & 0 & 0 & d_{36} \end{bmatrix}$$

$d_{14} = + \dfrac{B}{T-T_0}$, with $B = 8.67 \, x10^{-5} \, cgs$ units (Blinc & Zeks, 1974).

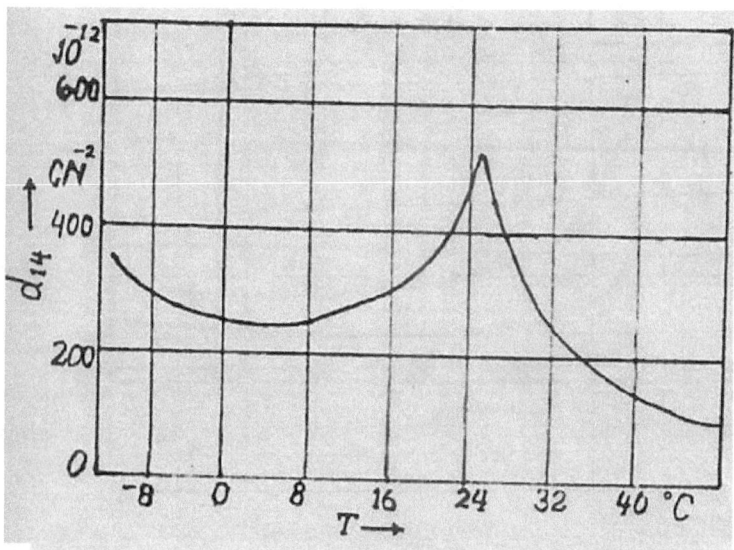

Fig 6.72 $NaKC_4H_4O_6 \cdot 4H_2O(R.S.)$, d_{14} vs T

Also $T_0 = T_C$.

Elastic compliance constants

$$\begin{bmatrix} s_{11} & s_{12} & s_{13} & 0 & 0 & 0 \\ 0 & s_{22} & s_{23} & 0 & 0 & 0 \\ 0 & 0 & s_{33} & 0 & 0 & 0 \\ 0 & 0 & 0 & s_{44} & 0 & 0 \\ 0 & 0 & 0 & 0 & s_{55} & 0 \\ 0 & 0 & 0 & 0 & 0 & s_{66} \end{bmatrix}$$

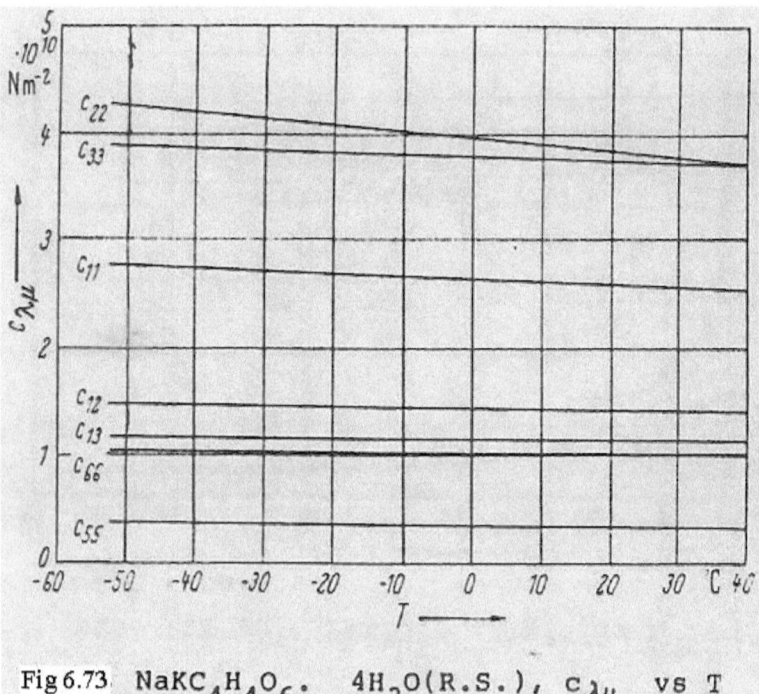

Fig 6.73 $NaKC_4H_4O_6 \cdot 4H_2O(R.S.)$, $c_{\lambda\mu}$ vs T

In the case of TGS

$$TGS: \begin{array}{|c c c|} \hline \text{Monoclinic} & T_C = +49.4\ °C & \text{Monoclinic} \\ \hline 2_1/m & \longleftarrow \longrightarrow & P2_1 \\ \hline \end{array}$$

In the paraelectric phase the crystal is centro-symmetric and is not piezoelectric. Below the Curie temperature, piezoelectric coefficient tensor is

$$\begin{bmatrix} 0 & 0 & 0 & d_{14} & 0 & d_{16} \\ d_{21} & d_{22} & d_{23} & 0 & d_{25} & 0 \\ 0 & 0 & 0 & d_{34} & 0 & d_{36} \end{bmatrix}$$

The temperature variation of these eight independent constants are reorted by Hellwege & Hellwege, Vol. III/2, 1969.

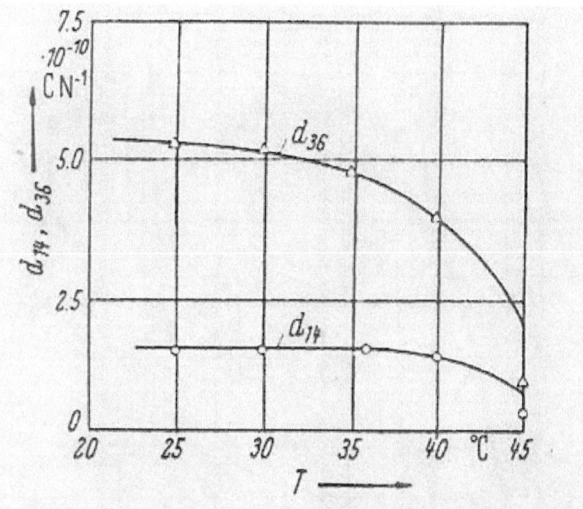

Fig 6.74 $(NH_2CH_2COOH)_3 \cdot H_2SO_4 \cdot$ (TGS),

The elastic constants are given by the tensor

$$
\begin{bmatrix}
s_{11} & s_{12} & s_{13} & 0 & 0 & 0 \\
0 & s_{22} & s_{23} & 0 & 0 & 0 \\
0 & 0 & s_{33} & 0 & 0 & 0 \\
0 & 0 & 0 & s_{44} & 0 & 0 \\
0 & 0 & 0 & 0 & s_{55} & 0 \\
0 & 0 & 0 & 0 & 0 & s_{66}
\end{bmatrix}
.
$$

The temperature dependence of the elastic coefficients are available in Hellwege & Hellwege, III/2, 1969..

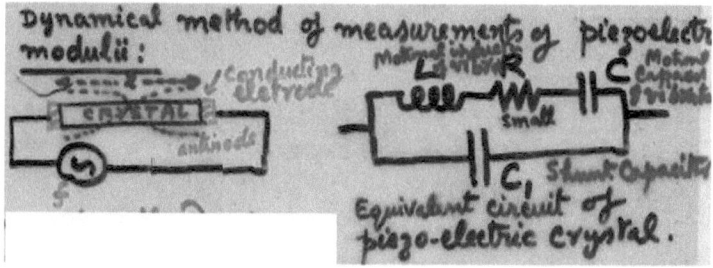

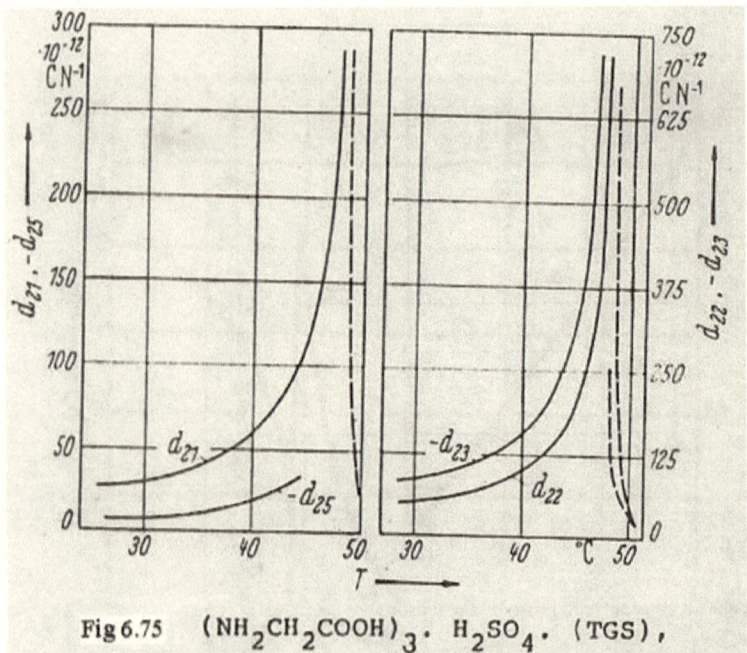

Fig 6.75 $(NH_2CH_2COOH)_3 \cdot H_2SO_4 \cdot (TGS)$,

$d_{21}, -d_{25} \quad d_{22}, -d_{23}$ vs T

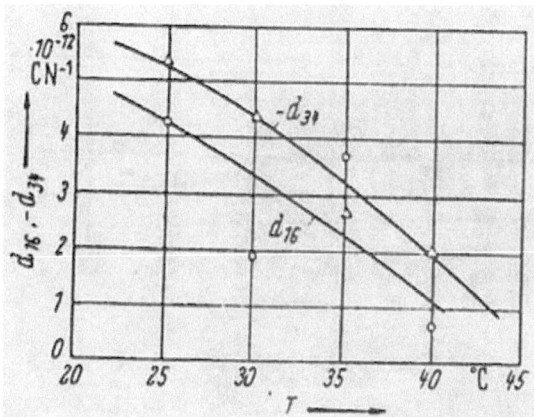

Fig 6.76 $(NH_2CH_2COOH)_3 \cdot H_2SO_4 \cdot (TGS)$,
d_{16}, $-d_{34}$ vs T

Wavelength λ of the mechanical eave in the crystal bar is

$$v = f\lambda$$

v =velocity of elastic wave.

f_0 = Natural frequency of longitudinal vibration of crystal,

When in resonance $f = f_0$, and $\lambda = 2\ell$.

$$\therefore v = 2\ell f_0$$

s =elastic compliance, $s = \dfrac{1}{\text{Youngs modulus}}$, ρ = density of crystal,

$$v = \frac{1}{\sqrt{\rho s}},$$

If the crystal bar is cut with $\ell \parallel X$ axis,

$$s_{11} = \frac{1}{\rho v^2} = \frac{1}{\rho 4\ell^2 f_0^2}$$

$$\boxed{s_{11} = \frac{1}{\rho 4\ell^2 f_0^2}}$$

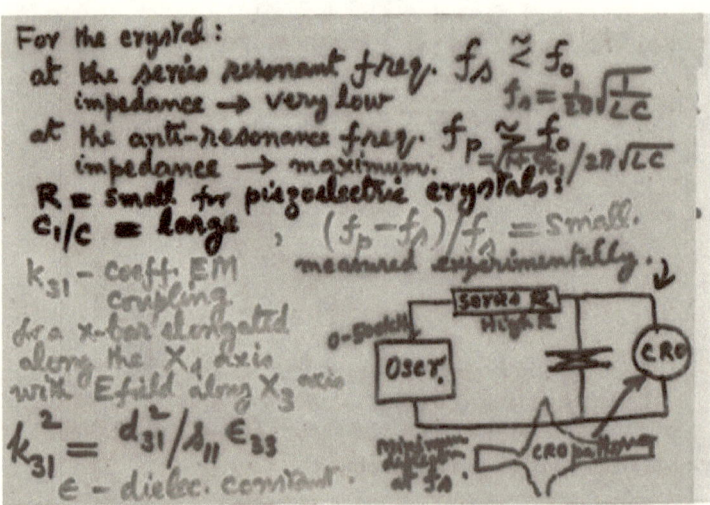

Fig 6.77   Dynamical Method of Measuring Piezoelectric modulii

Example

PbZrO$_3$ ceramic sample with $\ell = 51.04$ mm and breadth 6.20 mm, at room temperature, has thickness 2.02 mm and $\rho = 7.7$ $gm.cm^{-3}$. $f_s = 29.10$ kHz, $f_p = 30.13$ kHz

$$v = 2 f_s \ell = 298.800 \ cm.s^{-1}.$$

$$\rho s_{11} = 1/v^2; \ \text{gives} \ s_{11} = 14.5 x10^{-12} m^2 N^{-1}.$$

$$k_{31}{}^2 = \frac{\pi}{4} \frac{\Delta f}{f_s} \ (= \frac{d_{31}{}^2}{s_{11} \epsilon_{33}}); \ \kappa = 1400 = \frac{\epsilon'}{\epsilon_o};$$

$$k_{31} = \frac{\pi}{2} \sqrt{2.54 x10^{-2}} \approx 0.301;$$

$$d_{31} = 0.301 / \sqrt{(14.5 x10^{-12}) \frac{1400}{36\pi 10^9}} = 130 x10^{-22} MKS.$$

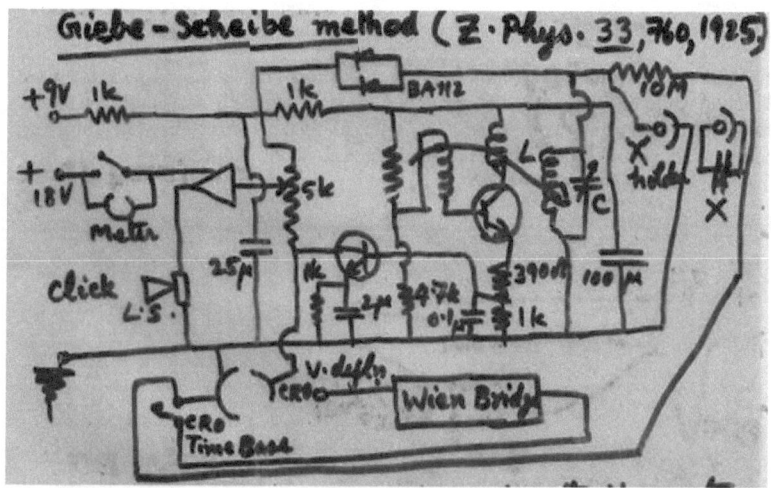

The linear combinations of the six adiabatic 3rd order elastic constants in cubic crystals of $SrTiO_3$ were determined by using the ultrasonic 2nd harmonic generation technique by James & Arnold (1969).

6.6 Raman Spectral analyses and Ferroelectric Transitions

In this section some of the experimental data available from the literature are reproduced.

Porto's notation of scattering geometry

It is a way to indicate the configuration of the Raman scattering experiment. This notation expresses the orientation of the crystal with respect to the polarization of the laser in both the excitation and analyzing directions. The notation of Porto, for Raman scattering processes, consists of four letters:

A(BC)D

$A \equiv$ The direction of the *propagation of the incident light* (k_i).

$B \equiv$ The direction of the *polarization of the incident light* (E_i).

$C \equiv$ The direction of the *polarization of the scattered light* (E_s).

$D \equiv$ The direction of the *propagation of the scattered light* (k_s).

Example: Y(XZ)X

In this case we have that the direction of the propagation of the *incident light* is along the
Y direction while the direction of the *scattered light* is along the X direction. The
direction of the polarization of the *incident light* is along the X direction while the
direction of the polarization of the *scattered light* is along the Z direction.

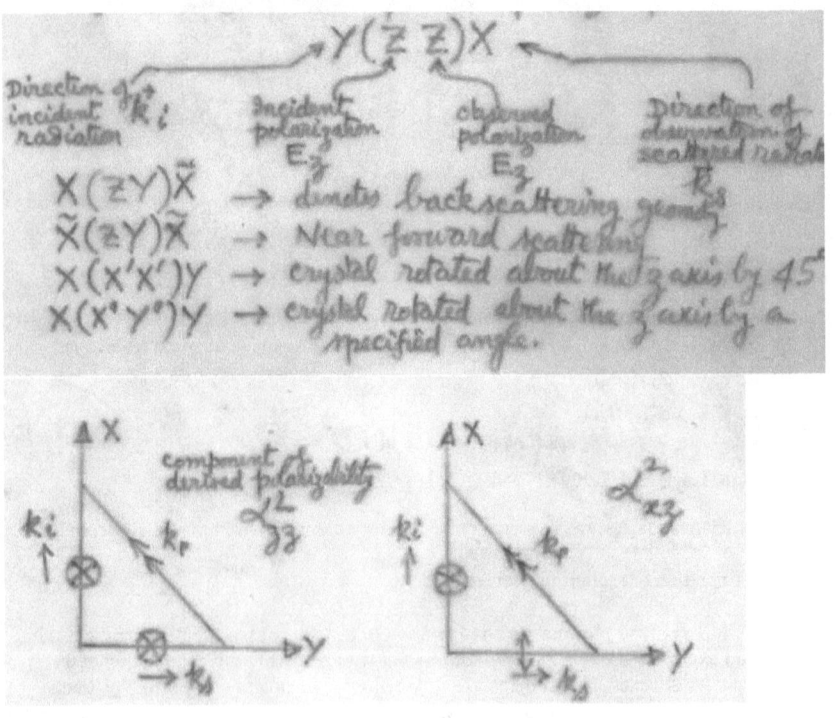

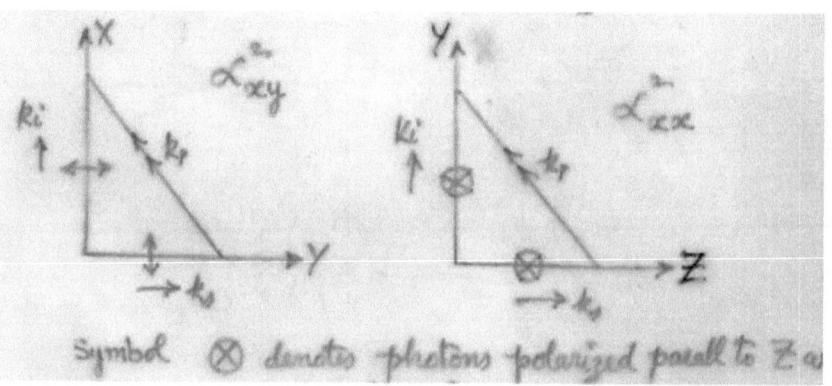

Symbol ⊗ denotes photons polarized parall to Z a

Barium Titanate

BaTiO$_3$:

Hexagonal		Cubic		Tetragonal		Orthorhombic		Trigonal
C6$_3$/mmc	$\xleftarrow{1460°C}$	Pm3m	$\xleftarrow{T_C = 120°C}$	P4mm, P$_S$‖ [001]	$\xleftarrow{T_C = 5.0°C}$	Amm2, P$_S$‖ [110]	$\xleftarrow{T_C = -90°C}$	R3m, P$_S$‖ [111]

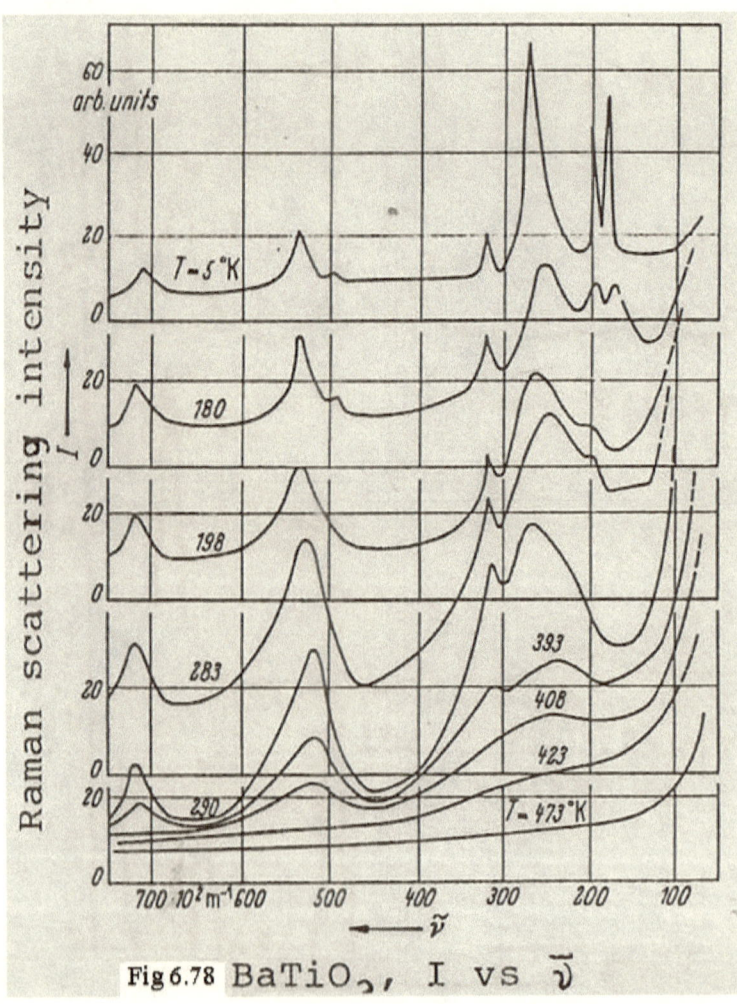

Fig 6.78 BaTiO₃, I vs $\bar{\nu}$

SbSI:	Tetragonal Point Group, P*nam*	$T_C = -15.7°C$	Orthorhombic Point Group P*na*2₁	$T_C = -36.6°C$	Monoclinic Point Group 2

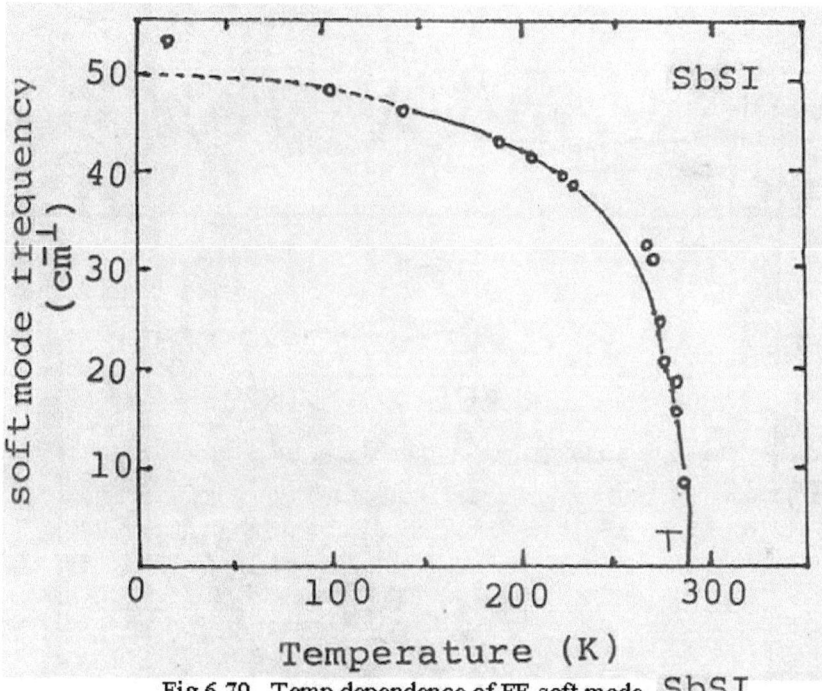

Fig 6.79 Temp dependence of FE soft mode SbSI

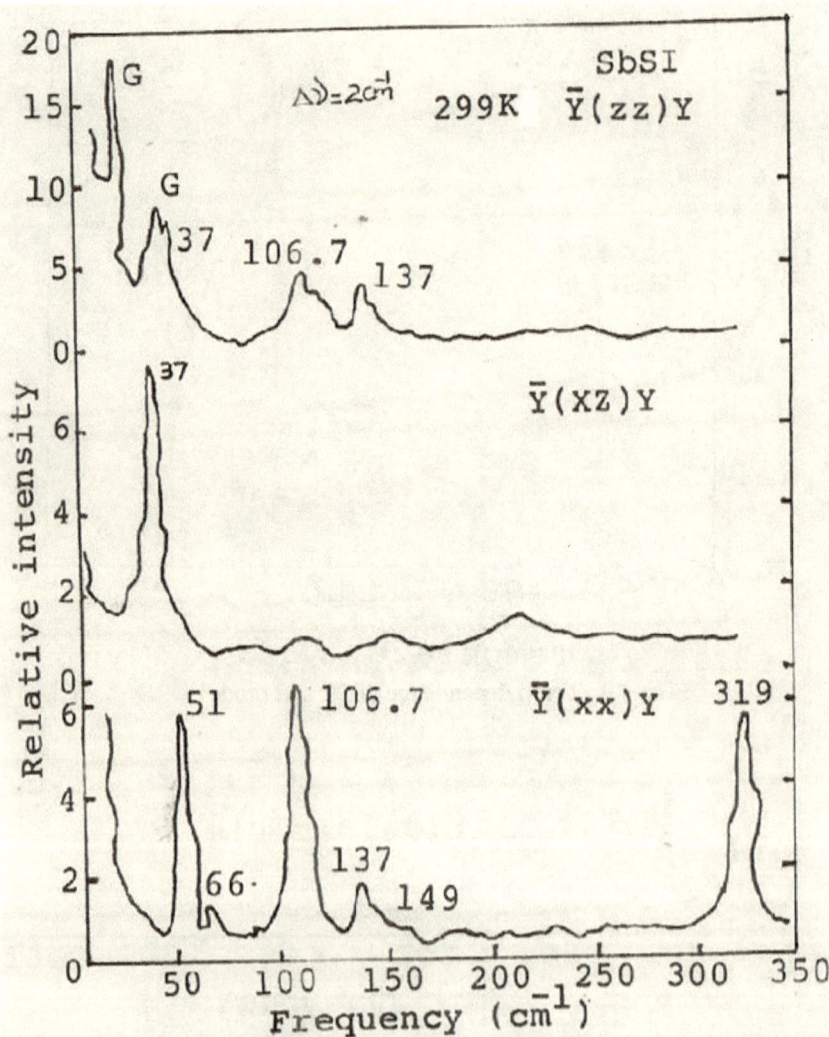

Fig 6.80 Raman spectrum of SbSI in the paraelectric phase.

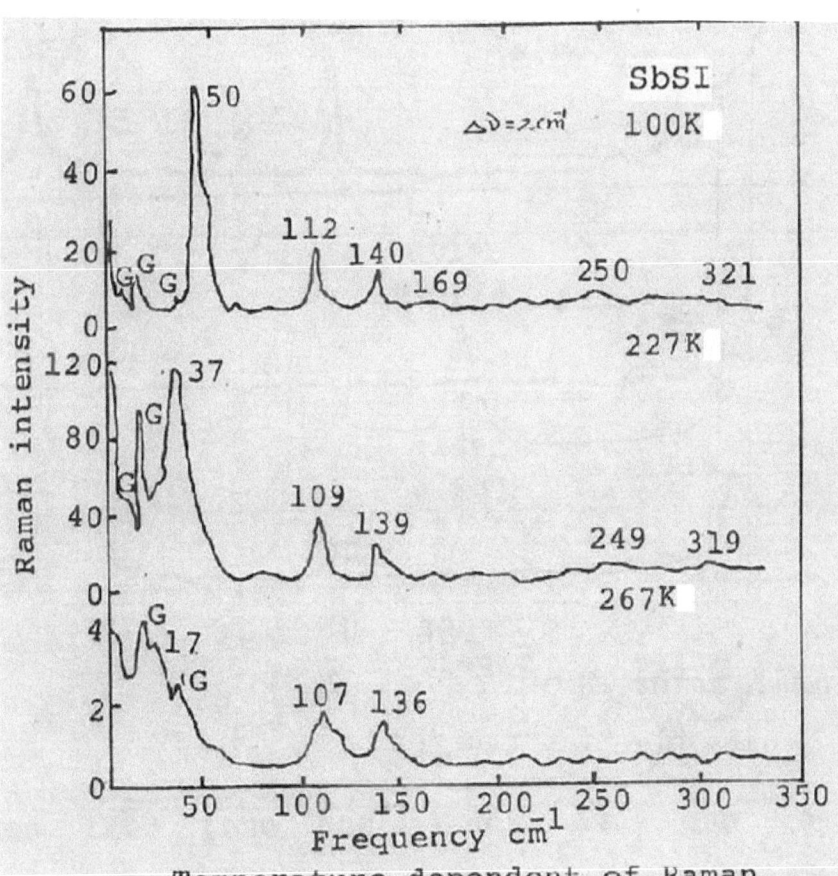

Fig 6.81., Temperature dependent of Raman spectrum of SbSI in the ferroelectric phase

KDP:	Tetragonal	$T_C = 123K$	Orthorhombic
	Point Group, $D_{2d} - I\bar{4}2d$	$\longleftrightarrow$	Point Group $Fdd2$

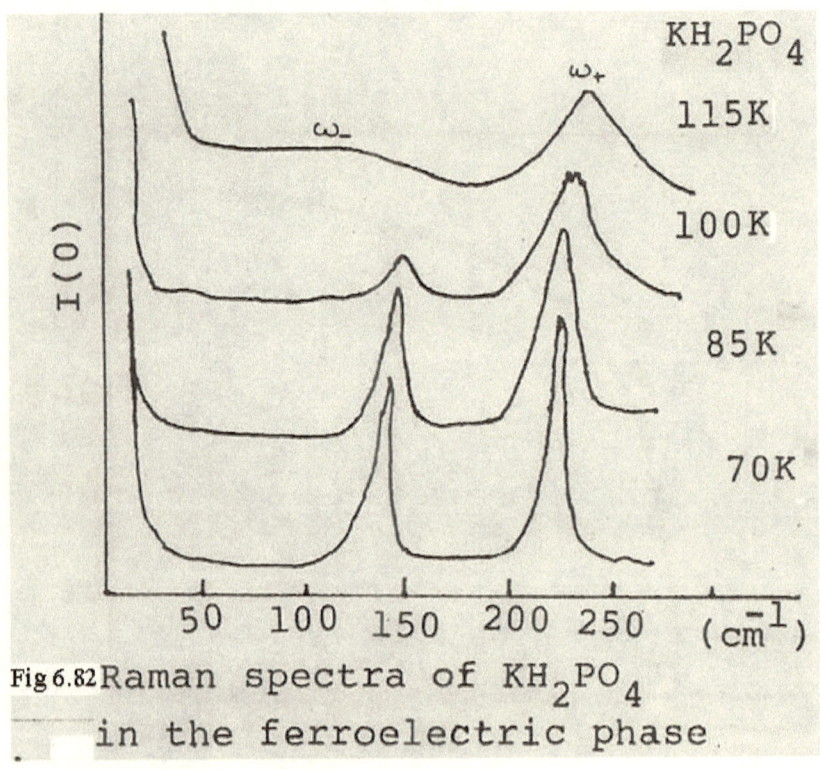

Fig 6.82 Raman spectra of KH_2PO_4 in the ferroelectric phase.

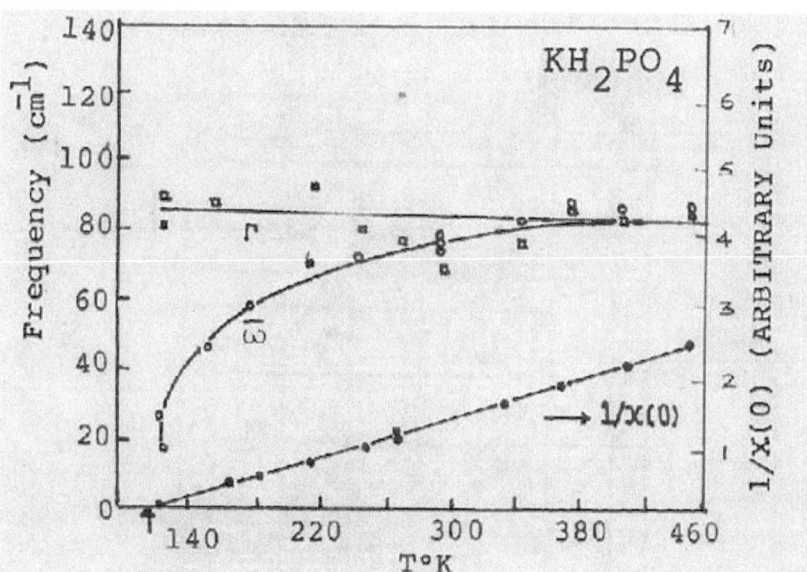

Fig 6.83 Temperature dependence of the soft mode parameters in KH_2PO_4

NaNO$_2$:	$T_C = 164.7°C$	Orthorhombic Point Group, - Immm	$T_C = 163.8°C$	Orthorhombic Point Group Im2m	$T_C = -105.0°C$

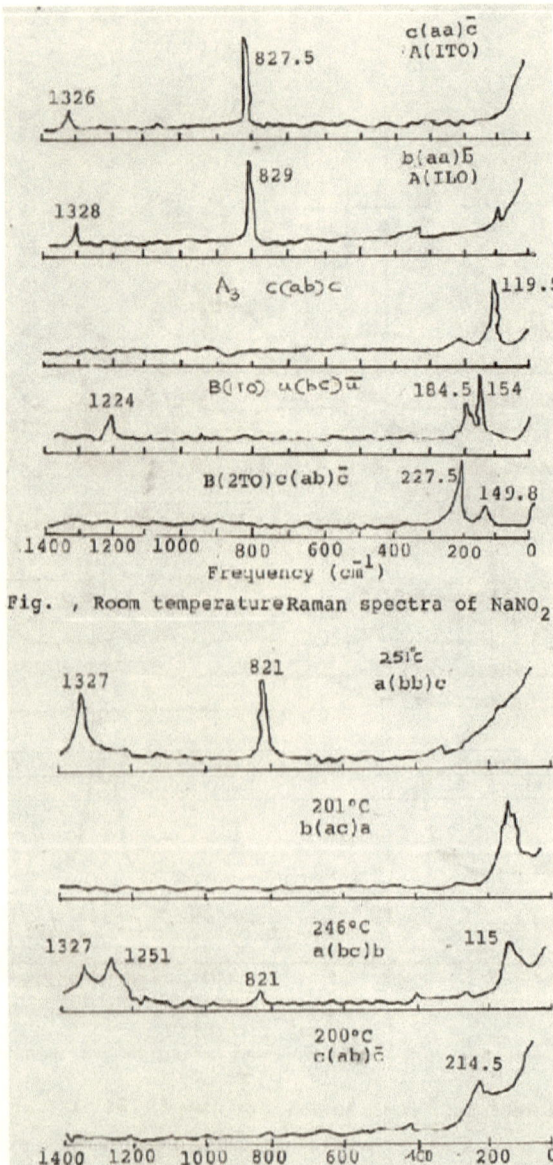

Fig. , Room temperature Raman spectra of NaNO$_2$

Fig 6.84 Raman spectra of NaNO$_2$ in the paraelectric phase

$Li_2Ge_7O_{15}$: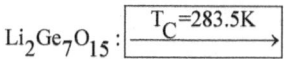

The Raman spectrum of Lithium hepta germanate by Wada et al., 1981 (Fig.6.85).The soft mode at $36.3 \ cm^{-1}$ at 21K decreases as the temperature decreases. Above the T_C=283.5K this mode ceases to be Raman active. The frequencies of the low lying modes as function of temperature is displayed in Fig 6. . The mode at $109.2 \ cm^{-1}$ at 293.5 K splits into three modes $107.6 \ cm^{-1}$, $110.6 \ cm^{-1}$, and $117.4 \ cm^{-1}$ at 106.5 K as the temperature is lowered. This is indicative of a superstructure in the crystal below the T_C. This suggest a prototype phase exits at the higher temperature range in the crystal.

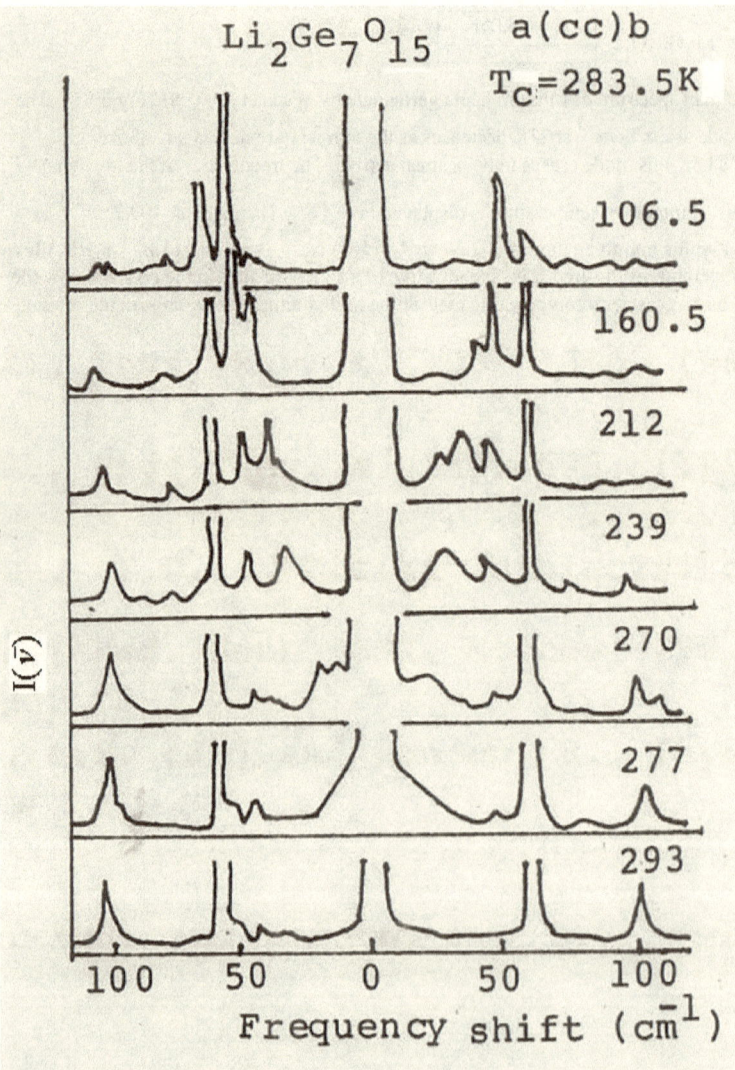

Fig 6.85 Raman spectra of $Li_2Ge_7O_{15}$

(Wada, M., *et al.*, 1981)

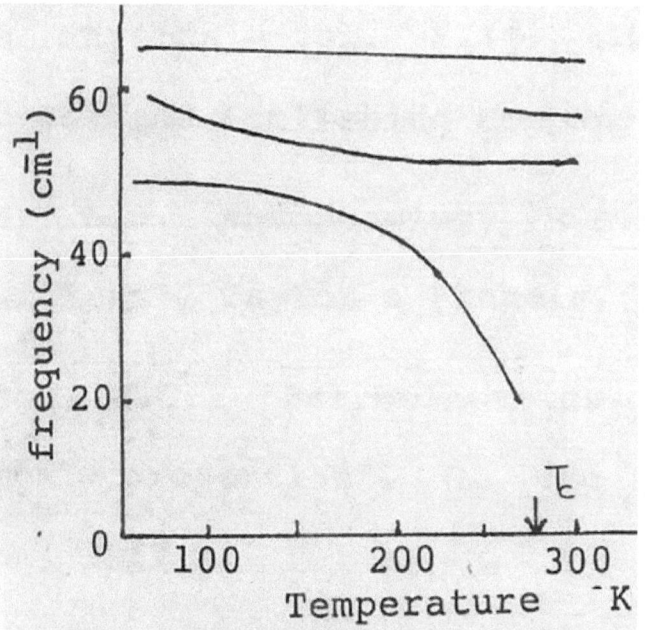

Fig 6.86 Frequency of the low lying modes in $Li_2Ge_7O_{15}$

LiNbO$_3$

LiNbO$_3$:	Tigonal Point Group, - R$\bar{3}$c	$\xleftarrow{\quad T_C = 1210K \quad}$	Trigonal Point Group R3c

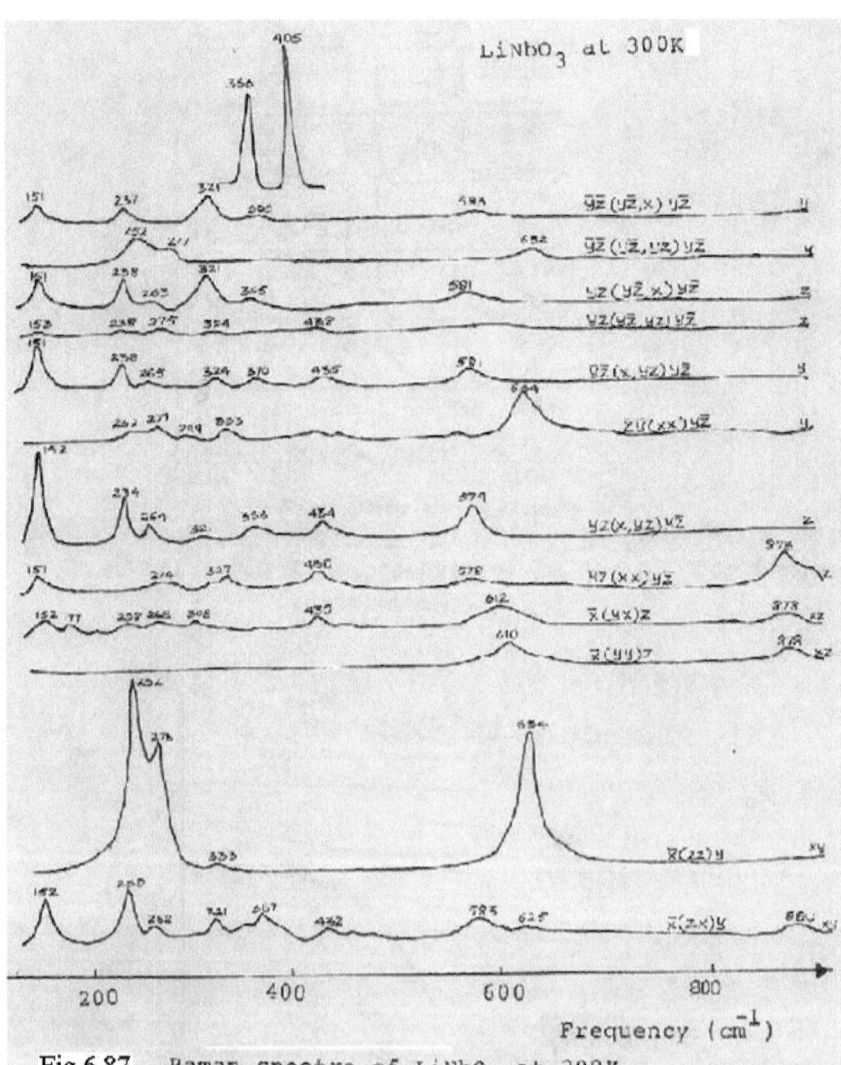

Fig 6.87   Raman spectra of LiNbO₃ at 300K

(Barker & Loudon, 1967)

Raman spectra of Incommensurate Transitions and Ferroelastic LiCsSO$_4$

$LiCsSO_4$:

Normal phase	$T = 202 K$	Ferroelastic
Orthorhombic, $D_{2h}^{16}-Pcmn$, $Z=4$	Incommensurate	Monoclinic, $C_{2h}^5-P2_1/n$, $Z=4$

Polarized Raman spectra were recorded for the $z(x\,x)y$ and $z(x\,z)y$ geometries.

Measurements were made in the range 30 L to 290K, at 25 different temperatures. Using SPEX 1402 double grating Spectrometer, with an Ar-ion laser operating at 514.5 nm as the exciting line striking the sample at 25 mW, with spectral resolution 1.5 cm^{-1}. A Displex ^{4}He closed cycle refrigerating system was used as a cryogenics.

A typical spectrum recorded (G. Morell, S. Devanarayanan & RS Katiyar,1990), shown in Fig 6.88, Fig 6.89 & Fig 6.90 indicates the quality of the instrument for all measurements.

(Morell, Devanarayanan & Katiyar, 1990, unpublished data)

Fig 6.88

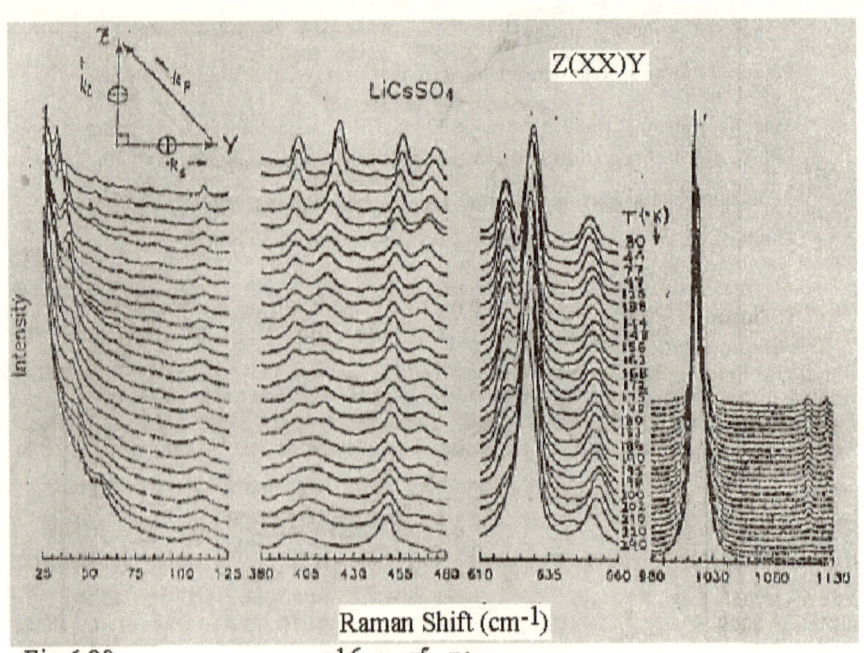

Raman Shift (cm⁻¹)

Fig 6.89 D_{2h}^{16} & C_{2h}^{5} Phases

Devanarayanan *et al.*, Proc. Kerala Acad. Sci., 2, 9 (1991)

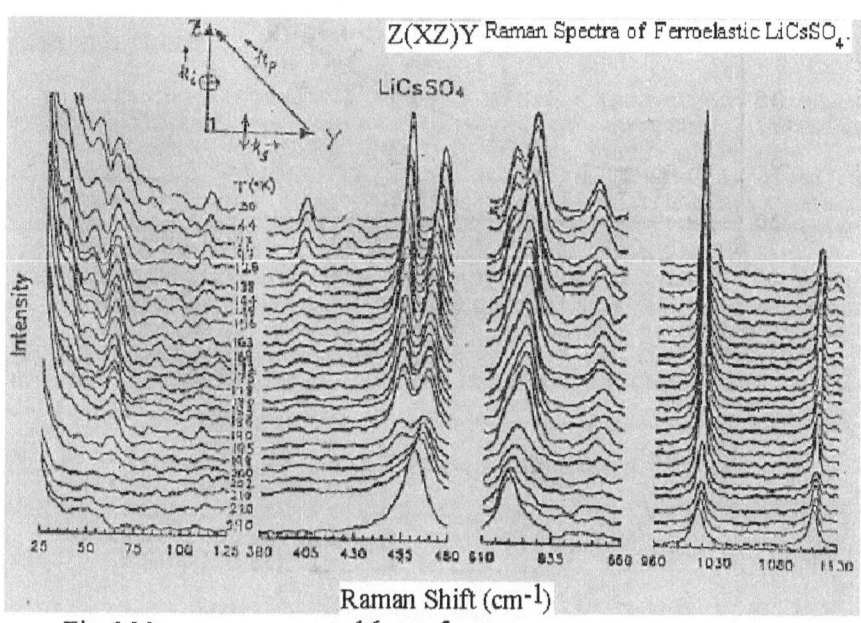

Fig 6.90 D_{2h}^{16} & C_{2h}^{5} Phases

Devanarayanan *et al*., Proc. Kerala Acad. Sci., **2**, 9 (1991)

In the D_{2h}^{16} phase, as per Factor group theory, there are 84 vibrational modes in the crystal. These are the 36 internal modes and 48 lattice (external) modes. Of the lattice modes 12 are libratory types (of SO_4) referred to as R' and 36 translatory (T'). The lattice modes are Raman phonons, as listed in Table 6.3. and are IR (irreducible representation) A_g, B_{1g}, B_{2g}, and B_{3g} species in the appropriate scattering geometry. In the low temperature C_{2h}^{5} phase only A_g and B_g species are present (Table 6.). The asymmetric stretching mode $\nu_1 = 1016\ cm^{-1}$, symmetrical bending mode $\nu_2 = 465\ cm^{-1}$, totally symmetric stretching mode (S-O) $\nu_3 = 1110\ cm^{-1}$ and asymmetrical bending $\nu_4 = 627\ cm^{-1}$ are the only observes phonons.

The intensity of $\nu_3 = 1110\ cm^{-1}$ indicates the phase change in the crystal at $T_i = 202K$ and again at $T_C = 160K$ (Fig 6.91).

$v_2 = 465\ cm^{-1}$ splits into $456\ cm^{-1}$ and $465\ cm^{-1}$.(Fig 6.92).

Phonons $627\ cm^{-1}$ and $620\ cm^{-1}$ are highly polarized and temperature dependent (Fig 6.93).

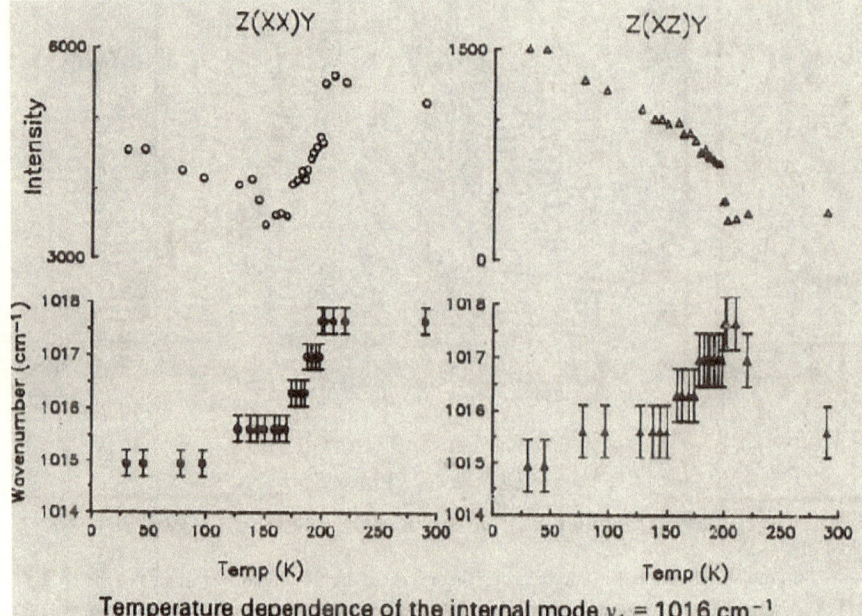

Temperature dependence of the internal mode $v_1 = 1016\ cm^{-1}$

Fig. 6.91 having *xx* and *xz* polarizations through the T_i and the T_C
(Morell, Devanarayanan & Katiyar, 1991)

Table   Raman-active phonons in LiCsSO$_4$

Crystal phase	Mode species	Lattice vibrations Acoustic n_{ac}	Optical $n(T')$	Librational $n(R')$	Internal modes of SO$_4^{2-}$ n_{vib}	Components
Normal (N) phase,	A_g	0	6	1	6	xx,yy,zz
Pcmn $(D_{2h}^{16} > T_i$	B_{1g}	0	3	2	3	xy
	B_{2g}	0	6	1	6	xz
	B_{3g}	0	3	2	3	yz
Commensurate (C) phase,	A_g	0	9	3	9	xx,yy,zz,xy
P2$_1$/n $(C_{2h}^5) < T_c$	B_g	0	9	3	9	xz,yz

The frequency and temperature dependence of these two modes are shown Fig. 6.93. The square of the $627\ cm^{-1}$ versus temperature plot is reproduced in Fig. 6.94. The plots show transition in the crystal at $T_i = 202K$. The square of the mode has a variation given by $(377856.5 \pm 41.6)\ T$. It reaches zero at $110K$, and thereafter is a constant till $30K$.

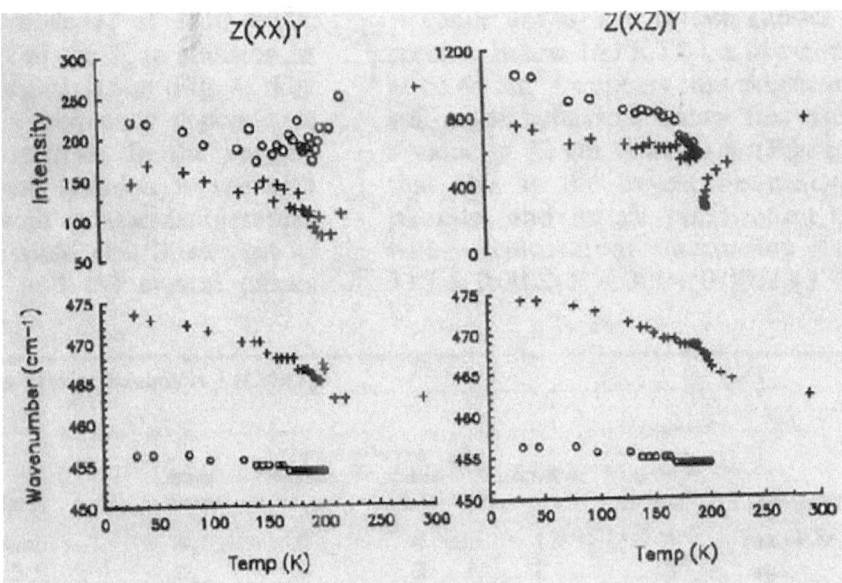

Fig. 6.92 Temperature-dependent internal mode v_2 at 465 cm^{-1} in the $z(xx)y$ and $z(xz)y$ scattering geometries.
(Morell, Devanarayanan & Katiyar, 1991)

Below $T_C = 160K$, a low energy lattice mode around $44\ cm^{-1}$ appears and is probably a soft mode that and whose value decreases to $32\ cm^{-1}$ at $30K$. This TO mode varies as $(31.7x0.0033(T-30) + 0.00023(T-30)^2$. This mode ps also an unlocking phase transition at $T_O = 30K$.

267

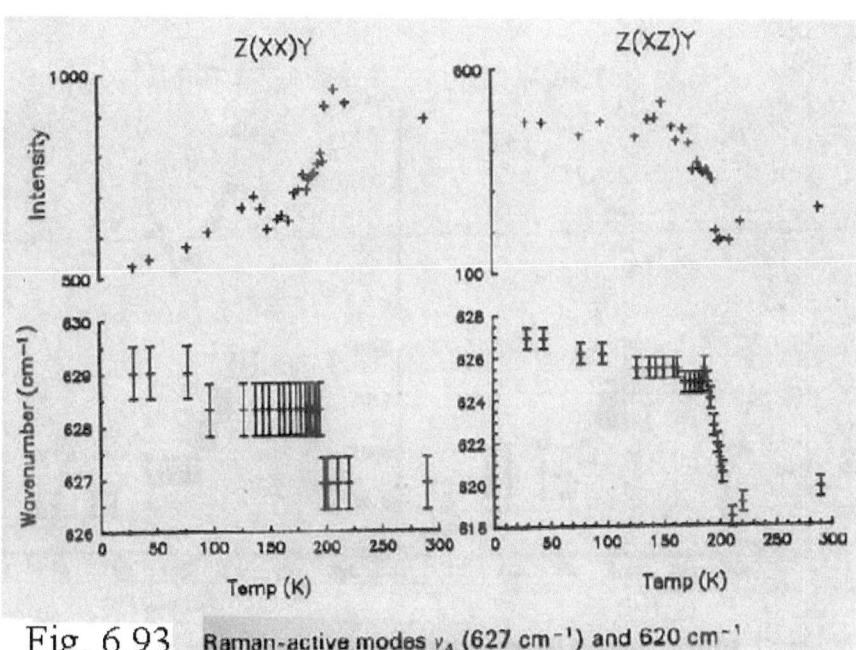

Fig. 6.93 Raman-active modes v_4 (627 cm^{-1}) and 620 cm^{-1}

temperature dependences of frequency and intensity in D_{2h} and C_{2h}^5
(Morell, Devanarayanan & Katiyar, 1991)

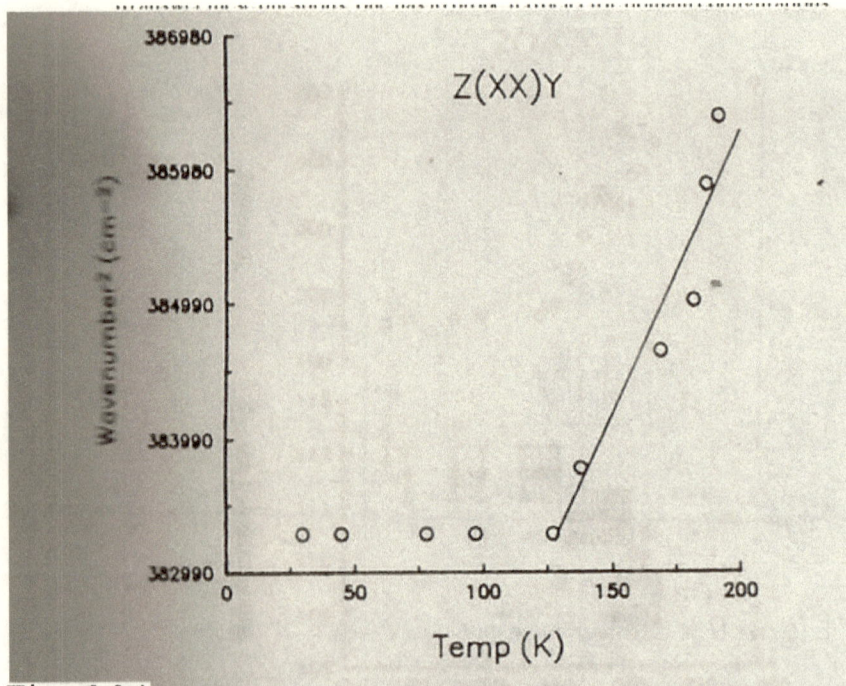

Fig. 6.94 Behaviour of the internal mode $v_4{}^2$ with temperature in ... ase in the $z(xx)y$ configuration. D_{2h}^{16}

(Morell, Devanarayanan & Katiyar, 1991)

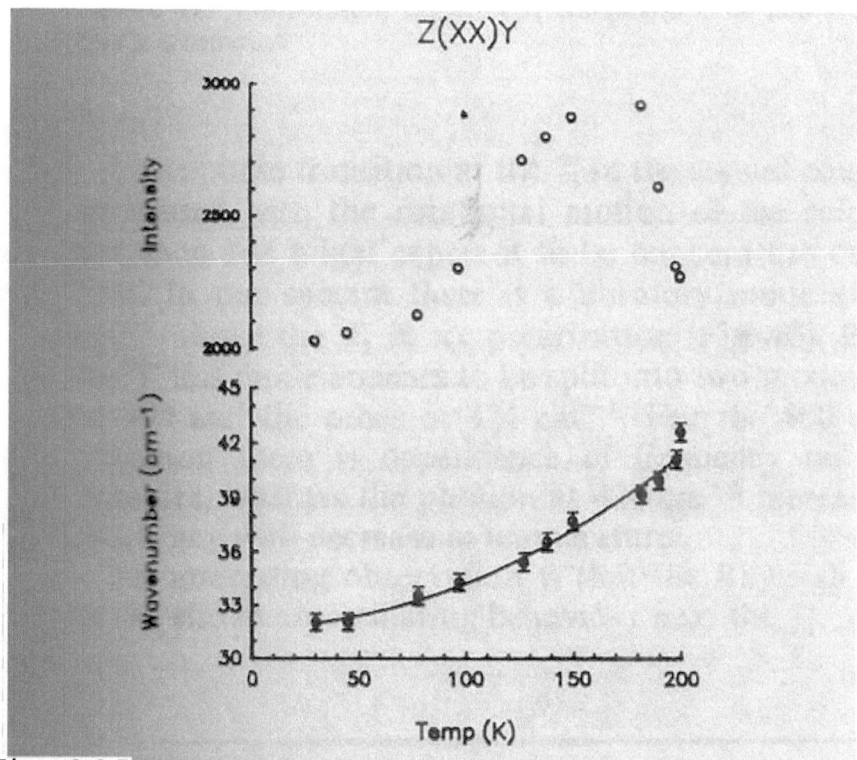

Fig. 6.95 The lowest frequency optical mode 44 cm⁻¹ in $z(xx)y$ geometry in the C_{2v} phase of LCS. (Morell, Devanarayanan & Katiyar, 1991)

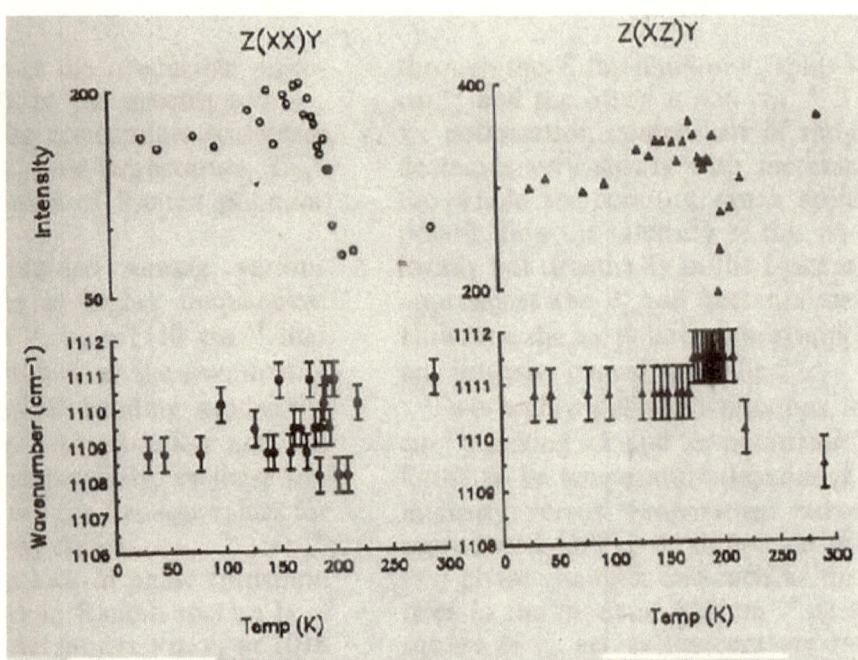

Fig. 6.96 Behaviour of *xx* and *xz* components of the phonon v_3 at 1110 cm^{-1} at the T_i and the T_c.

(Morell, Devanarayanan & Katiyar, 1991)

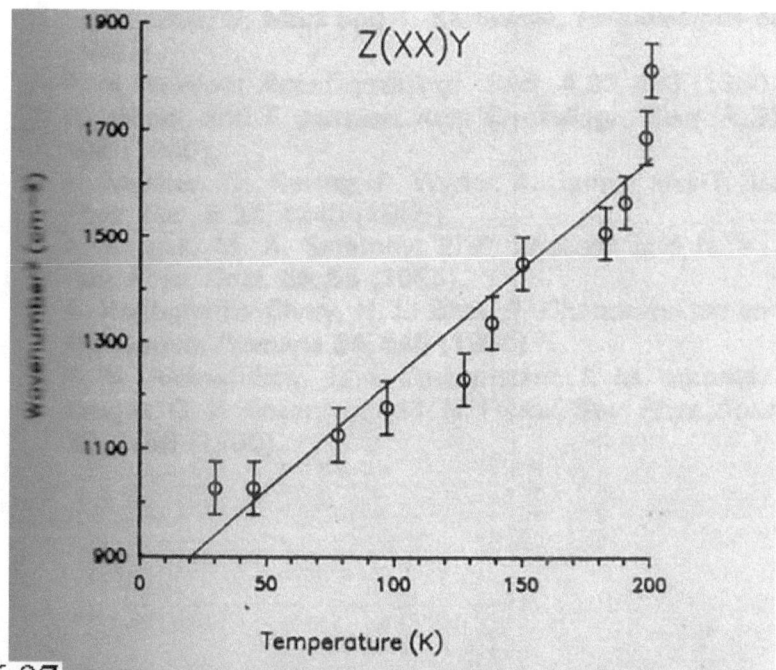

Fig. 6.97 Variation of square of frequency of the soft phonon in
perature in $z(xx)y$ scattering.

(Morell, Devanarayanan & Katiyar, 1991)

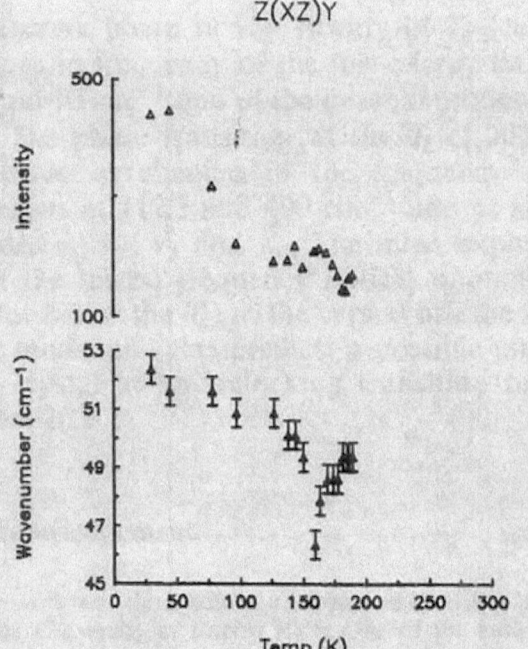

Fig. 6.98 Intensity versus temperature for the 50 cm⁻¹ mode with *xz* polarization showing the phase change at the T_c.

(Morell, Devanarayanan & Katiyar, 1991)

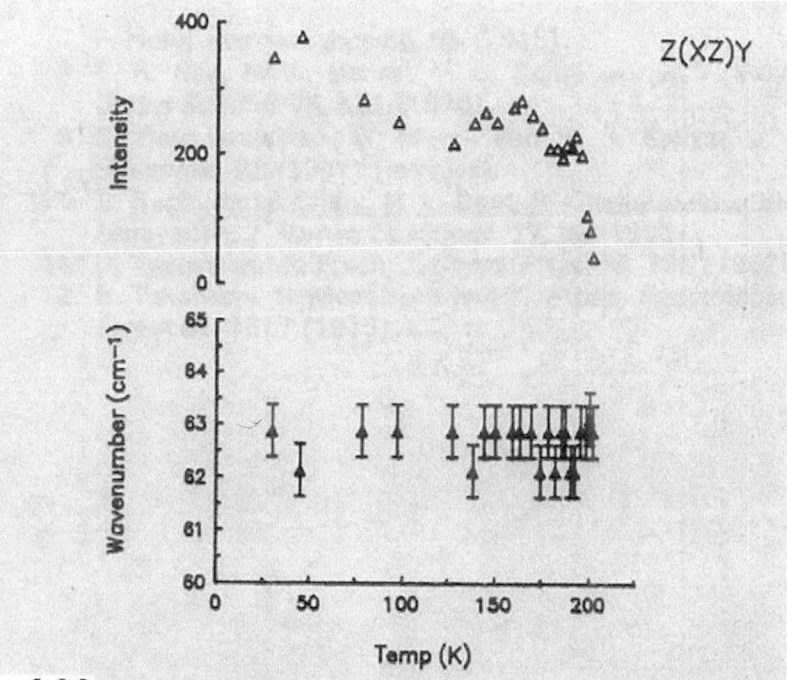

Fig. 6.99 Behaviour of the *xz* polarized lattice phonon at 62 cm^{-1} with temperature in the crystal.(Morell, Devanarayanan & Katiyar, 1991)

Thus

112.4 & 400 cm^{-1},N phase $D_{2h}^{16}-Pcmn$, $Z=4$	$T_i = 202K$ Incommensurate	465 cm^{-1}, Ferroelastic $C_{2h}^5-P2_1/n$, $Z=4$	$T_C = 160K$ 400 & 424 cm^{-1}	$T_Q = 20K$ 44 →32 cm^{-1}

KDP

(Peercy (1975, Bruce & Cowley, 1981) recorded the Raman spectra. The soft mode at 150 cm^{-1} in KDP is of resonant nature identifies as B_2 species. Though the soft mode decreases in the PE phase with temperature it drops suddenly at the T_C, indicates a second order transition. Just below the T_C in the FE phase the soft mode is A_1 which is a TA mode whose frequency attains zero, (Fig. 6.100).

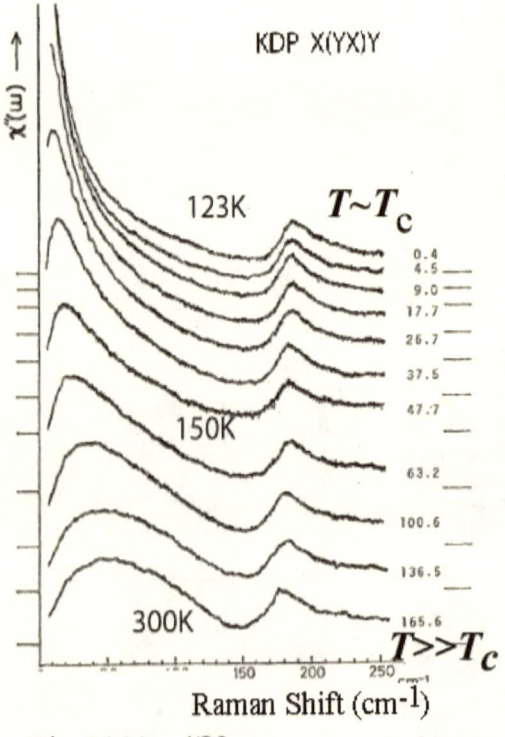

Fig 6.100   KDP

<u>Lead Zirconate titanate with lanthanum doping</u>, (PLZT) ceramic specimens

In situ Raman spectra focused on a fixed grain under various compressed stresses (Li Ping Liang, et al.,2016), were obtained for PLZT ceramic samples.

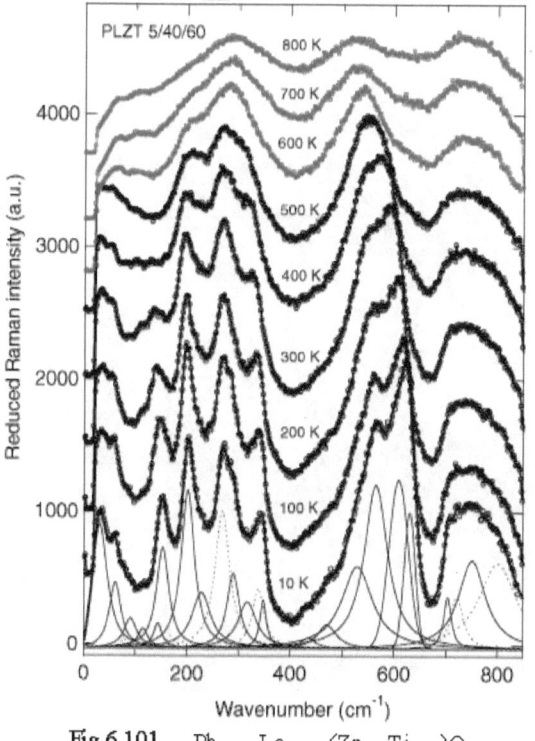

Fig 6.101 $Pb_{0.95}La_{0.05}(Zr_{0.4}Ti_{0.6})O_3$

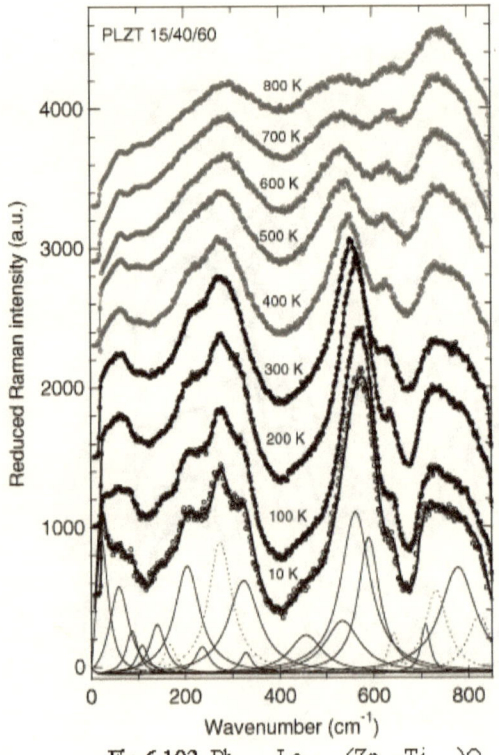

Fig 6.102 $Pb_{0.85}La_{0.15}(Zr_{0.4}Ti_{0.6})O_3$

K_2ZnBr_4

In this sample the soft mode, symmetry class A, in Raman spectra indicates displacive type contribution. (T. Britz, & Unruh, 1996)

$K_2ZnI_4\ (T_C = 272\ K)$

Polarized Raman spectra between 220 and 310K (Jochum, & Unruh, 1998), used Jobin Yuvon triple spectrometer and CCD MCA detector, Ar 514.32 *nm*, laser, studied. $x(yy)\hat{x}$, $x(zy)\hat{x}$, and $x(zz)\hat{x}$ geometries. Mode A_u^1 appeared only below T_C, in $z(xz)\hat{x}$ and is so A_u species above T_C. The second one is B_g species of PE phase.

6.7 Mossbauer spectroscopic Studies
(Mossbauer fraction, isomer shift, quadrupole splitting and hyperfine field)

Canner (1969). investigated Mossbauer effect studies on ferroelectric phase transition in $PbZrO_3$-$PbTiO_3$-$BiFeO_3$ ternary system.

Gleason & Walker (1969), studied Mossbauer effect of Potassium Ferrocyanide trihydrate and Ferric ammonium sulphate dodecahydrate. The recoil free fraction f and isomer shifts were as function of temperature have been studied. In FAS, there appeared a minimum in f at T_C , and no change in isomer shift. The behaviour of f is in consistent with a description of the transition in FAS in terms of the soft mode (TO) , i.e., $\omega_{TO} \rightarrow 0$, at T_C .

For KFCT no anomalous behavior of Mossbauer parameters was detected. This lack of any anomaly is consistent with an order-disorder model of KFCT.

The Mossbauer spectra of the system $(PbTiO_3)_{0.95}(BiFeO_3)_{0.05}$ solid solution have been studied from RT to 585 $°C$ (Yagnik, et al.,1969). From the isomer shift of 0.491 ± 0.02 $mm.s^{-1}$ at RT the iconicity is found to be 60%. A discontinuity in the plot of isomer shift vs. T at the T_C is observed. The quadrupole splitting at RT is 0.295 ± 0.02 $mm.s^{-1}$. A dip in the Mossbauer fraction is observed at the T_C

6.8. Magnetic Resonance (NMR, NQR)

It is always useful to peruse the Table 6.2 while applying the various Magnetic resonance techniques to investigate ferroelectric materials (Blinc, 1966).

Table 6.2

Various observable quantities in Magnetic Resonce and information available

I $\boxed{\text{NMR:}}$ $\boxed{H = H_z + H_{dd} + H_Q}$

Observable quantity	Information derived
1) Positiion of resonance and chemical shift	Electronic structure, change in chemical bonding at T_C
2) Shape and second moment of resonant line	Crystal structure, molecular motion
3) Quadrupole splitting of spectrum	EFG tensor, Crystal structure, molecular motion
4) Spin-lattice relaxation time, t_1	Dynamics of nuclear motion
5) Temp behaviour spectrum on going through T_C	Favourable cases: Displacement of order-disorder nature of transition as well as I or II order transition.

II. $\boxed{\text{EPR:}}$ $\boxed{H = H_{z,e} + H_F + H_{HF} + H_{Z,N} + H_{Q,N}}$

1) Positiion of resonance lines and chemical shift radiation damage centres	g-tensor
2) Number of components and magnitude of splitting	Fine and hyperfine coupling tensors describing crystalline field effects & interaction with nuclei
3) Spin-lattice relaxation time, t_l and line width	Dynamics of electron spins & informations on phonons.
4) Change in intensity of Spectra of radiation damage centres	Kinetics of defect motion resulting in recombination of damage centres.

$\underline{CsH_2AsO_4}$

Gupta (1969) studied quadrupole coupling parameter of As^{75} and Cs^{133} in a single crystal of CsH_2AsO_4 which has T_C at $143K$. As^{75} Resonance could not be observed below 153K. The temperature coefficient of As^{75} quadrupole coupling constant exhibits an anomalous behaviour. Cs^{133} resonance was seen in the FE phase also. The average of the two EFG tensors corresponding to the two physically non-equivalent Cs sites inn ferroelectric state is exactly equal to the EFG tensor in the paraelectric phase.

In CsH_2AsO_4 Blinc & Mali (1969) have studied EFG terms at the Cs^{133} sites, as function of T in the PE and FE phases. $|e^2Qq/h| = 371\ kHz$, $\eta = 0$ above T_C, whereas at I=123K, it is $|e^2Qq/h| = 342\ kHz$, $\eta = 0.84$.

$NaNO_2$

NMR study of the phase transition in $NaNO_2$ by Betsuyaku (1969) by studying EFG at the Na site both in the PE and FE phases. He found (i) the EFG varies linearly with the square of the long range order parameter determined by the X-ray study in the FE phase,(ii) a small but abrupt change, at T_C in the field gradient indicatin a first order phase transition, (iii) slight anomalies exist at 168 and 180 C in the temp variation of EFG, (iv) the EFG decreases linearly with increasing temperature between 168-175C , and 180-230C. The unusual behaviour of Na^{23} EFG have close relations with molecular rotation of $(NO_2)^-$ ions and the structural changes in the rystal which causes the phase transition.

6.9   Effect of Hydrostatic Pressure

The major part of this Chapter has shown that a crystalline material can be forced to undergo a structural phase change by changing its temperature. It has been known that the properties of a substance are in general change with its temperature. In the case of a ferroelectric material, phenomenon of ferroelectricity can be made to occur when the temperature of the material is brought below its T_C. Merz (1950) and Samara (1966, 1971) have shown that the physical properties of materials undergo variation at the influence external pressure on the material. It is known that deep under the earth at high temperatures and high hydrostatic pressure crystals of diamond are formed from Carbon.

In past three decades studies have been performed for the elucidation of the fundamental aspects of ferroelectricity by means of Raman spectra.

Whereas the atomic spacing variation to 1% is difficult to achieve in a material with change of its temperature alone (in the absence of a phase change), 10% changes in atomic spacing can be effected by applying pressure. This change in atomic spacing with pressure may produce drastic effects on the physical properties of a solid. The equipment for high pressure studies utilize either a piston & cylinder technique providing a cubic press or a tungsten carbide anvil method (Sanjurjo et al., 1983a, 1983b), Jayaraman (1983) and Sikka et al., (1989)..

Like the transition temperature T_C, there corresponds a transition pressure P_C at which the crystal changes from one phase to another. To determine the P_C: a) In First order phase change Raman modes disappear from the spectrum, whereas b) two different modes coalesce to the same frequency mode at P_C .in the second order transition. A shift of the Curie point under pressure depends on the increase or decrease of the polarization by the pressure (Stewart , 1967).

The effect of pressure and temperature dependence studies on Raman spectra of a few typical ferroelectric materials, such as KFCT, KDP, $BaTiO_3$, $PbTiO_3$, SbSI, $NaNO_2$, $LiNbO_3$, $Li_2Ge_7O_{15}$.

KFCT

Dielectric constant *versus* temperature in KFCT crystals along the FE axis [101] was investigated under hydrostatic pressures up to 5600 $kg.cm^{-2}$. It was found that $\frac{dT_C}{dP} = 2.6 \times 10^{-3} \, {}^{\circ}C.cm^{-2}.kg^{-1}$, and $T_C = -24.9^{\circ}C$, at atm.pressure.

Table 6.3

			$\frac{dT_C}{dP} \times 10^3 K^{-1}.cm^{-2}.kg^{-1}$
1) NH_4HSO_4	$P2_1/c \xleftrightarrow{-3^{\circ}C} Pc$	$P_S \,\|\, [001]$	+14
2) Rochelle salt	$P2_12_12_1 \xleftrightarrow{+24^{\circ}C} P2_1$	$P_S \,\|\, [100]$	10.4 & 11.0
3) TGS	$P2_1/m \xleftrightarrow{+49^{\circ}C} P2_1$	$P_S \,\|\, [010]$	+2.6 & +2.85 to $1.16 \times 10^{-4} P$
4) TGSe	$P2_1/m \xleftrightarrow{+22^{\circ}C} P2_1$	$P_S \,\|\, [010]$	+3.8
5) TGFB	$P2_1/n \xleftrightarrow{+70^{\circ}C} P2_1$	$P_S \,\|\, [010]$	+2.5
6) KDP	$I\bar{4}2d \xleftrightarrow{-150^{\circ}C} Fdd$	$P_S \,\|\, [001]$	-4.52
7) DKDP	$I\bar{4}2d \xleftrightarrow{-60^{\circ}C} Fdd$	$P_S \,\|\, [001]$	-2.63
8) KFCT	$C2c \xleftrightarrow{-24^{\circ}C} Cc$	$P_S \,\|\, [10\bar{1}]$	+2.6
9) KFCT	$I4_1a \xleftrightarrow{-55^{\circ}C} Cc$	$P_S \,\|\, [10\bar{1}]$	+2.5
10) KFCT	$C2/c \xleftrightarrow{+24^{\circ}C} Cc$	$P_S \,\|\, [10\bar{1}]$	+2.2 - 2.4

KDP

Raman spectral studies show that the soft mode frequency varies with temperature as well as with pressure as shown in Fig 6.103. In KDP the soft mode seems to have B_2

symmetry. The data show that the spectral region from -250 to +250 cm^{-1} contains both the soft mode and coupled mode of the lattice. The soft mode is over damped at moderate pressures and appeared to be centred at $\omega = 0$. Up to high pressures, say 5 *kbars*, the soft mode appears to be under damped. As T is increased towards T_C, damping increases. At T_C the soft mode gets over damped. His means the soft mode in KDP which is over damped at atmospheric pressure becomes under damped for pressures more than 6 kbar at $296K$.

At $T \ll T_C$, the Raman spectrum consists of two well defined modes, *viz.*,

$\omega^- = 150 \ cm^{-1}$ and $\omega^+ = 225 \ cm^{-1}$. Both the modes are found to decrease in frequency with increase of pressure. ω^- is the soft mode, the line shape of which changes near T_C. As T is increased ω^- shows a non-linear dependency on pressure, whereas all other modes display normal P dependence. Effect of P on T_C is seen in Fig 6.105. Wada *et al.*, 1981 showed that T_C decreases with increase of P. Therefore the soft mode and any mode coupled to it should soften with pressure. ω^- is the soft mode and for $T < T_C$ and that ω^- and ω^+ remain coupled even in the ferroelectric phase.

In perovskite materials the soft mode is very sensitive to changes in pressure. For example in $PbTiO_3$ the transition is a first order one at the T_C at atmospheric pressure. When pressure is applied this transition ceases to be of first order, but becomes one of second order. Sanjurjo *et al.* (1983) recorded the Raman spectra of $PbTiO_3$. At 300K and at low pressures the transition if of first order.

At its high temperature phase $PbTiO_3$ has 12 optic modes. A the pressure changes these modes transform as $(3T_{1u} + T_{2u})$ where T_{2u} is the silent mode. At very low temperatures, the T_{1u} splits into two modes: A_1 and E. The lowest frequency modes are E(1TO) and T(1TO), which are soft modes vanishing at the P_C. When the pressure reaches P_C there is transition changing from fist one to second. The modes T_{2u} and T_{3u} coalesce to a single one. Modes disappear at the T_C and no Raman intensity is seen above P_C.

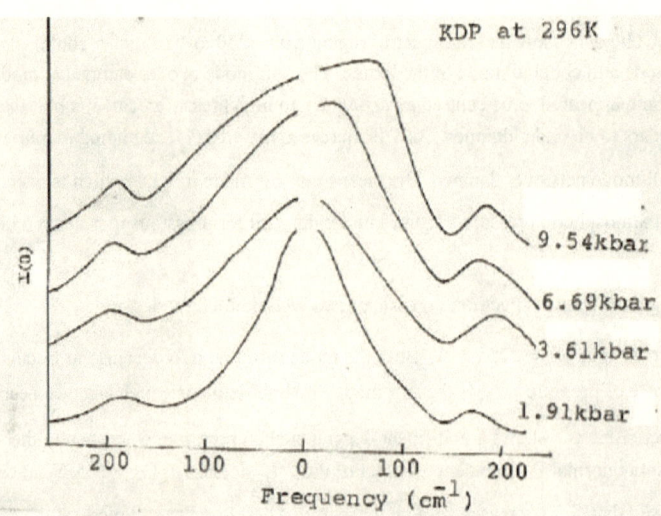

Fig 6.103 Soft mode Raman spectra of KDP for various pressures at 219K

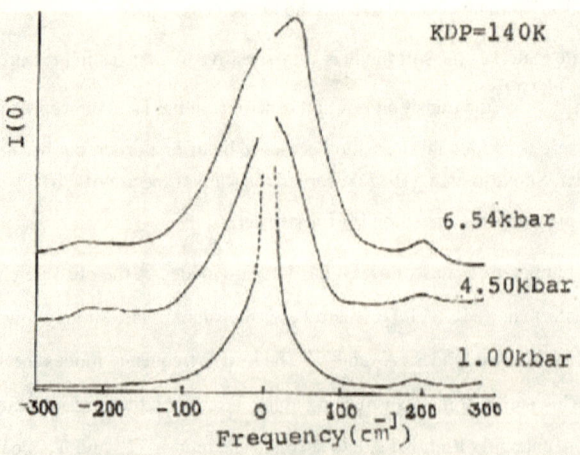

Fig 6.104 Soft mode spectra of KDP for different pressures at 140K.

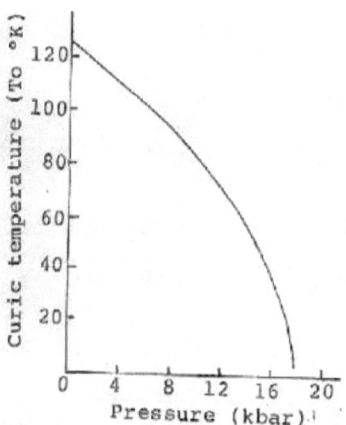

Fig 6.105 Variation of transition temperature with pressures in KH_2PO_4.

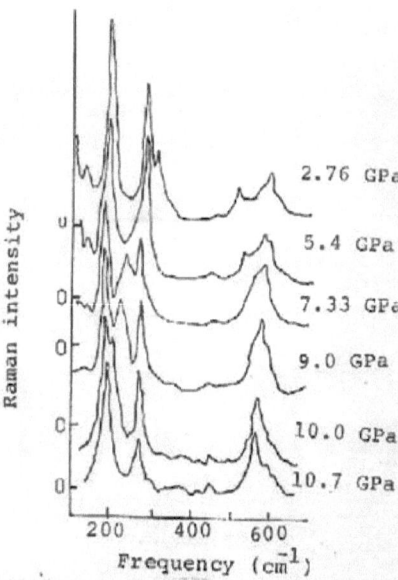

Fig 6.106 Raman spectra of high frequency modes at pressures in $PbTiO_3$

Lead Titanate, PbTiO$_3$

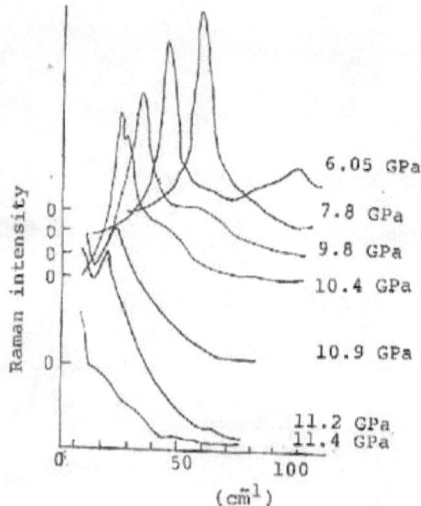

Fig 6.107 Raman spectra of soft phonons at various pressures in PbTiO$_3$

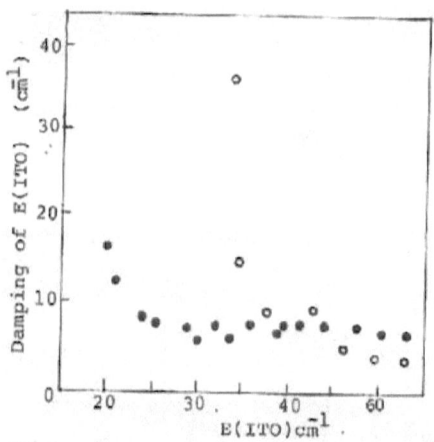

Fig 6.108 Damping constant versus soft mode frequency for PbTiO$_3$

For BaTiO$_3$ Raman studies by Sanjurjo *et al.* (1983)show that at T_C or P_C ,all (3A$_1$+3E) modes as well as (B$_1$+ E) modes are silent and disappear above P_C . A(1TO) is mode is under-damped, whereas E(1TO) is highly over-damped at low pressure. As pressure is increased E(1TO) is under-damped at the P_C . Both these modes are observed in the spectrum.

In both BaTiO$_3$ and PbTiO$_3$ the $(T_C - T_o)$ tends to zero as pressure increases. This is is indicative of the transition changes from first to second order.

In SbSI , as pressure increases the phase transition changes from first order to second order at the pressure induced critical point P_C (Peercy, 1975). When the pressure in increased in this crystal the transition temperature is decreased at the rate $\frac{dT_c}{dP} = -40K.(kbar)^{-1}$. The transition is terminated at $\geq 9.54\ kbar$. Raman spectra at 119K are redrawn in Fig.6.109. At low pressures the soft mode appears to be coupled with an optic mode of the lattice near $32 \sim 42\ cm^{-1}$. The transition at $T \leq 235$ K and $P \geq 1.4\ kbar$ the transition appears to be os second order.

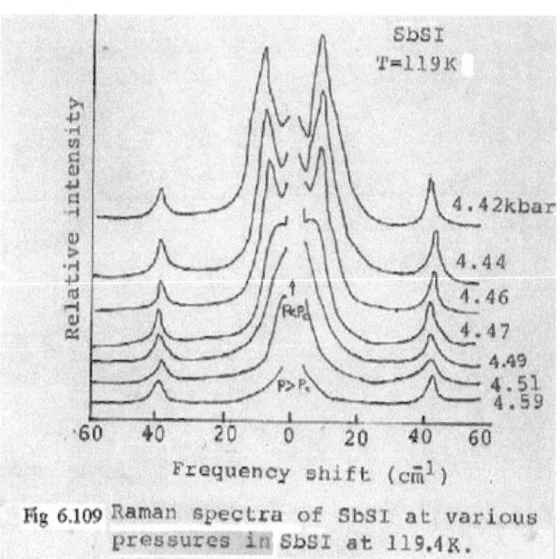

Fig 6.109 Raman spectra of SbSI at various pressures in SbSI at 119.4K.

In the case of $Li_2Ge_7O_{15}$ (Lithium hepta-germanate) the frequency of soft mode at $36.3\ cm^{-1}$ at 212K is found to decrease with decrease of temperature, It is seen that $\frac{dT_c}{dP} = 14.6K.(kbar)^{-1}$. Fig 6.111 exhibits the variation of T_C with pressure (Wada *et al.*, 1981). The soft mode is $34.8\ cm^{-1}$ at 3.4kbar goes to zero as P_C is approached from above. Fig 6.114 shows the variation of low lying modes with pressure in this material.

Raman studies on $LiNbO_3$ at room temperature and pressures 1 bar to 80 kbar (Mendes Felho et al., 1984). Only TO phonons were studied.

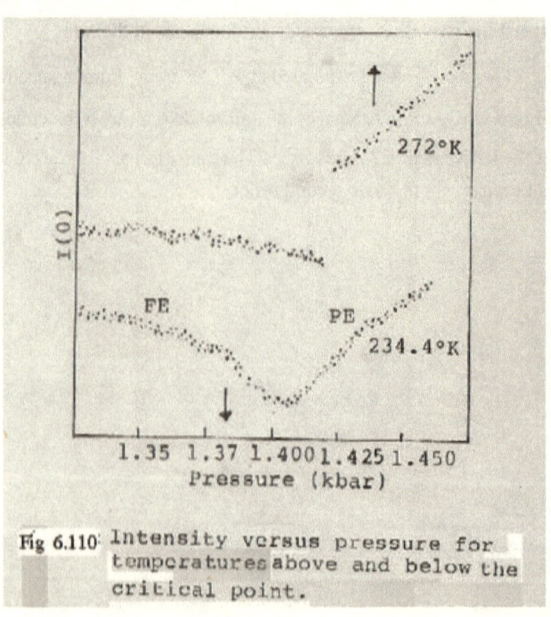

Fig 6.110 Intensity versus pressure for temperatures above and below the critical point.

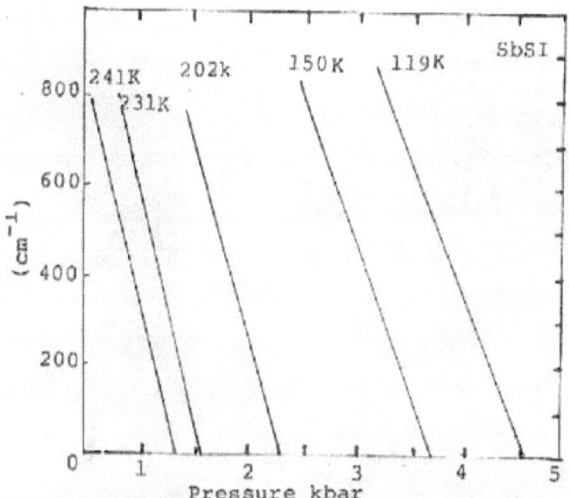

Fig 6.111 Pressure dependence of soft mode
in SbSI.

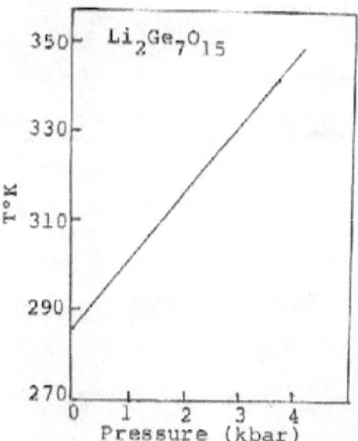

Fig 6.112 Variation of Tc with Pressure
in $Li_2Ge_7O_{15}$.

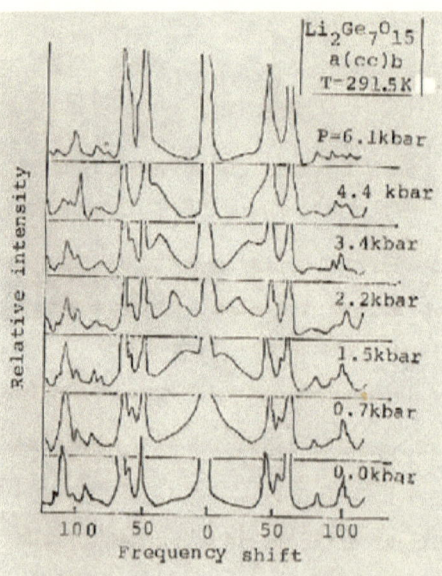

Fig 6.113 Raman spectra of $Li_2Ge_7O_{15}$ under various pressures.

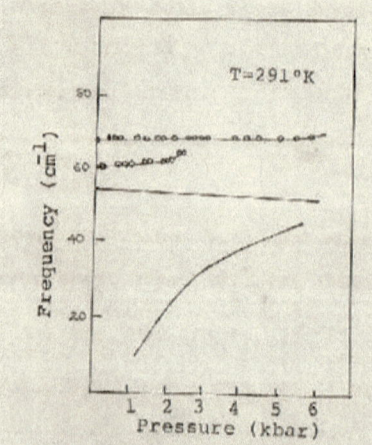

Fig 6.114 Frequency at peak position of low lying modes of $Li_2Ge_7O_{15}$ as functions of pressure

%&%&%&%&%&%&%&%&%

Chapter 7

List of Ferrolectric (FE) and

Antiferroelectric (AFE) Materials

Chapter 7

List of Ferrolectric (FE) and

Antiferroelectric (AFE) Materials

"The important thing in science is not so much to obtain new facts as to discover new ways of thinking about them". ~William Lawrence Bragg

A ferroelectric is defined as a material that shows spontaneous and reversible dielectric polarization in a temperature range".

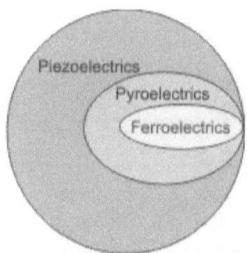

Most Solid Solutions are not included. The question mark indicates that Ferro electricity (FE), antiferroelectricity (AFE) or the Curie temperature (T_C) of the compound has not been definitely established till 1990. T_C is shown in bold italic characters.

Material	Chemical Composition	Transition Temp ($°C$)	Point groups PE	FE ($\vec{P}_s$)	Year of discovery

(1) FERROELECTRIC MATERIALS:

Perovskite (ABO_3) Type (Isomorphism) Oxides :

$[A^{2+} B^{1+} O_3]$ (A^{2+} = Ba, Sr, Rb, Cs, K, Cs, Ag, Na, Cd)

$(B^{1+} = \text{Ti, Nb, Ta Ge, I})$:

Barium Titanate	$BaTiO_3$		$C6_3 / mmc$	
		1460	Pm3m	
		120	P4mm [001]	1945
		5	Amm2 [110]	
	-90		R3m [111]	(FE -

Ferroelastic, Toledano *et al* 1983)(Displacive FE)

Strontium Titanate	$SrTiO_3$		Pm3m	1956
		(-163)	(4mm)	
		(-208)	(mm2)	
		(-238)	(3m)	

(Purely Ferroelastic, Toledano *et al*, 1983)

Potassium Niobate	$KNbO_3$		Pm3m	
		434	4mm [001] 1949	
		210	Bmm2 [110]	
		-33	3m [111]	

Lead Titanate	$PbTiO_3$		Pm3m	
		490	P4mm	1950
(Makedonite)		-100		
		-150		

Cadmium Titanate?	$CdTiO_3$	960	$Pc2_1n$		1950
		(-218)	Trigonal		
Sodium Niobate	$NaNbO_3$	638	F4/mmb	`	1951
		575	Ccmm		
		520	Pnmm		
		360	Pbma	Mono [001]	
		-100	F3c		
(AFE at RT, 1955)					
Sodium Tantalate	$NaTaO_3$	630			1949
		550			
		475	$Pc2_1n$		
Rubidium Tantalate	$RbTaO_3$	*247*			1951
Potassium Tantalate?	$KTaO_3$		Pm3m		1949, 2000
		-260			
Potassium Iodate	KIO_3	*212*	TrigonalTrigonal		1961
		70			
		-10, -18			
		-190			
Cesium Germanium	$CsGeCl_3$		Pm3m		1965
Chloride		*155*	R3		

Silver Niobate?	AgNbO$_3$	340	cubic	Tetragonal	1958
		240		Orthorhombic	
		50			
Silver Tantalate?	AgTaO$_3$	485			1958
.		*370*			
Silver Vanadate?	AgVO$_3$	170 ~ 180			1960
Bismuth Ferrite?	BiFeO$_3$	*850*		R3*c*	1962
		~575			
		370		R3*m*	

(AFM at < 370°C)

Dysprosium Chromite?	DyCrO$_3$	*575*	
Holmium Chromite?	HoCrO$_3$	*495*	
Ytterbium Chromite?	YbCrO$_3$	*515*	
Lutetium Chromite?	LuCrO$_3$	*440*	
Praseodymium Chromite?	PrCrO$_3$	*630*	
Yttrium Chromite?	YCrO$_3$	*550*	
Lithium Columbate?	LiCbO$_3$	?	1949
Lithium Galliumate?	LiGaO$_3$	*97*	1949

Complex Perovskite type Oxides :

Potassium Bismuth Titanate $Pm3m$ 1959

$(K_{0.5}, Bi_{0.5})TiO_3$ 377

270 Tetragonal

($270°C < AFE < at\ 377°C$) (FE at RT)

Sodium Bismuth Titanate? $(Na_{0.5}, Bi_{0.5})TiO_3$ $Pm3m$ 1959

317

~200 rhombohedral

(FE at RT; AFE at $>200°C$)

Lead Cobalt Tungstate $Pb(Co_{0.5}, W_{0.5})O_3$ cubic 1963

Or, Pb_2CoWO_6 32 orthorhom

-190

-264

(AFE at $<32°C$)(FE at $<-190°C$ (weakly FM at $<-264°C$)

Lead Scandium Niobate $Pb(Sc_{0.5}Nb_{0.5})O_3$ 90 Cubic Tetragonal

Lead Iron Niobate $Pb(Fe_{0.5}Nb_{0.5})O_3$ 114 $Pm3m$ rhombohedral 1958

-130

(FE at T $<114°C$, AFM at T $<-130°C$)

Lead Indium Niobate? $Pb(In_{0.5}Nb_{0.5})O_3$ 90 Cubic Monoclinic 1965

Lead Scandium Tantalate $Pb(Sc_{0.5}Ta_{0.5})O_3$ 26 Cubic Tetragonal 1959

Lead Iron Tantalate $Pb(Fe_{0.5}Ta_{0.5})O_3$ -30 $Pm3m$ Rhombohedral 1959

Or, Pb_2FeTaO_6 -130

(AFM at $<-130°C$) (FE at T$<30°C$)

Lead Magnesium Niobate	$Pb(Mg_{1/3}Nb_{2/3})O_3$	*-8*	Pm3m ?			1958
Lead Zincate Niobate	$Pb(Zn_{1/3}Nb_{2/3})O_3$	*140*	Cubic ?			1960
Lead Cobaltate Niobate	$Pb(Co_{1/3}Nb_{2/3})O_3$	*-98*	Pm3m	?		1960
Lead Nickel Niobate	$Pb(Ni_{1/3}Nb_{2/3})O_3$	*-120*	Pm3m	?		1958
Lead Magnesium Tantalate $Pb(Mg_{1/3}Ta_{2/3})O_3$		*-98*	Pm3m ?			1960
Lead Cobalt Tantalate $Pb(Co_{1/3}Ta_{2/3})O_3$		*-140*	Pm3m ?			1960
Lead Cadmium Niobate $Pb(Cd_{1/3}Nb_{2/3})O_3$		*270*	?			
Lead Nickel Tantalate $Pb(Ni_{1/3}Ta_{2/3})O_3$		*-180*	Pm3m ?			1960
Lead Ferrum Tungstate $Pb(Fe_{2/3}W_{1/3})O_3$		363	Cubic	Cubic		1959
		--95		?		
Lead Scandium Tungstate? $Pb(Sc_{2/3}W_{1/3})O_3$		*-12*	?			
Sodium Bismuth Titanate $(Na_{1/2}Bi_{1/2})TiO_3$		*320*	?			1961
		318				

(AFE at T < 320°C) (FE above 318°C)

Strontium Sodium Bismuth Titanate		*240*	?	FE		1974
$Sr_x(Na_{1/2}Bi_{1/2})_{1-x}TiO_3$		*150*		AFE		

(AFE at T < 150°C) (FE below 240°C)

$SrMgF_4$			?			2002

Solid Solutions with Perovskite Oxides as end members :

Sodium Niobate--Potassium -Niobate   ?

     $NaNbO_3$-$NaTaO_3$?

Potassium Tantalate- Niobate   60 ~ 140   ?

     $KNbO_3$-$KTaO_3$

Potassium Tantalate- Niobate (KTN)   ?

     $KTa_{0.35}Nb_{0.65}O_3$

Barium Strontium Titanate $(Ba_xSr_{1-x})TiO_3$ *~145*   ?   1954

     ($x = 0.38, 0.32, 0.26, 0.20$)

Lead Strontium Zirconate $(Ba_x,Sr_{1-x})ZrO_3$?

Lead Zirconate Titanate $Pb(Zr_x,Ti_{1-x})O_3$?   FE

Sodium Niobate Vanadate $Na(Nb_{0.7},V_{0.3})O_3$?

Barium Copper Tungstate?   *1200*

     $Ba(Cu_{0.5},W_{0.5})O_3$

Strontium Copper Tungstate?   *920*

     $Sr(Cu_{0.5},W_{0.5})O_3$

Barium Copper Tantalate?   *470*

     $Ba(Cu_{1/3},Ta_{2/3})O_3$

Strontium Copper Tantalate?   *1250*

     $Sr(Cu_{1/3},Ta_{2/3})O_3$

Barium Copper Niobate?   *380*

     $Ba(Cu_{1/3},Nb_{2/3})O_3$

Strontium Copper Niobate?   *390*

$Sr(Cu_{1/3},Nb_{2/3})O_3$

Barium Bismuth Oxide? $BaBiO_{2.8}$	370		
Barium Bismuth Oxide? $BaBiO_3$	*340*		1971
Barium Bismuth Niobate?	*420*		1965
$Ba(Bi_{0.5},Nb_{0.5})O_3$			
Barium Bismuth Tantalate?	*410*		
$Ba(Bi_{0.5},Ta_{0.5})O_3$			
Barium Bismuth Vanadate?	*320*		
$Ba(Bi_{0.5},V_{0.5})O_3$			
Barium Bismuth -Tungstate?	*>400*		
$Ba(Bi_{2/3},W_{1/3})O_3$			
Barium Bismuth Molybdate?	500		
$Ba(Bi_{0.5},Mo_{0.5})O_3$			
Barium Bismuth Molybdate?	*260*		
$Ba(Bi_{2/3},Mo_{1/3})O_3$			
Thallium Zirconium Tungstate?	*310*		
$Tl(Zr_{0.5},W_{0.5})O_3$			
Cadmium Hafnate? $CdHfO_3$	*605*		
Cadmium Iron Niobate?	*450*		
$Cd(Fe_{0.5},Nb_{0.5})O_3$			
Cadmium Scandium Niobate?	*430*		

$Cd(Sc_{0.5},Nb_{0.5})O_3$

Cadmium Magnesium Niobate? **295**

$Cd(Mg_{1/3},Nb_{2/3})O_3$

Lead Chromium Niobate? **5**

$Pb(Cr_{1/2},Nb_{1/2})O_3$

Lead Lithium Scandium Tungstate? **-90**

$Pb(Li_{0.25}Sc_{0.25},W_{0.5})O_3$

Lead Lithium Iron Tungstate? **-70**

$Pb(Li_{0.25}Co_{0.25},Fe_{0.5})O_3$

Lead Lithium Cobalt Tungstate? **30**

$Pb(Li_{0.25}Co_{0.25},W_{0.5})O_3$

Lead Lithium Indium Tungstate? **-5**

$Pb(Li_{0.25}In_{0.25},W_{0.5})O_3$

Lead Lithium Yttrium Tungstate? **5**

$Pb(Li_{0.25}Y_{0.25},W_{0.5})O_3$

Lead Lithium Terbium Tungstate? **20**

$Pb(Li_{0.25}Tb_{0.25},W_{0.5})O_3$

Lead Lithium Ytterbium Tungstate? **-40**

$Pb(Li_{0.25}Yb_{0.25},W_{0.5})O_3$

Lead Lithium Holmium Tungstate? **0**

$Pb(Li_{0.25}Ho_{0.25},W_{0.5})O_3$

Lead Lithium Gadolinium Tungstate? **0**

$$Pb(Li_{0.25}Gd_{0.25},W_{0.5})O_3$$

Lead Lithium Praseodymium Tungstate?   *-18*

$$Pb(Li_{0.25}Pr_{0.25},W_{0.5})O_3$$

Lead Lithium Lanthanum Tungstate?   *-20*

$$Pb(Li_{0.25}La_{0.25},W_{0.5})O_3$$

Lead Lithium Samarium Tungstate?   *-15*

$$Pb(Li_{0.25}Sm_{0.25},W_{0.5})O_3$$

Lead Sodium Yttrium Tungstate?   *120*

$$Pb(Na_{0.25}Y_{0.25},W_{0.5})O_3$$

Lead Sodium Holmium Tungstate?   *150*

$$Pb(Na_{0.25}Ho_{0.25},W_{0.5})O_3$$

Lead Lithium Zirconium Tungstate?   *-10*

$$Pb(Li_{1/4},Zr_{1/4},W_{0.5})O_3$$

Lead Lithium Zirconium Tungstate?   -30

$$Pb(Li_{1/3},Zr_{1/6},W_{0.5})O_3$$

Lead Cadmium Niobium Tungstate?   *495*

$$Pb(Cd_{1/4},Nb_{1/4},W_{0.5})O_3$$

Lead Scandium Niobium Tungstate?   *<-170*

$$Pb(Sc_{5/9},Nb_{1/3},W_{1/9})O_3$$

Lead Scandium Chromium Niobate?   *57*

$$Pb(Sc_{1/4},Cr_{1/4},Nb_{1/2})O_3$$

Lead Cadmium Manganese Niobate?   *20*

$$Pb(Cd_{1/4},Mn_{1/4},Nb_{1/2})O_3$$

Lead Magnesium Manganese Tungstate?   *>200*

$$Pb(Mg_{1/4},Mn_{1/4},W_{1/2})O_3$$

Lead Cadmium Manganese Tungstate?   *>250*

$$Pb(Cd_{1/4},Mn_{1/4},W_{1/2})O_3$$

Lead Cobalt Manganese Tungstate?   *≈110*

$$Pb(Co_{1/4},Mn_{1/4},W_{1/2})O_3$$

Lead Nickel Manganese Tungstate?   *≈100*

$$Pb(Ni_{1/4},Mn_{1/4},W_{1/2})O_3$$

Lead Nickel Manganese Niobate?   *-25*

$$Pb(Ni_{1/4},Mn_{1/4},Nb_{1/2})O_3$$

Lead Cobalt Manganese Niobate?   *-20*

$$Pb(Co_{1/4},Mn_{1/4},Nb_{1/2})O_3$$

Lead Magnesium Manganese Niobate?   *0*

$$Pb(Mg_{1/4},Mn_{1/4},Nb_{1/2})O_3$$

Lead Zinc Manganese Niobate?   *30*

$$Pb(Zn_{1/4},Mn_{1/4},Nb_{1/2})O_3$$

Lead Lithium Niobium Tungstate?   *-45*

$$Pb(Li_{1/3},Nb_{1/3},W_{1/3})O_3$$

Lead Magnesium Manganese Tantalate?   *-135*

$$Pb(Mg_{1/4},Mn_{1/4},Ta_{1/2})O_3$$

Lead Nickel Manganese Tantalate?   *-180*

$$Pb(Ni_{1/4}, Mn_{1/4}, Ta_{1/2})O_3$$

$\boxed{(A^{2+}A^{3+}B^{1+})TeO_6, \textit{or } (A^{2+}A^{2+}B^{1+})TeO_6}$]   1974

$(A^{2+} = Pb, Ba, Sr, Bi, La)(B^{1+} = Li, Na)$

Barium Bismuth Lithium Tellurate $BaBiLiTeO_6$ *140*

Barium Bismuth Sodium Tellurate $BaBiNaTeO_6$ *370*

Barium Lanthanum Lithium Tellurate $BaLaLiTeO_6$ *400*

Barium Lanthanum Sodium Tellurate   *520*

$$Cd(Mg_{1/3}, Nb_{2/3})O_3$$

Lead Bismuth Lithium Tellurate $PbBiLiTeO_6$ *350*

Lead Bismuth Sodium Tellurate $PbBiNaTeO_6$ *600*

Lead Lanthanum Lithium Tellurate $PbLaLiTe_6$ *630*

Lead Lanthanum Sodium Tellurate $PbLaNaTe_6$ *760*

Strontium Bismuth Lithium Tellurate $SrBiLiTeO_6$ *430*

Strontium Bismuth Sodium Tellurate $SrBiNaTeO_6$ *630*

Strontium Lanthanum Lithium Tellurate   *700*

$$SrLaLiTeO_6$$

Strontium Lanthanum Sodium Tellurate   *800*

$$SrLaNaTeO_6$$

$(NaBiB^{2+})TeO_6$ B^{2+} = Mg,Co, Zn, Mn, Cd, Ca			
Sodium Bismuth Magnesium Tellurate	*400*		
$NaBiMgTeO_6$			
Sodium Bismuth Cobalt Tellurate	*440*		
$NaBiCoTeO_6$			
Sodium Bismuth Cadmium Tellurate	*700*		
$NaBiCdTeO_6$			
Sodium Bismuth Calcium Tellurate	*760*		
$NaBiCaTeO_6$			
Barium Magnesium Tellurate Ba_2MgTeO_6			
Lead Cobalt Tellurate Pb_2CoTeO_6	250	FM	
	90	AFE	
	-60		FE
Lead Iron Tellurate $Pb(Fe_{2/3}Te_{1/3})O_3$	*600*		
Barium Tellurate	$BaTeO_3$	*120 ~ 150*	
Lead Tellurate	$PbTeO_3$	*> 400*	
Cadmium Tellurate	$CdTeO_3$	*> 400*	

Bismuth Manganate?	$BiMnO_3$	>500		
Tungsten Trioxide	WO_3	1230	Tetragonal	1949

910	P4/*mmm*,	
710	P4/*mnb*	
330	P2$_1$/*n*	
17	P1	
-40		P*c* (FE)

Lithium Niobate family Compounds :

Lithium Niobate	LiNbO$_3$	*1210*	R$\bar{3}$c R3c	1949

(Best Electro-optic)

Lithium Tantalate	LiTaO$_3$	*665*	" "	1949
Lithium Iron Tantalum Oxyfluoride		*580*		

Li(Fe$_{1/2}$Ta$_{1/2}$)O$_2$F

Manganites: (RMnO$_3$) family (Hexagonal phase for

(R = Ho, Er, Tm, Yb, Lu, or Y are FE and FM ordered)

(R = La, Ce, Pr, Nd, Sm, Eu, Gd, Tb or Dy , have Orthohombic phase show only FM ordering)BERTAN T, F., FORRA T, F., & PAN G, P.: C. R. Acad. Sci. **256**, (1963) 1958; Acta Crystallogr. **16**, (1963) 957

Yttrium Manganite YMnO$_3$	≈ *640*	P6$_3$/*mcm* P6$_3$*cm*	1963
	-46	AFM	

(Hexagonal crystal transforms to Perovskite Pbnm at high pressures when heated)

Erbium Manganite	ErMnO$_3$	≈ *560*	P6$_3$*cm*	1963

Holmium Manganite	$HoMnO_3$	≈ 600	$P6_3cm$	1963
Thulium Manganite	$TmMnO_3$	> 300	$P6_3cm$	1963
Ytterbium Manganite	$YbMnO_3$	≈ 720	$P6_3cm$	1963
Lutetium Manganite	$LuMnO_3$	> 300	$P6_3cm$	1963
Bismuth Manganite	$BiMnO_3$	> 500		
	$Y_{1-x}Bi_xMnO_3$ $(0 < x < 0.1)$	$550 \sim 1000$		1974

Tungsten- Bronze type Oxides :

Lead Meta Niobate	$PbNb_2O_3$		$P4/mbm$	1953
		560	$Cmm2$	
Lead Meta Tantalate	$PbTa_2O_3$		Orthorhombic	1954
		265	Ortho-	
		150		
	$Ca_xBi_{1-x}Nb_2O_6$ $(x = 0.20\ to\ 0.40)$			
	$(Ba_{1-x}Sr_x)Nb_2O_6$ $(x = 0.25\ to\ 0.75)$			
	$x = 0.75$	75	$P4bm$	1968

Fresnoite	$Ba_2Si_2TiO_8$		$P4bm$	1976
Lanthanum doped PLZT				1972

$$(Pb_{0.88}La_{0.12})(Zr_{0.4}Ti_{0.6})O_3$$

Bismuth Barium Titanium Iron Oxide   ?   1971

(FE - AFM) $Bi_5Ba_4Ti_3Fe_5O_{27}$

Ts-TS-19 type   ?   FE   1972

_____ - _____ - _____ -

$\boxed{A_4^{1+}A_2^{3+}B_{10}^{5+}O_{30}}$:

$(A_4^{1+} = K)\,(A_2^{3+} = Bi, La)\,(B_{10}^{5+} = Nb)$

Potassium Bismuth Niobate $K_2BiNb_5O_{15}$ *350*   1963

Potassium Lanthanum Niobate   -120

$\qquad K_2LaNb_5O_{15}$

_____ - _____ - _____ -

$\boxed{A_2^{1+}A_4^{3+}B_{10}^{5+}O_{30}}$: $(A_2^{1+} = K, Li, Na)\,(A_4^{3+} = Sr, Ba, Pb)\,(B_{10}^{5+} = Nb)$

Rubidium Strontium Niobate   139

$\qquad RbSr_2Nb_5O_{15}$

Potassium Strontium Niobate $KSr_2Nb_5O_{15}$ *156*   P4*bm*   1967

Potassium Strontium Niobate   120

$\qquad KSr_{4.5}Nb_{10}O_{38}$

Potassium Strontium Niobium Oxyfluoride   -178

$\qquad K_2Sr_2Nb_5O_{14}F$

Sodium Strontium Niobate $NaSr_2Nb_5O_{15}$ 270

Sodium Barium Niobate $NaBa_2Nb_5O_{15}$	560	*4mm*	1967
	300	*mm2*	

(Good electro-optic crystal)

Lithium Potassium Strontium Niobate   145

     $LiKSr_4Nb_{10}O_{30}$

Lithium Sodium Strontium Niobate   260

     $LiNaSr_4Nb_{10}O_{30}$

Potassium Barium Niobate $KBa_2Nb_5O_{15}$ *373*		1967

(Potassium, Strontium) Niobate

$(KNb_2)_{1-x}(SrNb_2)_xO_6$ $x = 0.40$	*167*		1967
$x = 0.75$	*160*		1967

Lithium Barium Niobate $LiBa_2Nb_5O_{15}$ *586*		
Potassium Lead Niobate $KPb_2Nb_5O_{15}$ *374*		1969
Rubidium Strontium Niobate $RbSr_2Nb_5O_{15}$ *139*		1967

Barium Magnesium Niobate   -25

     $Ba_9MgNb_{14}O_{45}$

Strontium Magnesium Niobate   10

     $Sr_9MgNb_{14}O_{45}$

Barium Strontium Niobate   75

     $Ba_{1.3}Sr_{3.7}Nb_{10}O_{30}$

Barium Strontium Niobate $Ba_2Sr_3Nb_{10}O_{30}$ 78

Barium Strontium Niobate   132

$$Ba_{2.7}Sr_{2.3}Nb_{10}O_{30}$$

Potassium Lithium Niobate ?   430

$$K_3Li_2Nb_5O_{15}$$

Potassium Lithium Tantalate   266

$$K_3Li_2Ta_5O_{15}$$

Barium Sodium Lanthanum Niobate ?   -25

$$Ba_2Na_3LaNb_{10}O_{30}$$

Barium Sodium Lanthanum Niobate?   -50

$$Ba_3NaLaNb_{10}O_{30}$$

Barium Sodium Europium Niobate ?   155

$$Ba_2Na_3EuNb_{10}O_{30}$$

Barium Sodium Gadolinium Niobate?   170

$$Ba_2Na_3GdNb_{10}O_{30}$$

Barium Sodium Lanthanum Niobate?   20

$$Ba_3Na\ GdNb_{10}O_{30}$$

Barium Sodium Dysprosium Niobate?   220

$$Ba_2Na_3DyNb_{10}O_{30}$$

Barium Sodium Yttrium Niobate?   220

$$Ba_2Na_3YNb_{10}O_{30}$$

Barium Sodium Yttrium Niobate?   145

$$Ba_3Na\ YNb_{10}O_{30}$$

Barium Titanium Niobate   245

$$Ba_4Ti_2YNb_8O_{30}$$

Barium Zirconium Niobate?   None

$$Ba\ Zr_{0.25}Nb_{1.5}O_{5.25}$$

Barium Iron Niobate $Ba_6FeNb_9O_{30}$ -140

Strontium Titanium Niobate $Sr_6Ti_2Nb_8O_{30}$ 130

_____-_____-_____-

$\boxed{A_4^{2+}A_2^{3+}B_2^{3+}B_8^{5+}O_{30}}$:

(A_4^{2+} = Ba, Sr) (A_2^{3+} = Nd, Sm, Gd, Yb) (B_2^{3+} = Fe), (B_{10}^{5+} = Nb)

Barium Neodymium Iron Niobate   ≈ 55   1964

$$Ba_4Nd_2Fe_2Nb_8O_{30}$$

Barium Gadolinium Iron Niobate   ≈ 130   1964

$$Ba_4Gd_2Fe_2Nb_8O_{30}$$

Barium Samarium Iron Niobate   ≈ 130   1964

$$Ba_4Sm_2Fe_2Nb_8O_{30}$$

Strontium Ytterbium Iron Niobate   ≈ 10   1962

$$Sr_4Yb_2Fe_2Nb_8O_{30}$$

_____-_____-_____

$\boxed{A_2^{2+}A_4^{3+}B_3^{3+}B_7^{5+}O_{30}}$

$:(A_2^{2+} =)\,(A_4^{3+} =)\,(B_3^{3+} =),\,(B_7^{5+} =)$

None

———————————- ———————————-·-

$\boxed{A_6^{2+} B_2^{4+} B_8^{5+} O_{30}}$: $(A_6^{2+} = Ba\ )\,(B_2^{4+} = Ti, Zr)\,(B_8^{5+} = Nb),$

Barium Titanium Niobate $Ba_6Ti_2Nb_8O_{30}$	?	?	1960
Barium Zirconium Niobate $BaZr_{0.25}Nb_{1.5}O_{5.25}$	?	Pba2	1960

———————————- ———————————-

$\boxed{A_6^{1+} B_4^{1+} C_7^{5+} D_3^{5+} O_{30}}$: $\quad (A_6^{1+} =)\,(B_4^{1+} =)\,(C_7^{5+} =),\,(D_3^{5+}\ )$

Potassium Lithium Niobate (KLN) ? $K_3Li_2Nb_5O_{15}$

(Completely filled Tungsten Bronze, KLN))

Potassium Lithium Niobate (KLN) ?	*430*	P4bm	1967
$\quad K_{0.6}Li_{0.4}NbO_3$			
Potassium Lithium Niobate (KLN) ?	*405*	Tetra-	1978
$\quad K_{2.89}Li_{1.55}Nb_{5.11}O_{15}$			
Potassium Bismuth Niobate (KLN) ?	*360*	Ortho-	1978
$\quad K_{1.92}Bi_{0.91}Nb_{6.07}O_{15}$			
Potassium Barium Niobate (KBN)	*392* ?	P4bm	1989
$\quad K_2Ba_4Nb_{10}O_{30}$			
Potassium Lithium Tantalum Niobate		Ietragonal	1970
$\quad$ (KLTN) $K_3Li_2(Ta_x,Nb_{1-x})O_{15}$			
$\qquad (x \le 0.5)$			

Potassium Lithium Tantalum Niobate		≈ 30	Ietragonal		1967

$$K_3Li_2(Ta_{0.6},Nb_{0.4})O_{15}$$

(KLTN)   ($x \geq 0.5$)

ABO$_4$ Type: : StibioTantalate family:

Stibio Tantalate	$SbTaO_4$	400	AFE	Ortho=FE	1970
Stibio Niobate	$SbNbO_4$	410	AFE	FE	1974
Bismuth Tantate	$BiTaO_4$	247	?	?	1974
Bismuth Niobate	$BiNbO_4$	247		?	1974
Bismuth Antimony Oxide? $BiSbO_4$		?			
Bismuth Vanadate	$SbVO_4$	255			1975
Antimony Thallium Niobium Oxide?		400			1970
$Sb(Ta,Nb)O_4$				–	
α-Lead Ortho Phosphate α-$Pb_3(PO_4)_2$		535	$R\bar{3}m$	C2/c	1973
Lead Ortho Vanadate $Pb_3(VO_4)_2$		100	R3m	$P2_1/n$		1980
Antimony Ortho Niobate $Sb(NbO_4)$		403			1972

Barium Bismuth Titanium Niobate?	-31
$Ba_4Bi_2Ti_4Nb_6O_{30}$	
Barium Bismuth Titanium Niobate?	15
$Ba_5BiTi_3Nb_7O_{30}$	

Strontium Bismuth Titanium Niobate?   10

$$Sr_5BiTi_3Nb_7O_{30}$$

Barium Lanthanum Titanium Niobate?   -130

$$Ba_5La_3Ti_5Nb_5O_{30}$$

Barium Lanthanum Titanium Niobate?   -80

$$Ba_4La_2Ti_4Nb_6O_{30}$$

Barium Lanthanum Titanium Niobate?   -80

$$Ba_5La_3Ti_5Nb_5O_{30}$$

Barium Lanthanum Titanium Niobate?   -55

$$Ba_5LaTi_3Nb_7O_{30}$$

Strontium Lanthanum Titanium Niobate?   -33

$$Sr_4La_2Ti_4Nb_6O_{30}$$

Strontium Lanthanum Titanium Niobate?   -7

$$Sr_5LaTi_3Nb_7O_{30}$$

Potassium Barium Titanium Niobate?   290

$$KBa_5TiNb_9O_{30}$$

Sodium Barium Titanium Niobate?   414

$$NaBa_5TiNb_9O_{30}$$

Potassium Strontium Titanium Niobate?   118

$$KSr_5TiNb_9O_{30}$$

Sodium Strontium Titanium Niobate?   157

$$NaSr_5TiNb_9O_{30}$$

Sodium Barium Tungsten Niobate?　　460

$Na_3Ba_3WNb_9O_{30}$

Sodium Barium Tungsten Niobate?　　365

$Na_4Ba_2W_2Nb_8O_{30}$

Potassium Strontium Tungsten Niobate?　　70

$K_3Sr_3WNb_9O_{30}$

Potassium Strontium Tungsten Niobate?　　65

$K_4Sr_2W_2Nb_8O_{30}$

———————————–————————————————–————

| Pyrochlore (Mineral Ca Na Nb_2O_6 F) type Compounds | $A_2B_2O_7$:

(A_2 = Cd, Ca, Ba, Pb, La, Sr, Sb)

(B_2 = Nb, Fe, Cr, Zn, Li, Bi, Ta, W, Mo, Ge, Ti, O)

Cadmium pyro Niobate　$Cd_2Nb_2O_7$　　*-88*　　Fd3m　*4mm*　　　　　1952

　　$[(1-x) Cd_2Nb_2, x Pb_2Nb_2]O_7$　　*-93 ~ -223*

　　　　($x \leq 0.2$)

(Coexistence of two FE states; Triple hysteresis observed)

Strontium pyro Niobate　$Sr_2Nb_2O_7$　　*1350*　*Cmcm*　*Pmc2₁*　　　1971

　　　　　　　　　　　　　　215

Calcium pyro Niobate $Ca_2Nb_2O_7$　　　?　　　　　　　　　　　　1974

Cadmium Niobium Oxysulfide $Cd_2Nb_2O_6S$ *282*

Cadmium Chromium Niobate Cd_2CrNbO_6 *-203*

Cadmium Iron Niobate Cd_2FeNbO_6 *-153*

Barium Zinc Tantalate? $Ba_2Zn_{4/3}Ta_{2/3}O_6$ 300

Lead Barium Lithium Niobate? ≈ 100

$Pb_{1.9}Ba_{0.1}Li_{0.5}Nb_{1.5}O_6$

Lead Bismuth Niobate? Pb_2BiNbO_6 *475* 1965

Lead Bismuth Tantalate? Pb_2BiTaO_6 *420* 1965

Lead Bismuth Tungstate? $Pb_2Bi_{4/3}W_{2/3}O_6$ *400* 1965

Lead Bismuth Molybdate ? *500* 1965

$Pb_2Bi_{4/3}Mo_{2/3}O_6$

Neodymium pyro Titanate $Nd_2Ti_2O_7$ *>500* $P2_1$ 1974

Lanthanum pyro Titanate $La_2Ti_2O_7$ *1500* $P2_1$ 1974

—————————-—————————-————————

| Layer Structure Oxides | :

Strontium Bismuth Tantalate? $SrBi_2Ta_2O_9$ *310* Tetra- Ortho- 1961

Strontium Bismuth Niobate? $SrBi_2Nb_2O_9$ 440

Barium Bismuth Niobate? $BaBi_2Nb_2O_9$ *200*

Barium Bismuth Tantalate? $BaBi_2Ta_2O_9$ *110*

Lead Bismuth Niobate? $PbBi_2Nb_2O_9$ *550*

Lead Bismuth Tantalate? $PbBi_2Ta_2O_9$ *430*

Calcium Bismuth Tantalate $CaBi_2Ta_2O_9$ *575* I4/ *mmm* F*mm*2 1960

Calcium Bismuth Niobate	$CaBi_2Nb_2O_9$	*625*	I4/*mmm* F*mm*2	1960
		577		
		547		
Bismuth Strontium Barium Tantalate		*200*	I4/*mmm* A2$_1$*am*	1973
$(Sr_{0.9}Ba_{0.1})Bi_2Ta_2O_9$				
$SrBi_2Ta_2O_9$		?	FE	1998
$Sr_{0.9}Bi_{2.1}Ta_2O_9$				1998
(Thin films for Ferroelectric Memories)				
$Sr_{0.7}Bi_{2.3}Ta_2O_9$				1998
(Thin films)				
Bismuth Tungstate? Bi_2WO_6		*935*	2*mm*	1972
Bismuth Niobium Oxyfluoride? Bi_2NbO_5F		*30*		1971
Bismuth Tantalum Oxyfluoride? Bi_2TaO_5F		*10*		
Bismuth Titanium Niobate? Bi_2TiNbO_9		*900–950*	I4/*mmm* F*mm*2	1960
		740		
Bismuth Titanium Tantalate? Bi_3TiTaO_9		*870*	I4/*mmm* F*mm*2	1962
Praseodymium Bismuth Iron Titanate		*740*		1974
$Pr_2Bi_4Ti_3Fe_2O_{18}$ (Magnetic FE)				
Lanthanum Bismuth Iron Titanate		*600*		1971
$La_2Bi_4Ti_3Fe_2O_{18}$				
Bismuth Titanate $Bi_4Ti_3O_{12}$		*675*	4/*mmm*	1961

(used in optical memory and light valves)   *700*   B$c2b$

Praseodymium Bismuth Titanate? $PrBi_3Ti_3O_{12}$ 400

Holmium Bismuth Titanate? $HoBi_3Ti_3O_{12}$ 440

Lanthanum Bismuth Titanate? $LaBi_3Ti_3O_{12}$ 315

Calcium Bismuth Titanate? $CaBi_3Ti_3O_{12}$ 790

Barium Bismuth Titanate? $BaBi_4Ti_4O_{15}$ 395

Lead Bismuth Titanate? $PbBi_4Ti_4O_{15}$ 570

Strontium Bismuth Titanate? $SrBi_4Ti_4O_{15}$ 330

Lanthanum Bismuth Iron Titanate? $LaBi_4Ti_4O_{15}$ 600

Praseodymium Bismuth Iron Titanate?   740

         $PrBi_4Ti_3FeO_{15}$

Barium Bismuth Titanate? $Ba_2Bi_4Ti_5O_{18}$ *325*   Tetra- Ortho-   1962

Lead Bismuth Titanate? $Pb_2Bi_4Ti_5O_{18}$ *310*   1962

Strontium Bismuth Titanate? $Sr_2Bi_4Ti_5O_{18}$ *285*   1962

Bismuth Titanate? $Bi_2Ti_4O_{11}$ 1200   C$2/m$

                                                   250   C$2/c$

Bismuth Iron Titanate? $Bi_5Ti_3FeO_{15}$ 750

Bismuth Iron Titanate? $Bi_6Ti_3FeO_{18}$ > 750

Bismuth Iron Titanate? $Bi_5Bi_4Ti_3Fe_5O_{27}$ 830

Calcium Bismuth Iron Titanate? $CaBi_5Ti_4FeO_{18}$ ≈ 770

Strontium Bismuth Iron Titanate? $SrBi_5Ti_4FeO_{18}$ ≈ 580

Lead Bismuth Iron Titanate ? $PbBi_5Ti_4FeO_{18}$ ≈ 550

Barium Bismuth Iron Titanate? $BaBi_5Ti_4FeO_{18}$ ≈ 570

DiLithium HeptoMolybdo-tetraGadolinate[7] *52*   1970

     $(Li_2Gd_4)(MoO_4)$ 0

                                  -26

| Barium Aluminium Oxy-Fluoride, $Ba\,Al_2\,F_4$ type Compounds | : |

Pure compound   Nil

Barium Lithium Aluminium Oxyfluoride   1960

     $BaLi_{2x}Al_{2-2x}F_{4x}O_{4-4x}$

(Solid Solution) $[(1-x)(BaAl_2F_4)\,x(BaLi_2F_4)]$

         $x = 0.15$ *140*   Hexagonal
   $x = 0.2$ 134   "

      $x = 0.3$ 153   "

Lead Tungsten OxyFluorate $Pb_5W_3O_9F_{10}$ *512*   ?   Tetrago-   1987

Lead Hafnate Titanate $PbHf_{0.9}Ti_{0.1}O_3$?   Rhombohedr FE   1999

      $Sr_2(FeF_6)_2,-NbO_2$?   2003

      $Sr_3(FeF_6)_2,-TlBO_2$?   2003

      $Sr_3(FeF_6)_2,-CrOF_3$?   2003

Barium Fluoride Ba M^{2+} F_4 type Compounds: (Simultaneously AFM - FE)

Barium Magnesium Fluoride BaMgF$_4$	-18	2*mm*	1969
(AFE – AFM - FE)	*-247*	AFM FE	
Barium Manganese Fluoride? BaMnF$_4$	None		
Barium Iron Fluoride? BaFeF$_4$	None		
Barium Cobalt Fluoride BaCoF$_4$	None	2*mm*	1969
Barium Nickel Fluoride BaNiF$_4$	None	2*mm*	1968
(FE – AFM simultaneously)			
Barium Zinc Fluoride BaZnF$_4$	None	2*mm*	1969

(A_3 MO_3 F_3) family:(A = K, Rb, Cs); (M = Mo, W)

Rb$_2$KMoO$_3$F$_3$	1980
Cs$_2$RbMoO$_3$F$_3$	1980
Na$_5$W$_3$O$_9$F$_5$	1981

Molybdates: Improper Ferroelectrics (FE - Ferroeleastic)

In improper FE, the values of birefruingence and electro-optic coeffs in the FE phase cannot be explained by classical description based on electro-optic effect.

_Gadolinium

Molybdate β - Gd$_2$(MoO$_4$)$_3$	42*m*		
(First FE – Ferroelastic)	*159*	P*ba*2	1966

Samarium Molybdate $Sm_2(MoO_4)_3$	*190*	?	Orthogonal	1967
Europium Molybdate $Eu_2(MoO_4)_3$	*161*	?	Orthogonal	1967
Praseodymium Molybdate $Pr_2(MoO_4)_3$	235			
Neodymium Molybdate $Nd_2(MoO_4)_3$	225			
Dysprosium Molybdate $Dy_2(MoO_4)_3$	145			
Holmium Molybdate $Ho_2(MoO_4)_3$	134			
Terbium Molybdate $\beta\text{-}Tb_2(MoO_4)_3$	*157*	?	Orthogonal	1967
Europium Gadolinium Molybdate $(Eu_{0.1}Gd_{0.9})_2(MoO_4)_3$	*157*	?	?	1967
Europium Gadolinium Molybdate $(Eu_{0.2}Gd_{0.8})_2(MoO_4)_3$	*160*	?	?	1967
Europium Terbium Molybdate $(Eu_{0.1}Tb_{0.9})_2(MoO_4)_3$	*159*	?	?	1967
Europium Terbium Molybdate $(Eu_{0.5}Tb_{0.5})_2(MoO_4)_3$	*170*	?	?	1967
Gadolinium Yttrium Molybdate $(Gd_{0.8}Y_{0.2})_2(MoO_4)_3$	*147*	?	?	1967
Gadolinium Neodymium Molybdate $(Gd_{0.8}Nd_{0.2})_2(MoO_4)_3$	*159*	?	?	1967
Gadolinium Terbium Molybdate $(Gd_{0.25}Tb_{0.75})_2(MoO_4)_3$	*157*	?	?	1967

Gadolinium Molybdate Tungstate   *148*   ?   ?   1967

$$Gd_2[(Mo,W)O_4]_3$$

—————————————— - ————————————— - ——————————— -

| Boracites: $(Me_3B_7O_{13}X)$ family | :   (X = halogen) (Me = Divalent)

Magnesium Boracite? $Mg_3B_7O_{13}Cl$ *265*   F43*c* Ortho-   1957

(Improper FE) (Simultaneously FE- Ferri-Magnetic)

Chromium Boracite? $Cr_3B_7O_{13}Cl$ *-18*   "   "   1974

Manganese Boracite? $Mn_3B_7O_{13}Cl$ *407*   1970

Iron Boracite? $Fe_3B_7O_{13}Cl$ *330*   1970

(FE-FM)   250

Cobalt Boracite $Co_3B_7O_{13}Cl$ *350*   1970
(FE-FM)   174

Nickel Boracite? $Ni_3B_7O_{13}Cl$ *333*   1964

Nickel Iodine Boracite $Ni_3B_7O_{13}I$ -153   F43c-   1966
                 -209   Pca

(Piezoelectric- PM at T > -153C; Piezoelectric-AFM –209 < T < -153 C)

(FE – weak FM < -209C)

Cobalt Iodine Boracite $Co_3B_7O_{13}I$ *- 81*   Cubic Ortho-   1970

Iron Iodine Boracite $Fe_3B_7O_{13}I$ 72   Cubic Ortho-   1970

                 -72   Tetrago-

Copper Boracite $Cu_3B_7O_{13}Cl$ *92*   Cubic Ortho-   1964
            -268.8

Zinc Boracite $Zn_3B_7O_{13}Cl$ **350**   1970
193

------ - ------------------- - --------------- - ------

Colemanite :

Colemanite $Ca_2B_6O_{11} \cdot 5H_2O$ ≈ -7 $P2_1/c$ 2   1956

Or $CaB_3(OH)_3 \cdot H_2O$

------------------- -- --------------- - ------------- -

Miscellaneous Oxides ::

Rubidium Tanatalate? $RbTaO_3$ 247

Sodium Vanadate?   380?

Silver Vanadate? $AgVO_3$ 170

                                        180

Lead Germanate $Pb_5Ge_3O_{11}$ –

(Or, $5PbO_3GeO_2$ 177   P6   P3   1971

(First FE in which optical rotation changes from left to right as $\vec{P}_S$ is reversed)

Stibio Tantalite $Sb(Ta,Nb)O_4$ 400

Strontium Tellurite $SrTeO_3$ 485

Strontium Niobate $Sr_2Nb_2O_7$

Lead Niobate ? $PbNbO_4O_{11}$

------------- - --------------- - ------------- - --------

Halides:

Hydrogen Chloride HCl	*- 174.6*	F*m3m* B*b*2$_1$*m*		1967
Deuterium Chloride DCl	*- 168*			1967
Hydrogen Bromide HBr	-160	B*b*cm		
	- 183	B*b*2$_1$*m*		1967
Deuterium Bromidec DBr	*- 152.4*			
Ammonium Chloride (NH$_4$)Cl	- 31	*m3m* *43m*		
Thallium Chloride? TlCl	?	Cubic ortho.		1968
Thallium Bromide? TlBr	?	Cubic Ortho.		1968
Thallium Iodide? TlI	?	Cubic Ortho.		1968

Aniline Hydro Bromide C$_6$H$_5$NH$_2$HBr *27.5*?   ?   1958

Inorganic (Antimony Sulfide Iodine type) Compound
other than Oxides: Semi conducting type, V - VI - VII Groups

:

:   (All Photoconductive)

Antimony Sulfide Iodide	SbSI	P*nam*	1962
	~15.7	P*na*2$_1$	
	-36.6	2	

(Photoconductive)

Antimony Sulfide Bromide	SbSBr	*-180*	P*nam* ?	1964
Antimony Selenide Iodide	SbSeI	23		
Bismuth Sulfide Bromide	BiSBr	*-170*	P*nam* ?	1964
Bismuth Sulfide Iodide	BiSI	-160		

Bismuth Trisulfide?	Bi_2S_3	≈ 50		
Antimony Trisulfide?	Sb_2S_3	26		
Antimony Pentaiodide?	SbI_5			
	$As_{0.1}SI\text{-}Sb_{0.9}SI$	*0*		1964
	$SbO_{0.2}S_{0.8}I$	*65*		1964
	$SbO_{0.05}S_{0.95}I$	?		1964
	$(SbSBr)_x(SbSI)_{1-x}$	?		1964
	$(SbSI)_x(SbSeI)_{1-x}$	?		1964
	$Sb_{0.1}Bi_{0.1}SI$	?		1964
Iron Sulfide	FeS	*137*		1969
	(AFM-FE)			
	$Fe_{1-x}S$			1970
Chalcocite	Cu_2S	?		1987
Chromium OrthoBeryllate	Cr_2BeO_4	?		1978
	(Magnetoferrolectricity)			

_____-_____-_____-_____-_

$A^{1+}B_2^{4+}(PO_4)_3$: (A = Li, Na, K, Rb, Cs,) (B = Th, U,)			

Sodium DiThorium TriPhosphate	?	C*c*	1969
$NaTh_2(PO_4)_3$			
(or Sodium TriPhosphate DiThorate)			
Sodium DiUranium TriPhosphate	?	?	1969

$NaU_2(PO_4)_3$

Sodium salt of De-Oxy Ribo Nucleic Acid (DNA) ?   ?   1960

(X)3 H (YO4)2 - Type:

Tri Ammonium Hydro Di-Sulphate			R3m	1976
$(NH_4)_3H(SO_4)_2$		140	A2/a	
		-8	?	
		-132		?
		-140		?
$(NH_4)_3H(SeO_4)_2$		2		1977
		-92		
$(NH_4)_3D(SO_4)_2$		-9.5		1977
		-24		
		-65		
		-100		

Nitrites ;

Sodium Nitrite	$NaNO_2$	164.7	Immm	1958
		163.8	Im2m	
		-105.0		
Potassium Nitrite	KNO_2	*47.4*	Pm3m Am	1966
		-13		

Silver Sodium di-Nitrite $AgNa(NO_2)_2$	*38*	F*ddd*	F*d2d*	1969

Nitrates :

Potassium Nitrate KNO_3		P*mn*2$_1$		1958
(Niter)	130	R3*m*		
	124	R3*m*		
	115	P*nma*		
Sodium Nitrate $NaNO_3$	?	R3*c*	R3*c*	1969
Ammonium Iodate $(NH_4)IO_3$	*85 ~ 95*	P*c*2$_1$*n*		1976

Potassium Dihydrogen Phosphate [KH_2PO_4] type Compounds :

Potassium dihydrogen Phosphate (KDP) KH_2PO_4	*- 150*	I$\bar{4}$2*d*	F*dd*2	1935
(Best Electro-optic material) (FE-Ferroelastic)				
Potassium dideuterium Phosphate (DKDP) KD_2PO_4	*- 60*	I$\bar{4}$2*d*	F*dd*2	1942
Rubidium dihydrogen Phosphate RbH_2PO_4	*- 126*	?		1945
Potassium dideuterium Phosphate RbD_2PO_4	*- 55*	?		1952
Cesium dihydrogen Phosphate CsH_2PO_4	*- 121*	P2$_1$/*m*	F*dd*2	1950
(Ferroelastic in the PE phase)				
Potassium dihydrogen Arsenate KH_2AsO_4	*- 176*	I$\bar{4}$2*d*	F*dd*2	1937

Potassium dideuterium Arsenate KD_2AsO_4 - *111*   1953

Rubidium dihydrogen Arsenate RbH_2AsO_4 - *162* ?   ?   1947

Rubidium dideuterium Arsenate RbD_2AsO_4 - *95*   1953

Cesium dihydrogen Arsenate CsH_2AsO_4 ~160   ? $P2_1/m$ 1953

                                                        - *130* $Fdd2$

Cesium dideuterium Arsenate CsD_2AsO_4 - *61*   1953

$$K(D_{0.88}H_{0.12})_2PO_4 \quad -64$$

‑‑

Ammonium Meta Phosphate   none $C2$ or Cm 1952

‑‑

$\boxed{A_2 \, B \, X_4 \qquad \text{Sulphates and Related Compounds}}$: (A = K, Rb, NH_4, Li,)

(B = S, Be, Se, Zn,) (X = O, F, Cl, Br)

Ammonium Sulphate $(NH_4)_2SO_4$ - *49.5* $Acam$ $Pna2_1$ 1956

                (Typical Improper FE)

Deuterated Ammonium Sulphate   - *49* $Acam$ $Pna2_1$ 1958

$$(ND_4)_2SO_4)$$

Ammonium Fluoroberyllate $(NH_4)_2BeF_4$ $Pnam$ (FE)   1957

                                                        -*188* $Pnma$ (AFE)

                                                        -175 $Pn2_1a$

Deuterated Ammonium Fluoroberyllate   - 94 $Acam$ $Pn2_1a$ 1958

$$(ND_4)_2BeF_4$$

Potassium Selenate K_2SeO_4	472	$P6_3/mmc$ $Pnam$		1969
	-143.5	$Pna2_1$		
	-180			
(Ferroelastic)				
Rubidium Zincate Bromide Rb_2ZnBr_4	77	$Fcmn$		1977
	-73	$Bc2_1n$		
Rubidium Zincate Chloride Rb_2ZnCl_4	77	$Fcmn$		1977
Rubidium Zincate Chloride Rb_2CoCl_4	-81	$2mm$		1986
Ammonium Lithium Sulphate $(NH_4)LiSO_4$	186.5	$Pnam$ $Pna2_1$		1977
	10	$2/m$		
	-245	?		
Lithium Cesium Sulphate $LiCsSO_4$		$Pcmn$ (Normal Phase)		
	-71	$Pcmn$?		
	-113	?	$P112_1/n$	
Below $T_I = -71°C$ Incommensurate Ferroelastic Phase	-253 ?			?

_____-_____-_____-_____-

Ammonium Bisulphate $(NH_4)HSO_4$	*-3*	$P2_1/c$ Pc	1958
	-119	$P1$	
Deuterated Ammonium Bisulphate	*-11*		
$(ND_4)DSO_4$	*-115*		
Rubidium Bisulphate $RbHSO_4$	*-15*	$P2_1/c$ Pc	1960

Lithium Hydrazinium Sulphate $LiN_2H_5SO_4$	none	$Pbn2_1$		1958
Lithium Hydrazinium FluoroBeryllate	none			1971
$LiN_2H_5BeF_4$				

Lithium Potassium Sulphate $LiKSO_4$	*163*	$P6_2$		1962
Lithium Rubidium Sulphate $LiRbSO_4$	*204*	?	?	1976
	202	?	?	
	188	?	?	
	166	?	?	

Double Ammonium Thallium	*-123*	Tetrag-		1963
$(NH_4)_4Tl_3(H_2AsO_4)$				
MonoArsenate (ATLAS)				

Langbeinites (Double Salts) $(M^{1+})_2 (M^{2+})_2 (SO_4)_3$:
$(M^{1+} = K, Rb, Tl, NH_4)$:
$(M^{2+} = Zn, Co, Mn, Mg, Cd)$

Diammonium dicadmium Sulphate	*-178*	$P2_13$	$P2_1$	1956
$(NH_4)_2Cd_2(SO_4)_3$				
DiThallium dicadmium Sulphate	*-145*	$P2_13$	$P2_1$	1972
$Tl_2Cd_2(SO_4)_3$	-153	P1		

	-181	$P2_12_12_1$		
DiRubidium dicadmium Sulphate	*-144*	$P2_13$ $P2_1$		1976
$Rb_2Cd_2(SO_4)_3$	-170			
DiPotassium diZinc Sulphate $K_2Zn_2(SO_4)_3$	*-135*	$P2_13$		1973
	-198			
DiPotassium diMagnesium Sulphate	*-72*	$P2_13$ $P2_12_12_1$		1978
$K_2Mg_2(SO_4)_3$				

-

Leconites :

Sodium Ammonium Sulphate dehydrate	*-173.9*	$P2_12_12_1$	$P2_1$?	1959
$Na(NH_4)(SO_4).2H_2O$	-181			
Sodium Ammonium Selenate dehydrate	*-97*	$P2_12_12_1$	$P2_1$	1970
$Na(NH_4)(SeO_4).2H_2O$				

-

Lead Monohydro Phosphate $PbHPO_4$	*37*	$P2/m$ Pc		1974
Lead MonoDeutero Phosphate $PbDPO_4$	*179*	$P2/m$ Pc		1974

-

Alums :

Methyl Ammonium Aluminium Sulphate.	*-96*	$P2_13$ $P2_1$		1956
DoDecahydrate $(CH_3)(NH_3)Al(SO_4)_2.12H_2O$ (MASD)				
Deuterated Methyl Ammonium	*-96*			1957

Aluminium Alum $(CD_3)(ND_3)Al(SO_4)_2.12D_2O$

Methyl Ammonium Gallium Alum	*-102*	1959
$(CH_3)(NH_3)Ga(SO_4)_2.12H_2O$		
Methyl Ammonium Chrome Alum	*-109*	1957
$(CH_3)(NH_3)Cr(SO_4)_2.12H_2O$		
Methyl Ammonium Iron Alum	*-104*	1957
$(CH_3)(NH_3)Fe(SO_4)_2.12H_2O$		
Methyl Ammonium Vanadium Alum	*-116*	1957
$(CH_3)(NH_3)V(SO_4)_2.12H_2O$		
Methyl Ammonium Indium Alum	*-109*	1957
$(CH_3)(NH_3)In(SO_4)_2.12H_2O$		
Methyl Ammonium Aluminium Selenate Alum	*-57*	1957
$(CH_3)(NH_3)Al(SeO_4)_2.12H_2O$		
Urea Chrome Alum	*-113*	1958
$CO(NH_2)_2HCr(SeO_4)_2.12H_2O$		
Ammonium Iron Alum	*-185*	1957
$(NH_4)Fe(SO_4)_2.12H_2O$		
Deuterated Ammonium Iron Alum	*-185*	1957
$(ND_4)Fe(SO_4)_2.12D_2O$		
Ammonium Indium Alum	*-146*	1957
$(NH_4)In(SO_4)_2.12H_2O$		
Ammonium Vanadium Alum	*-157*	1957

$(NH_4)V(SO_4)_2.12H_2O$

--- - --- - --- - --- -

| Di-Methylamine Aluminium Sulfate | *-123* | $2/m$ | m | 1988 |

$(CH_3)_2(NH_2)Al(SO_4)_2.6H_2O$

--- - --- - --- -

Guanidinium Compounds :

| Guanidinium Aluminium Sulphate | None | $P3_1m$ | 1955 |

Hexahydrate (GASH) $C(NH_2)_3Al(SO_4)_2.6H_2O$
   (First H-bonded FE discovered)

| Guanidinium Chromium Sulphate | None | $P3_1m$ | 1955 |

Hexahydrate (GCrSH) $C(NH_2)_3Cr(SO_4)_2.6H_2O$

| Guanidinium Gallium Sulphate | None | $P3_1m$ | 1955 |

Hexahydrate (GGaSH) $C(NH_2)_3Ga(SO_4)_2.6H_2O$ — Guanidinium

| Vanadium Sulphate Hexahydrate | None | $P3_1m$ | 1956 |

(GVSH) $C(NH_2)_3V(SO_4)_2.6H_2O$

| Guanidinium Aluminium Selenate | None | $P3_1m$ | 1955 |

Hexahydrate (GASeH) $C(NH_2)_3Al(SeO_4)_2.6H_2O$

| Guanidinium Chromium Selenate | None | $P3_1m$ | 1955 |
| Hexahydrate (GCrSeH) | | | |

$C(NH_2)_3Cr(SeO_4)_2.6H_2O$

| Guanidinium Gallium Selenate | None | $P3_1m$ | 1955 |
| Hexahydrate (GGaSeH) | | | |

$C(NH_2)_3Ga(SeO_4)_2.6H_2O$

Deuterated Guanidinium Aluminium	None	$P3_1m$		1955

Sulphate Hexahydrate $C(ND_2)_3Al(SO_4)_2.6D_2O$

_____-_____-_____-_____-_

Selenites:

Lithium Trihydrogen Selenite $LiH_3(SeO_3)_2$	None	Pn		1959
Sodium Trihydrogen Selenite $NaH_3(SeO_3)_2$	-79	$P2_1/a$	P1	1959
	-162		m	
Sodium Trideuterium Selenite $NaD_3(SeO_3)_2$	-2.5	$P2_1/n$	Pn	1968
	-15			
Potassium Trihydrogen Selenite $KH_3(SeO_3)_2$	-62	Pbcn	$P2_1/b$	1967
(Pure Ferroelastic)				
Potassium Trideuterium Selenite $KD_3(SeO_3)_2$	23.5	mmm		1968
Rubidium Trihydrogen Selenite $RbH_3(SeO_3)_2$	-115	$P2_12_12_1$	$P2_1$	1969
Rubidium Trideutero Selenite $RbD_3(SeO_3)_2$	-121	$P2_12_12_1$	$P2_1$	1975

_____-_____-_____-_____-

Potassium Cyanide [KFCT] s :

Potassium Hexa Cyano Ferrite Trihydrate	-24.5	C2/c	Cc	1959
$K_4Fe(CN)_6.3H_2O$				
Deuterated Potassium Ferro Cyanide	-18.0	C2/c	Cc	1960
Trihydrate $K_4Fe(CN)_6.3D_2O$				

Potassium Manganese Cyanide Trihydrate	*-40*			1960
$K_4Mn(CN)_6 \cdot 3H_2O$				
Potassium Ruthenium Cyanide Trihydrate	*-14.5*	C2/*c*	?	1960
$K_4Ru(CN)_6 \cdot 3H_2O$				
Deuterated Potassium Ruthenium Cyanide	*-7.3*			1960
Trihydrate $K_4Ru(CN)_6 \cdot 3D_2O$				
Potassium Osmium Cyanide Trihydrate	*-2.4*			1960
$K_4Os(CN)_6 \cdot 3H_2O$				
Deuterated Potassium Osmium Cyanide	*1.8*			1960
Trihydrate $K_4Os(CN)_6 \cdot 3D_2O$				

Triglycine Sulfate (L-Alanine) and Related Compounds :

Triglycine Sulfate (TGS)	*49.4*	$P2_1/m$	$P2_1$	1956
$(NH_2CH_2COOH)_3H_2(SO_4)$				
Deuterated Triglycine Sulfate	*60*			1961
$(ND_2CD_2COOD)_3D_2(SO_4)$				
Triglycine Selenate (TGSe)	*22*	$P2_1/m$	$P2_1$	1956
$(NH_2CH_2COOH)_3H_2(SeO_4)$				
$(ND_2CD_2COOD)_3D_2(SeO_4)$	*60*			1961
	?	?	?	
Triglycine Fluoroberyllate (TGFB)	*75*	$P2_1/m$	$P2_1$	1957

$(NH_2CH_2COOH)_3H_2(BeF_4)$

Deuterated Triglycine Fluoberyllate	*77*			1966

$(ND_2CD_2COOD)_3D_2(BeF_4)$

Di-Glycine Nitrate (DGN)	*-67*	$P2_1/a$	Pa	1958

$(NH_2CH_2COOH)_2.HNO_3$

Glycine Silver Nitrate	*-55*	$P2_1/a$ 2		1957

$(NH_2CH_2COOH).AgNO_3$

$(ND_2CD_2COOD).AgNO_3$?   1977

Diglycine Manganous Chloride Dihydrate	None	$P2_1$		1958

$NH_2(CH_2COOH)_2MnCl_2.2H_2O$

TrisSarcosine Calcium Chloride	. *−146*	$Pnma$	$Pn2_1a$	1962

$(CH_3NHCH_2COOH)_3,CaCl_2$

Semi- Carbazide Hydro Chloride	*21*			1972

$H_2NCONHNH_2HCl$

Tetra-Methyl Ammonium Tetra-	20	*Pmcn*		1978
Chloro Zincate $[N(CH_3)_4]_2.ZnCl_4$	6.0 ?	Ortho-		
	3.3	Monoclinic ?		
-92	Monoicl	?		
	-112	Ortho-	?	

Tetra-Methyl Ammonium Tetra- Iodo Zincate $[N(CH_3)_4]_2.ZnI_4$	*-63*	Pmcn P12$_1$		1987
DiPotassium TetraChloro Zincate K_2ZnCl_4	280	P6$_3$/mmc		1987
	130	Pnam P$na2_1$		
Troegerite $H(UO_2)AsO_4.4H_2O$	~20	Tetrag-		1974
Tanane (Nitroxy Di-Tetra	14.5	$\overline{4}2m$ $mm2$		1972
Methyl 2.2.6.6. piperidene Oxyle)				
Ammonium bis-Chloroacetate	?	FE		2000

TATRATES : Rochelle Salt and Related Compounds

Sodium Potassium Tartrate	*24*	P2$_1$2$_1$2 P2$_1$		1921
Tetrahydrate (Rochelle Salt) (RS)	*-18*			
$NaKC_4H_4O_6.4H_2O$	-61	P2$_1$2$_1$2		
(First FE to be discovered)				
Deuterated Rochelle Salt	*35*	P2$_1$2$_1$2	P2$_1$	1939
$NaKC_4D_4O_6.4D_2O$	*-22*	P2$_1$2$_1$2		
Ammonium Rochelle Salt	*-164*	P2$_1$2$_1$2	P2$_1$	1958
$NaNH_4C_4H_4O_6.4H_2O$				
Deuterated Ammonium Rochelle Salt	*-157*			1978
$NaND_4C_4H_4O_6.4D_2O$				

Solid Solutions of RS $Na[K_{1-x}(NH_4)_x]C_4H_4O_6 \cdot 4H_2O$ $\qquad$ 1932

$\qquad$ x = 0 – 0.03 $\qquad$?

$\qquad$ x = 0.03 – 0,18 $\qquad$?

$\qquad$ x = 0.18 – 0.88 $\qquad$?

$\qquad$ x = 0.88 –0.93 $\qquad$?

$\qquad$ x = 0.93 – 1.00 $\qquad$?

Lithium Ammonium Tartrate Monohydrate *-167* $P2_12_12$ $\qquad$ $P2_1$ $\qquad$ 1951

$\qquad$ $LiNH_4C_4H_4O_6 \cdot H_2O$ (LAT)

Lithium Thallium Tartrate Monohydrate *-263* $P2_12_12$ $\qquad$ $P2_1$ $\qquad$ 1951

$\qquad$ $LiTlC_4H_4O_6 \cdot H_2O$ (LTT)

Deuterated LT T $\qquad$ ~ *-263* 222 $\qquad$ 1969

$O[H,D(CHCOO)_2 \, LiTl](H,D)_2O$

Deuterated LAT $\qquad$? $\qquad$ 1958

| Metallic Solid (Alloys) | IV - VI Semiconductors | : |

Germanium Telluride? $\qquad$ GeTe $\qquad$ *354* Cubic Trigonal $\qquad$ 1966

$\qquad$ (First Diatomic FE discovered)

Tin Telluride SnTe $\qquad$ -27 Cubic Trigonal $\qquad$ 1966

$\qquad$ (p-type Semiconductor)

$\qquad$ x (GeTe) - (1 - x)(SnTe) $\qquad$? Cubic Trigonal $\qquad$ 1966

$\qquad$ x = 0.1, 0.3

Lead Telluride PbTe		*-203*		1970
Altaite	$Pb_{1-x}Sn_xTe$		$Fm3m$	1977
	$Pb_{1-x}Ge_xTe$			
Bismuthinite	Bi_2S_3		$Pna2_1$	1969
	$Pb_{0.94}Ge_{0.06}Te$	-113		1970
	$Pb_{0.9544}Ge_{0.0456}Te$	-143		
	$Pb_{0.972}Ge_{0.028}Te$	-173		
Vanadium Silicide ? V_3Si		-243	$Pm3n$ $P4mc$	1965
(Purely Ferroelastc)				
Alexandrite (Cr-chrysoberyl) Cr_2BeO_4			$Pnam$	1978
(Magneto-ferroelectric)				

Alkali Tungsten Bronze (Superconducting)$(M_x W_{1-x}) O_3$:
$(M_x W_{1-x}) O_3$: M = Na, K; $x > 0.5$?

Cassiterite	SnO_2	$P4_2/mnm$	1961
Chalcocite	Cu_2S	$P2_1/c$	1987
Chalcostibite	$CuSbS_2$	$Pna2_1$	1976
Lewisite (Ti-roméite)?	$FeCa_2Sb_2O_7$	$Fd3m$	1971
Nitrokalite ?	$Ca(H_2O)(NO_3)_2$	$P2_1/c$	1971
Polianite-Pyrolusite	MnO_2	$P4_2/mmm$	1960

Pyrrhotite	$Fe_{1-x}S$	$P6_3/mmc$	1970,1971
Russelite	Bi_2WO_6	I42d	1969
Rutile	TiO_2	$P4_2/mmm$	1959
Stibiotantilite or stibiocolumbite		$Pna2_1$	1971
	$Sb(Ta,Nb)O_4$		
Stibnite or antimonite Sb_2S_3		$Pna2_1$	1967
Troilite	FeS	P62c	1969
Water (Hexagonal ice) H_2O		$P6_3/mmc$	1964

Complex Organic Compounds :

Thiourea	$SC(NH_2)_2$	-150	$Pb2_1m$	1956
-		*-104*		
		-97		
		-93		
		-71	Pbnm	
Deuterated Thiourea	$SC(ND_2)_2$	-88		
		-81		
		-77		
		-60		

Tetra Methyl Ammonium Tri Chloro	None
Mercurate $N(CH_3)_4HgCl_3$	
Tetra Methyl Ammonium Tri Bromo	None

Mercurate $N(CH_3)_4HgBr_3$

Tetra MethylPhosphonium Tri Bromo	None			
Mercurate $P(CH_3)_4HgBr_3$				
Tetra Methyl Ammonium Tri-Iodo	None			
Mercurate $N(CH_3)_4HgI_3$				

Di Calcium Strontium Propionate	*8.4*	$P4_12_12$	$P4_1$	1957
$Ca_2Sr(CH_3CH_2COO)_6$ or $Ca_2Sr(Et\,CO_2)_6$ -68.8				
DiCalcium Lead Propionate $Ca_2Pb(Et\,CO_2)_6$ *60*				1973
	-81.5			

Ammonium Mono ChloroAcertate	*-150*	$C2_1/c$ C2		1957
$(CH_2Cl\,COO)NH_4$				
Ammonium Di ChloroAcertate	*.-145*	C2/c Cc		1966
$NH_4H(Cl\,COO\,CH_2)_2$				
Ammonium Deutero bis-monochloro Acetate. *-143*				1976
$ND_4D(Cl\,COO\,CH_2)_2$				

Tetra-Methyl Ammonium Trichloro	None	$P2_1$		1960
Mecurate $N(CH_3)_4HgCl_3$				
Tetra-Methyl Ammonium Tribromo	None	$P2_1$		1962

Mecurate $N(CH_3)_4HgBr_3$

Tetra-Methyl Phosphonium TriBromo	None	1962
Mecurate $P(CH_3)_4HgBr_3$		
Tetra-Methyl Ammonium Tri-Iodo	None	1962
Mecurate $N(CH_3)_4HgI_3$		

$P(CH_3)_4SbCl_6$ Crystal		1998	
(TMACAB) Mixed Crystals	Ferroelastic	2000	
$(CH_3)_3]_3Sb_{2(1-x)}Bi_{2x}Cl_9$			
$N(CH_3)_4H(Cl_3CCOO)_2$	FE	2000	
$[NH_2(CH_3)_2]_3Sb_2Br_9$	FE	2000	
$[C(NH_2)_3]_3Bi_2Br_9$	Ferroelastic	1999	
(d-MABA) Crystal $(CH_3ND_3)_3]_3Sb_2Br_9$	FE	2001	
$(CH_3NH_3)_3]_3Sb_2I_9$			1998
$(C_5H_5NH)_5Bi_2Cl_{11}$	FE	2001	
$[NH_3CH_3]_5Bi_2Cl_{11}$	FE	1998	
Ferroic (DMACA) $[NH_2(CH_3)_2]_3Sb_2Cl_9$	FE	1998	
$P(CH_3)_4SbCl_6$		1999	
Two Relaxators Dielectric		2000	
$[C_5H_{10}NH_2]SbCl_6(C_5H_{10}NH_2)Cl$			

$[C_5H_5NH]_5Bi_2CBr_{11}$ FE   2001

(d-MABA)Crystals $(CH_3ND_3)_3Sb_2Br_9$ 2001

$(C_5H_5NH)BiCl_4$ 2001

$(CH_3NH_3)_5Bi_2Cl_{11}$,   3001

$(C_5H_{10}NH_2)SbCl_6$,   2001

(Ferroelastic)

_____-_____-_____-_____-

Poly Vinylidene Fluoride (PVF$_2$) ? $[CH_2CHF]_n$ 1972

Poly Vinyl Fluoride (PVF)? $[CH_2CHF]_n$ 1972

_____-_____-_____-_____-_

| Cyano-SPINEL Type $[K_2M(CN)_4]$ (M = Zn, Cd, Hg) |

$Rb_2Hg(CN)_4$.   *398*   Cubic R3*c*   1978

$K_2Hg(CN)_4$.   *125*   Cubic R3*c*   1978

_____-_____-_____-_____-_

| Rare Earth Oxides (Re P$_5$ O$_{34}$): (Re = La, Tb, Pr, Nd, Ce, Sm, Eu, Gd) |

KTP   KTiOPO$_4$?   ?   1992

$Sn_2P_2Se_6$ 1993

Bismuth Germanate, $Bi_{12}GeO_{20}$

Lithium Hepta Germanate $Li_2Ge_7O_{15}$ 283.5

A_2BX_4-type FE, A = K,Tl ; B = Co, Zn;X = I :

Tl_2ZnI_4 209 K, $P2_1/m$ $Z = 2$, FE $\| b$ - axis $P2_1$, $Z = 2$

         1985

K_2ZnBr_4 272 K   1985

LIQUID CRYSTALS :

p- Azoxyanisole(Nematic) $C_{14}H_{14}N_2O_3$ 134   1966

         116

p- Azoxyphenetole (Nematic) $C_{16}H_{18}N_2O_3$ 168   1965

         137

p- Azoxybenzoic Acid (Nematic) $C_{11}H_{14}O_3$?   1966

p- Methoxy Cinnamic Acid (Nematic) $C_{10}H_{10}O_3$?   1966

Azobenzene   (Nematic) $C_{12}H_{18}N_2$?   1966

Azoxybenzene(Nematic) $C_{12}H_{18}N_2O$?   1966

2-Methyl butyle p-[p-decycloxy   ?   1975

         Benzlidene amino] cinnamate

Cu-Phthalocyanine (organic Semicond.) CuPc *10*   1974

         H_2Pc   *5*   1974

         D_2Pc   *-12*   1974

         $9H_2$Pc-1FePc   *18*   1974

New ferroelectric inorganic materials predicted in point group 4mm	SC Abrahams 996	
$Ce_5B_2C_6$	1996	
$NbTe_4$	1996	
$Na_3Nb_{12}O_{31}F$	1996	
Ca_2FeO_3Cl	1996	
$K_4CuV_5O_{15}Cl$	1996	
$TlBO_2$	1996	
$CrOF_3$		1996
$PbTeO_3$	1996	
$VO(HPO_3)(H_2O).3H_2O$	1996	
$MgB_2O(OH_6)$	1996	
β-boron	1996	
$CuBi_2O_4$	1999	
$WOBr_4$	1999	
Na_8PtO_6	1999	
SbF_2Cl_3	1999	
$Ba_{12}Ti_8O_{16}$	1999	
$Ni[SC(NH_2)_2]_4Cl_2$	1999	

$Ca_2SiO_3Cl_2$	1999
Mineral Caratiite	1999
NbAS	1999
β-NbO_2	1999
Ag_3BiO_3	1999

Stillwellite family of Ferroelectrics

$LaBGeO_5$	2003
Cu_2BaGeS_4	2003
$Fe_3(Fe,Si)O_4(OH)_5$	2003
Se_4S_5	2003
$K_2HCr_2AsO_{10}$	2003
IV-$RbNO_3$	2003
$Rb_2Sc(NO_3)_5$	2003
Na_3ReO_5	2003
$Nd_{14}(GeO_4)_2(BO_3)_6O_8$	2003
$CsHgCl_3$	2003
$Ba_2Cu_2AlF_{11}$	2003
KYF_4	2003
$SrS_2O_6 \cdot 4H_2O$	2003

$Cu_3PbTeO_6(OH)_2$	2003	
$ReH(CO)_4$	2003	
$Ni_2(NH_3)_9Mo(CN)_8 \cdot 2H_2O$		2003
$Ca_{1.89}Ta_{1.80}Sm_{0.16}Ti_{0.10}O_7$ -6T polytype	2003	
β-$LaBSiO_5$	2003	
$PbBAsO_5$	2003	
$BaBAsO_5$	2003	
$LiAsCu_{0.93}$	2000	
Na_2UF_6	2000	
$BiTeI$	2000	
$BaGe_4O_9$	2000	
α-UMo_2O_8	2000	
Cu_2SiO_3	2000	
$Co(IO_3)_2$	2000	
$Sr_7Al_{12}O_{25}$	2000	
KSn_2F_5	2000	
$YbIn_2S_4$	2000	
$Na_5CrF_2(PO_4)_2$	2000	
$Sn(ClO_2)_2(ClO_4)_6$	2000	

Eu_3BWO_9 2000

$Li(H_2O)_4B(OH)_4).2H_2O$ 2000

$Mn_3V_{1/2}(SiO_4)O(OH)_2$ 2000

$Ca_6(Si_2O_7)(OH)_6$ 2000

$Na_{6.9(2)}[Al_{5.6(1)}Si_{6.4(1)}O_{24}]S_2O_3)_{1.0(1)}.2H_2O$ 2000

$BaCa_2In_6O_{12}$ 2000

$Ni(H_2O)_6[Sb(OH)_6]_2$ 2000

$Sr_4Cr_3O_9$ 2000

$Cu_5O_2(VO_4)_2.CuCl_2$ 2000

Anhydrous ferroelectric aminoguani-dinium(2+)   1998
hexafluorozirconate (Ferroelectric)

AFE Compounds :

Perovskite Type Compounds.

:Sodium Niobate	$NaNbO_3$		P*m3m*	1955
		640		
		562		
		354	P*bma*	
		-200	Monoclinic	
Lead Zirconate	$PbZrO_3$	*230*	P*m3m* P*ba2*	1951

Lead Hafnate	$PbHfO_3$	215		$Pbam$	1953
		163			
AFE at low Temp.					1998
Bismuth Ferrite?	$BiFeO_3$	850	Cubic		1961
		~575			
		370		$R3m$	
Cadmium Titanate?	$CdTiO_3$	960	?	$Pc2_1n$	1970
		110		Trigonal	
		-218			
Cadmium Hafnate?	$CdHfO_3$	605			
Cesium Lead Chloride?	$CsPbCl_3$	47			

Potassium Bismuth Titanate?	410	$Pm3m$		1959
$(K_{0.5}Bi_{0.5})TiO_3$	*270*	Tetrag-		
Sodium Bismuth Titanate?	320	$Pm3m$		1959
$(Na_{0.5}Bi_{0.5})TiO_3$	200			
Lead Magnesium Tungstate	*38*	Cubic	$C222_1$	1959
$Pb(Mg_{0.5}W_{0.5})O_3$				
Lead Cadmium Tungstate?	420	Cubic	Monocl	1963
$Pb(Cd_{0.5}W_{0.5})O_3$	120			
Lead Manganese Tungstate?	150			1964
$Pb(Mn_{0.5}W_{0.5})O_3$				
Lead Manganese Tungstate?	150			

$Pb(Mn_{2/3}W_{1/3})O_3$

Lead pyro Niobate? $Pb_2Nb_2O_7$	15.4	1953
Lead Cobalt Tungstate $Pb(Co_{0.5}W_{0.5})O_3$	32	1963
	-190 ~ -200	
	-263	
Lead Manganese Rhenate?	120	1964
$Pb(Mg_{0.5}Re_{0.5})O_3$	-170	
Lead Indium Niobate? $Pb(In_{0.5}Nb_{0.5})O_3$	90	
Lead Ytterbium Niobate	310	1963
$Pb(Yb_{0.5}Nb_{0.5})O_3$		
Lead Holmium Niobate?	240	1965
$Pb(Ho_{0.5}Nb_{0.5})O_3$		
Lead Lutetium Niobate?	270	1958
$Pb(Lu_{0.5}Nb_{0.5})O_3$		
Lead Ytterbium Tantalate?	285	1964
$Pb(Yb_{0.5}Ta_{0.5})O_3$		
Lead Lutetium Tantalate?	280	1958
$Pb(Lu_{0.5}Ta_{0.5})O_3$		
Lead Manganese Tungstate?	200	1965
$Pb(Mn_{2/3}W_{1/3})O_3$		
Lead Zirconate Titanate $Pb(Ti_{1-x}Zr_x)O_3$		1952
$x = 0.5 \sim 0.6$		
Lead Gallium Niobate? Pb_2GaNbO_6	100	

Lead Bismuth Niobate ? Pb_2BiNbO_6	-235			
Lead Cadmium Manganese Niobate? $Pb_2Cd_{0.5}Mn_{0.5}NbO_6$	-237			
Lead Cadmium Titanium Tantalate? $Pb_2Cd_{0.5}Ti_{0.5}TaO_6$	-253			
Lead Lithium Niobium Tungstate	110		$Pb(Li_{1/3}Mn_{1/3}W_{1/3})O_3$	
Lead Cadmium Manganese Tungstate	300		$Pb(Cd_{1/3}Mn_{1/3}W_{1/3})O_3$	
Lead Cadmium Tungstate? $Pb(Cd_{4/9}Nb_{2/9}W_{1/3})O_3$	?		1965	
Cadmium Scandium Niobate $Cd(Sc_{1/2}Nb_{1/2})O_3$	-203			
Cadmium Scandium Niobate	450		1971	
$Cd(Fe_{1/2}Nb_{1/2})O_3$	-225			

Pyrochlore type Oxides :

Ammonium Dihydrogen Phosphate $NH_4H_2PO_4$	*-125*	$I\bar{4}2d$ $P2_12_12_1$		1937 (ADP)
Deuterated Ammonium Phosphate $ND_4D_2PO_4$	*-31*	" "		1952 (ADP)
Ammonium Dihydrogen Arsenate (ADA) $NH_4H_2AsO_4$	*-57*	" "		1937
Deuterated Ammonium Arsenate $ND_4D_2AsO_4$	*31*	" "		1953

Ammonium Fluoro-Beryllate   *-97* $Ac2_1a$ *Acam*   1958

$(NH_4)_2BeF_4$

Selenites :

Cesium Trihydrogen Selenite?   *-128*   P1   P1   1962
     $CsH_3(SeO_3)_3$

Lithium TriDeutero Selenite?   None   ?   1968
     $LiD_3(SeO_3)_3$

Miscellaneous Materials :

Sodium Nitrite $NaNO_2$ 164.7   *Immm*   ?   1963

     163.4   Im2m

     -105   ?

Rubidium Nitrate $RbNO_3$ 291   Cubic   ?   1963

     219   $Pa3$

     164   $P3_12$

Sodium Azide NaN_3 *20* $R\bar{3}m$ C2/*m*   1986

Silver Tri-hydrogen Per-iodate   *-28*   R3*m*   Hexag-   1943

     $Ag_2H_3IO_6$

Silver Tri-deuterium Per-iodate   *12*   "   "   1956

$Ag_2D_3IO_6$

Ammonium Trihydrogen Periodate	*-20*	"	"	1943

$(NH_4)_2H_3IO_6$

DeuteroAmmonium Trideuterium	*-7*		1955
Periodate $(ND_4)_2D_3IO_6$			
Hexafluoro Phosphate Salt?	*-45*	C4/*mmm* Ortho	1959
$(NH_4)PF_6.NH_4F$	-101		
Lead Ortho Vanadate $Pb_3V_2O_8$	*100*	R$\bar{3}$m	1965
	0	?	
Lead Silicate Pb_4SiO_6	*155*		1965
Cesium Plumbo Chloride $CsPbCl_3$	*47*	P*m*3*m* P4/*mmm*	1969
	41.5	P*mmm*	
	37	P2/*m*	
Cesium Cobalt Chloride $CsCoCl_3$	?		
Cesium Plumbo Bromide $CsPbBr_3$		P*m*3*m*	
	130	P*mbm*	
	88	P*mbn*	
Cesium Strontium Chloride $CsSrCl_3$	113		1976
	108		
	89		
Lead Germanate $4PbO.GeO_2$	*67*		1978
(Or, Pb_4GeO_6)			

Stannous Chloride Dihydrate	*-65*	$P2_1/c$	1976
$SnCl_2.2H_2O$	-54		

(AFE, since the trademark for AFE T_N = 219.446 K < T_M = 219.093K)

Rubidium Lithium Sulphate $RbLiSO_4$	*204*	Pmcn	1975
	202	$P2_1/c$	
	188	P11n	
	166	P112$_1$	

Ammonium Lithium Sulphate NH_4LiSO_4 ?

Cupric Formate Tetrahydrate ?	*-38*	$P2_1/a$	?	1965
$Cu(HCOO)_2.4H_2O$ (AFM)	*-260*	?AFM		
Cupric Formate Tetra-Deuerate?	*-27.3*	$P2_1/a$	?	
$Cu(HCOO)_2.4D_2O$	-257	AFM		

Ammonium Nitrate NH_4NO_3	?		1976
Low Quartz? SiO_2			1950
Neodymium OrthoNiobate $NdNb_4$	807		
Gadolinium OrthoNiobate? $GdNb_4$	757		

Barium Manganese Fluoride $BaMnF_4$	-18	?	?	1969
	-247			
Barium Manganese Fluoride $BaCoF_4$	none	?	?	1974
Barium Manganese Fluoride $BaFeF_4$	none	?	?	1974

Potassium Manganese Fluoride KMnF$_4$	-89	?	?		1969
Potassium Cobalt Fluoride KCoF$_4$	-107	?	?		1970
[N(CH$_3$)$_4$]$_2$CoCl$_4$	20				
	7.1				
	4.6				
	3.0				
	-81				
	-151				

%&%&%&%&%&%&%&%&%

APPENDIX A

Values of Physical Constants

APPENDIX A

Values of Physical Constants

[Ref: *CODATA*, 1998; *J. Phys. Chem. Ref. Data*, Vol. 28 (6) 1999; *Rev. Mod Phys. 72*,(2) 2000]

Planck constant	$h = 6.6256 \times 10^{-34} J - s$;
	$\hbar = (h/2\pi) = 1.054 \times 10^{-34} J - s$
Permittivity of free space	$\varepsilon_0 = 8.8542 \times 10^{-12} F\ m^{-1}$;
	$\varepsilon_0 = 8.8542 \times 10^{-12} C^2 N^{-1}\ m^{-2}$
The Coulomb constant	$k = (1/4\pi\ \varepsilon_0) = 8.9875 \times 10^9\ Nm^2 C^{-2}$
	$k = (1/4\pi\ \varepsilon_0) = 8.9875 \times 10^9\ F^{-1}m$
Permeability of free space	$\mu_0 = 4\pi \times 10^{-7} T\ m\ A^{-1}$,
	$\mu_0 = 4\pi \times 10^{-7} N\ A^{-2}$.
Gravitational constant	$G = 6.67 \times 10^{-11}\ Nm^2 kg^{-2}$
Speed of light in Free space	$c = 2.997925 \times 10^8\ ms^{-1}$
Boltzmann constant	$k_B = 1.3805 \times 10^{-23}\ JK^{-1}$
Avogadro's constant	$N_A = 6.0225 \times 10^{23}\ mol^{-1}$
Stefan's constant	$\sigma = 5.67 \times 10^{-8} Wm^{-2} K^{-4}$,
Universal Gas constant	$R = 8.314\ J\ K^{-1} mol^{-1}$
Electronic Charge	$e = 1.6021 \times 10^{-19} C$
Electron Rest mass	$m_e = 9.1094 \times 10^{-31}\ kg$
	$m_e = 5.4858 \times 10^{-4} u = 0.5110\ MeV$
Electron Volt	$1\ eV = 1.6021 \times 10^{-19} J$
Proton Rest mass	$M_p = 1.6726 \times 10^{-27} kg$;

$$M_p = 1.007276 \, u = 938.272 \, MeV$$

Mass of Neutron $\qquad M_n = 1.67493 \, x \, 10^{-27} \, kg$

$$M_n = 1.008665 \, u = 939.57 \, MeV$$

Mass of Deuteron $\qquad M_d = 2.013553 \, u$

Mass of Alpha particle $\qquad M_\alpha = 4.002602 \, u$

Mass of H-atom $\qquad M_H = 1.67493 \, x \, 10^{-27} \, kg \; ;$

$$M(^1_1H) = 1.007825 \, u$$

Mass of Tritium $\qquad M(^3_1H) = 3.016049 \, u$

Mass of He $\qquad M(^4_2He) = 4.00387 \, u$

Mass of Nitrogen $\qquad M(^{14}_7N) = 14.00752 \, u$

Mass of Oxygen $\qquad M(^{17}_8O) = 17.00453 \, u$

Mass of U-238 $\qquad M(^{238}_{92}U) = 238.650784 \, u$

Mass of Th 234 $\qquad M(^{234}_{90}Th) = 234.043593 \, u$

Mass of Th-232 $\qquad M(^{232}_{90}Th) = 232.038124 \, u$

Mass of Ra-228 $\qquad M(^{228}_{88}Ra) = 228.031139 \, u$

Unified atomic mass unit $\qquad 1u = \frac{1}{12}(\text{mass of } ^{12}_6C) = 931 \, MeV/c^2$

hc $\qquad \hbar c = hc/2\pi = 1.973 \, x \, 10^{-11} \, MeV - cm$

Molar volume at 1 atm, 0°C $\qquad = 2.2414 \, x \, 10^{-2} \, m^3 mol^{-1}$

Specific electronic charge, $\qquad e/m_e = 1.7588 \, x \, 10^{11} \, Ckg^{-1}$

Rydberg constant, $\qquad R_H = 1.09737 \, x \, 10^7 \, m^{-1}$

Bohr first Radius, $\qquad a_0 = \hbar^2/m_e e^2 = 0.529167 \, x \, 10^{-10} \, m$

$$a_0 = \hbar^2/m_e e^2 = 0.5292 \, nm$$

Radius of Proton $\qquad r_0 = 1.2 \, x \, 10^{-15} \, m$

Fine Structure constant $\qquad \alpha = e^2/hc = 7.2973 \, x \, 10^{-3} = 1/137.036$

Bohr magneton, $\qquad \mu_B = 9.2740 \, x \, 10^{-24} \, A \, m^2 (= JT^{-1})$

$$\mu_B = 5.788383 \, x \, 10^{-11} \, MeVT^{-1}$$

Nuclear magneton, $\qquad 1 \, \mu_N = 1 \, nm = 5.9505 \, x \, 10^{-27} \, A \, m^2$

$$\mu_N = 1 \, nm = 3.152452 \, x \, 10^{-14} \, MeV \, T^{-1}$$

Magnetic dipole moment, electron $\quad \mu_e = 1.0011597 \, \mu_B$

Proton $\quad \mu_p = 2.792847 \, \mu_N$

Neutron $\quad \mu_n = -1.910426 \, \mu_N$

Barn $\qquad 1 \, b = 1 \, x \, 10^{-24} \, cm^{-2} = 100 \, fm^2$

Hydrogen Ground state $\qquad R_y = 13.6057eV$.

Calorie $\qquad 1 \, Cal = 4.1840 \, J$

Tesla (the unit of magnetic field)	$1\,T = 1.0\,Wb\,m^{-2} = 10^4\,G$
Atomic weight	$= N\,M_n + Z\,(M_p + m_e)$ - (binding energy)
Year in seconds	$1\,yr = 3.16 \times 10^7 s$
Atm. Pressure	$1\,atm = 1.01 \times 10^5 N\,m^{-2}$
Acceleration due to gravity	$1\,g = 9.81\,ms^{-2}$
Unit for energy absorption, *rad*	$1\,rad = 0.1\,J/kg$
Curie *Ci)	$1\,Ci = 3.7 \times 10^{10}\,s^{-1}$

&&*&*&*&*&

APPENDIX B
LIST OF ELEMENTS

APPENDIX B
List of Elements

LIST OF ELEMENTS

ATOMIC NUMBER	SYMBOL	NAME	ATOMIC WEIGHT	ATOMIC NUMBER	SYMBOL	NAME	ATOMIC WEIGHT
0	n	neutron	————	52	Te	tellurium	127.60
1	H	hydrogen	1.0079	53	I	iodine	126.9045
2	He	helium	4.00260	54	Xe	xenon	131.30
3	Li	lithium	6.941	55	Cs	cesium	132.9054
4	Be	beryllium	9.01218	56	Ba	barium	137.34
5	B	boron	10.81	57	La	lanthanum	138.9055
6	C	carbon	12.011	58	Ce	cerium	140.12
7	N	nitrogen	14.0067	59	Pr	praseodymium	140.9077
8	O	oxygen	15.9994	60	Nd	neodymium	144.24
9	F	fluorine	18.99840	61	Pm	promethium	————
10	Ne	neon	20.179	62	Sm	samarium	150.4
11	Na	sodium	22.98977	63	Eu	europium	151.96
12	Mg	magnesium	24.305	64	Gd	gadolinium	157.25
13	Al	aluminum	26.98154	65	Tb	terbium	158.9254
14	Si	silicon	28.086	66	Dy	dysprosium	162.50
15	P	phosphorus	30.97376	67	Ho	holmium	164.9304
16	S	sulfur	32.06	68	Er	erbium	167.26
17	Cl	chlorine	35.453	69	Tm	thulium	168.9342
18	Ar	argon	39.948	70	Yb	ytterbium	173.04
19	K	potassium	39.098	71	Lu	lutetium	174.97
20	Ca	calcium	40.08	72	Hf	hafnium	178.49
21	Sc	scandium	44.9559	73	Ta	tantalum	180.9479
22	Ti	titanium	47.90	74	W	tungsten	183.85
23	V	vanadium	50.9414	75	Re	rhenium	186.2
24	Cr	chromium	51.996	76	Os	osmium	190.2
25	Mn	manganese	54.9380	77	Ir	iridium	192.22
26	Fe	iron	55.847	78	Pt	platinum	195.09
27	Co	cobalt	59.9332	79	Au	gold	196.9665
28	Ni	nickel	58.71	80	Hg	mercury	200.59
29	Cu	copper	63.546	81	Tl	thallium	204.37
30	Zn	zinc	65.38	82	Pb	lead	207.2
31	Ga	gallium	69.72	83	Bi	bismuth	208.9804
32	Ge	germanium	72.59	84	Po	polonium	————
33	As	arsenic	74.9216	85	At	astatine	————
34	Se	selenium	78.96	86	Rn	radon	————
35	Br	bromine	79.904	87	Fr	francium	————
36	Kr	krypton	83.80	88	Ra	radium	————
37	Rb	rubidium	85.4678	89	Ac	actinium	————
38	Sr	strontium	87.62	90	Th	thorium	232.0381
39	Y	yttrium	88.9059	91	Pa	protactinium	————
40	Zr	zirconium	91.22	92	U	uranium	238.029
41	Nb	niobium	92.9064	93	Np	neptunium	————
42	Mo	molybdenum	95.94	94	Pu	plutonium	————
43	Te	technetium	————	95	Am	americium	————
44	Ru	ruthenium	101.07	96	Cm	curium	————
45	Rh	rhodium	102.9055	97	Bk	berkelium	————
46	Pd	palladium	106.4	98	Cf	californium	————
47	Ag	silver	107.868	99	Es	einsteinium	————
48	Cd	cadmium	112.40	100	Fm	fermium	————
49	In	indium	114.82	101	Md	mendelevium	————
50	Sn	tin	118.69	102	No	nobelium	————
51	Sb	antimony	121.75	103	Lr	lawrencium	————

BIBLIOGRAPHY

BIBLIOGRAPHY

1) Aizu, K. (1964), J. Phys. Soc. Japan,,19, 915.

2) Aizu, K. (1964), Phys. Rev., A 133, 1350.

3) Aizu, K. (1965), Phys. Rev., A 140, 590.

4) Aizu, K. (1969), J. Phys. Soc. Japan, 27, (1969).

5) Aizu, K. (1970) J. Phys. Soc. Japan, 28 (3), 706-16., , 28 (3), 717.

6) Anderson, P.W.,(1959), "Fizika Dielektrikov" (Edtd, BI Skanavi), (Akad. Nauk. SSSR) p 290.).

7) Artl, G and NA Pertsev (1991), J. Appl. Phys., 70, 2283-89.

8) Auciello, O. *et al.*,(1998), "The Physics of Ferroelectric Memories", *Physics Today*, pp 22-27, July 1998 issue.

9) Azaroff, LU and Donahue, R J (1969) "Laboratory Experiments in X-ray Crystallography" (MGH, NY).

10) Balkanski, M. and MK Teng, (1969), "Physics of the Solid State" (AP), pp 289-328.

11) Barker, AS.(Jr.) & R. Loudon (1967), Phys. Rev.., 158, 433.

12) Bell, AJ., (1994), Proc. 1994 Int. Symposium on Appl. Ferroelectrics, (1994).

13) Betsuyaku, H (1969), J. Phys. Soc. Japan, 27, 1485-500.

14)

15) Bhagavantam, S. (1967) "Crystal Symmetry and Physical Properties" (AP, London, NY).

16) Blinc, R. (1966), Proc. Intnl. Meeting on Ferrolectricity, Vol II, pp 333-57.

17) Blinc, R., A. Jovanovic, A. Levstik & A. Prelesnic (1965), J. Phys. Chem. Solids, 26, 1359.

18) Blinc R & M Mali, (1969), Sol. St. Commun. 7(19),1413- 15.

19) Blinc, R. & Zeks, (1974), "Soft Modes in Ferroelectrics & Anti-ferroelectrics" (North Holland, Amsterdam) 1974.

20) Born, Max, and Huang (1969), "Dynamical Theory of Crystalline Lattices" (Oxford), 1969

21) Borchardt-Ott, Walter,"Crystallography" (Springer-Verlag, Berlin).

22) Bötcher , C. J. F. (1973), *Dielectrics and Static Fields*, Vol. 1, 2nd edn, (Elsevier Scientific Publishing Company, Amsterdam).

23) Bötcher, C. J. F. and P. Bordewijk (1978), *Dielectrics in Time Dependent Fields*, Vol. 2, 2nd edn, (Elsevier Scientific Publishing Company, Amsterdam).

24) Boiko, AA. & LT. Dhat (1969), Kristallografiya, 14 (5), 825-28.

25) Britz, TC.& HG Unruh, (1996) Ferroelectrics, 185, 151.

26) Brice, J.C., (1965), The Growth of Crystals rom the Melt";Vol. 5, Series of Monographs on Selected Topics in Solid State Phyics, Edtd., E.P. Wohlfarth (North Holland, Amsterdam) 1965.

27) Bruce, A.D., & Cowley, R.A., (1981), "Structural Phase Transitions", (Taylor & Francis, London) 1981.

28) Burfoot, Jack C.(1967) "Ferroelectrics: An Introduction to the Physical Principles" (D. Van Nostrand, London) 1967.

29) Burfoot, Jack C. and Taylor, G.W.(1971), "Polar Dielectrics & Their Applications" (MacMillan).

30) Burger, MJ (1970), "Contemporary Crystallography" (MGH, NY)

31) Burger, M.J.,(1956) , "Elementary Crystallography", (John Wiley, NY) 1956

32) Busch, G. and P. Scherrer (1935), Naturwiss., $\underline{23}$, 737.

33) Canner, James P. (1969). Ph.D. Dissertation, "Mossbauer effect studies on ferroelectric phase transition $PbZrO_3$-$PbTiO_3$-$BiFeO_3$ ternary system" (Univ Missouri –Rolla, Curtis Laws Wilson Library) May 1969.

34) Cady, Walter Guyton., (1965), "Piezoelectricity" (An Introduction to the Theory and Applications of Electro-mechanical Phenomena in Crystals), New Revised Edition, Vol. 1 and Vol. II (Dover, New York, 1964).

35) Chadderton, (1965), "Radiation Damage in Crystals"(Methuen, London) 1965.

36) Chapman, D. (1966) Science Journal, pp 33-36.

37) Cochran, W. (1960), Advances in Physics, $\underline{9}$, 387.

38) Cohen, RE (1992) Nature, $\underline{358}$, (6382) 136-38.

39) Connolly, TF. & Turner, Errett, (1970), "Ferroelectric Materials and Ferroelectricity" (IFI / Plenum, NY), 1970.

40) Cross, LE., (1987) Ferroelectrics, $\underline{76}$ (3-4) 241-67.

41) Cross, LE.,(1993), Ferroelectric Ceramics-Tutorial Reviews, Theory, Processing and Applications, (N. Setter, and E. L. Colla, ed.) 1 (Birkhauser Verlag, Basel, 1993).

42) Daniel, VV., (1967) *Dielectric Relaxation*, (Academic Press, London).

43) Devanarayanan, S., (2013), "Quantum Chemistry", (SciTech, Chennai) 2013.

44) Devanarayanan, S. (1969), "Ferroelectric Crystals And Their Properties (Investigations on Thermal Expansion of Ferroelectrics)" Ph.D. Thesis (Indian Institute of Science, Bangalore) Feb 1969.

45) Devanarayanan, S. (1979), A Course of Lectures on "Ferroelectricity in Crystals" – Summer Institute of Physics for College Teachers, by UGC, at Univ. Kerala, Kariavattom, April 23 – May 31, 1979. .(<Reseachgate.net>).

46) Devanarayanan, S., (1995), "Ferroelectricity – Fundamentals and Theory", Four Lectures – Refresher Course in Physics, Calicut University, Kozhikode. 20 -21, Dec 1995 (<Reseachgate.net>).

47) Devanarayanan, S., (1999), "Phase Transitions",(UGC sponsored Refresher Course in Physics) Department of Physics, CUSAT, Kochi -682202, India.: 15 February 1999. (<Reseachgate.net>).

48) Devanarayanan, S..(1999), "Ferroelectric Materials – A Review", A course of three lectures at the Workshop on Research Activities with Ion Beams at NSC Sponsored by NUCLEAR SCIENCE CENTRE, Delhi, at Department of Physics, University of Kerala, Thiruvananthapuram, India, Feb. 1999. .(<Reseachgate.net>).

49) Devanarayanan, S.(1968), "Automatic Electronic System to Measure Interference Fringe Motion" (Get-together on Instrumentation, 22-23 Feb 1968, CISL, I.I.Sc., Bangalore, India.(<Reseachgate.net>).

50) Devanarayanan, S. (1968), "Low Temperature Vacuum Stage for Microscope" (Get-together on Instrumentation, 22-23 Feb 1968, CISL, I.I.Sc., Bangalore, India.(Reseachgate.net)

51) Devanarayanan, (1964) "Hot Stage for Polarization Microscope" (Indian Institute of Science, Bangalore, 1964). (Private Communication).

52) Devanarayanan, (1966), "A 3-Terminal Capacitor for dielectric constants" (Indian Institute of Science, Bangalore, 1966). (unpublished data).

53) Devanarayanan, S., Morrell, G, and Katiyar, RS., (1991).,Proc. Kerala Acad. Sciences, Thiruvananthapuram, Kerala, India, Vol 11, 09-13 (1991).

54) Devanarayanan, S. and Easwaran, KRK, (1966).Proc. Indian Acad. Sciences, A$\underline{64}$, (3) 173-84.

55) Devanarayanan, S. & PS Narayanan, (1967).Ind. J. Pure Appl. Phys., $\underline{5}$ (3) 104-05.

56) Devanarayanan, S. & PS Narayanan, (1968a). Ind. J. Pure Appl. Phys., $\underline{6}$ (10) 542-45.

57) Devanarayanan, S. & PS Narayanan, (1968b).Ind. J. Pure Appl. Phys., $\underline{6}$ (12) 714-15.

58) Devonshire, A.F., (1949), Phil. Mag., $\underline{40}$, 1040.

59) Devonshire, AF., (1964), Rep. Progr. Phys., $\underline{27}$, 1.

60) Diamont et al., (1957), Rev. Sci. Instr., $\underline{28}$, 30.

61) Dias, CJ, and DK Das-Gupta (1993), J. Appl. Phys., $\underline{74}$ (10) 6817.

62) Distler, GI, VP Kontantinova, YM Gerasimov and GA Tolmacheva (1968) Nature, $\underline{218}$, 762.

63) Dvorak, V., A. Fouskova & P. Glogar (Editors) (1966), "Proc. Intnl. Meeting on Ferroelectricity", Prague, Czechoslovakia.

64) Elwell, D and Scheel, H I., (1975), "Crystal Growth from High Temperature Solutions", (AP, London).

65) Fatuzzo, E. and W.J. Merz (1967), "Ferroelectricity", {in Selected Topics in Solid State Physics, Vol. VII, edtd., E.P. Wohlfarth}, (Wiley, NY), pp 171-89.

66) Fouskava, A. (1969), J. Phys. Soc. Japan, $\underline{27}$, 1699.

67) Forsbergh (Jr.) PW (1949), Phys. Rev., 76, 1187-1201.

68) Forsbergh (Jr.) PW.(1956), "Piezoelectricity, Electrostriction and Ferroelectricty" (In Handbuch der Physik) XVII, pp 264-392 – Dielektrika (Springer-Verlag, Berlin,).

69) Fredericks, (1971), Phys. Rev., B4, 911.

70) Froehlich, H.,(1958), "Theory of Dielectrics", (Clarendon, Oxford) 1958.

71) Ganesan, S., (1962), Acta Crystallographica , 15, 81(1962).

72) Giebe, E. and A. Scheibe, (1925), Zeit. Physik, , 33, 760.

73) Glass, AM (1969), J. Appl. Phys., 40 (12) 4699.-4713.

74) Gleason, Tj and JC Walker, (1969), Phys. Rev., 188 (2) 893-98,; Phys. Rev. Lett., 23 (19) A 13.

75) Grindley, J. (1970), "An Introduction to the Phenomenological Theory of Ferroelectricity" (Pergamon Press, NY) 1970, pp 270.

76) Gupta, LC, ()1969), J. Phys. Soc. Japan, 27, 1229.

77) Hablutzel, J. (1939), Helv. Phys. Acta, 12, 489.

78) Harrop, P.J.,(1972) "Dielectrics", (Butterworth) 1972.

79) Hippel, Arthur R. von (Ed.), (1954), "Dielectric Materials and Applications" Papers by Twenty-two Contributors (Jointly by the Technology Press of MIT and John Wiley & Sons, NY. Pp 30-40.Chapman & Hall Ltd, London) 1954

80) Hippel, Arthur R. von (Ed.),(1966), "Dielectric Materials and Applications" (The MIT Press, Cambridge) 1966.

81) Hippel, A von, (1950), Rev. Modern Physics, 22, 221, 1950

82) Hill, Nora E., Worth E. Vaughan, A. H. Price, and Mansel Davies (1969) *Dielectric Properties and Molecular Behaviour*, (van Nostrand Reinhold Company Ltd., London).

83) Hippel, A. R. von (1954) , *Dielectrics and Waves*, (Chapman & Hall, London).

84) Hooton, JA and WJ, Merz (1955) Phys. Rev., 98, 409.

85) Jaffe, B. Cook (Jr.) WR, and Jaffe, N.(1971), "Piezoelectric Ceramics" (Academic Press, London, NY) 1971.

86) James, E. & RT Arnold (1969), J. Appl. Phys., 40 (12) , 4806-11.

87) Jayaraman, A. (1983), Sci. Amer., 250, 42.; Rev. Mod. Phys., 55, 65.

88) Jaynes, ET.,(1953),"Ferroelectricity" (Princeton Univ Press, NJ)1953.

89) Jochum, M. and HG Unruh, (1998) Eur. Phys. J., B 5, 163-68.

90) Jona, F. & Shirane, G.(1962), "Ferroelectric Crystals" (Pergamon , London, 1962).

91) Kadanoff, LP., *et al.*(1967), Reviews on Modern Physics, 39, 395 (1967).

92) Kanzig, Werner, (1957),"Ferro-electrics and anti-ferroelectrics" (Solid State Physics Series - IV Edited Seitz & Turnbull) Academic press, 1957.

93) Kay, HF., (1948), Acta Crystall., 1, 229-37.

94) Kay, HF., (1969), Mater. Sci. Res. , 4, 206-28.

95) Kay, HF., & P. Vousden, (1949), Phil. Mag., 7, 1019-40.

96) Kalinin, SV, A. Rar and S. Jasse (2006), IEEE Trans. Ultrason. Ferroelec. Freq. Control, 53 (12) 2226-52.

97) Kell, RC. (1963), Brit. J. Appl. Phys., 14, 249-55.

98) Kittel, Charles, (1996),"Introduction to Solid State Physics", 7th Edition, (John Wiley, Singapore, NewYork) 1996.

99) Klug, H.P. and Alexander, L.E., (1970), "X-Ray Diffraction Procedures: for Crystalline and Amorphous Materials" 6th Print (John Wiley, NY) 1970

100)   Krasnikova, A.Yu., and IN Polandov, (1970) Soviet Phys. Solid State, 11, (7), 1421.

101)   Krishnan, RS., PS. Narayanan, & S. Devanarayanan, (1970), J. Phys. Soc. Japan,(Suppl.) 28, 163-65.

102)   Krishnan, RS., PS. Narayanan, & S. Devanarayanan, (1970), J. Phys. Soc. Japan,(Suppl.) 28, 166.

103)   Krishnan, RS, R. Srinivasan & S. Devanarayanan, (1979), "Thermal Expansion of Crystals" [Vol. 12, Intnl. Series on Science of Solid State, Ed. BR Pamplin] (Pergamon, Oxford) 1979.

104)   Krishnan, R.S. (1989, 1992. 1994, 1998), Source Book on Raman Effect, Vol. 1 (1928 - 57), Vol. II ((1958 - 70), Vol. III (19971 -74), Vol. IV (1975 - 96) , (Publ. Inf. Directorate / NISCOM, CSIR, New Delhi).

105)   Kunitomo, M., T. Terao and T. Hashi, (1970) Phys. Lett., 31A, 14.

106)   Landauer, R., DR Young, and ME Drougard (1956) , J. Appl. Phys., 27, 752.

107)   Lang, SB.,(1974), "Sourcebook of Pyroelectricity" (Gordon and Breach, New York, 1974).

108)   Liang, Li Pin, Xuan Cheng && Ying Zhang, (2016), Key Engineering Materials, Vol. 680, 25-29)

109)   Lee, Tu and Ilhan Aksay (2001), Crstal Growth & Design, 1 (5) 401-19.

110)   Lines, ME. & A.M. Glass.,(1996), Principles and Applications of Ferroelectrics and Related Materials.(Oxford University Press),1996.

111)   Loge, RE and Z. Suo (1996), Acta Mater.,33, 3429-38.

112)   Lovell, MC., Avery, J. & Vernon, MW.,(1981), "Physical Properties of Materials", (Van Nostrand, NY) 1981.

113)   Marta Deri, (1966),"Ferroelectric Ceramics", (Maclaren and Sons Ltd., London) 1966.

114)   Martin, Hans Joachim,(1964), "Die Ferroelektrika" (Geest and Portig, Leipzig) (Vol. 15, "Techniche-Physikaliche Monographen "), pp 551, 1964.

115)

116)   Mason, Warren P.,(1950), "Piezoelectric Crystals and Their Applications to Ultrasonics" (D, van Nostrand, N.Y.) 1950.

117)   Mathias, BT., and Hippel, von A. (1948), Phys. Rev,. 73, 1378-84.

118)   Megaw, Helen D., (1957), "Ferroelectricity in Crystals", (Methuen, London) 1957.
119)   McCammon & Work (1965), Rev. Sci. Instr., 36, 1169-73.
120)   Mendes-Felho, J., V. Lemos & F. Cardeira (1984), J. Raman Spectrosc., 15 (Dec),3(
121)   Merz, W.I.(1950) Phys. Rev., 78, 52.
122)   Merz, WJ. (1953), Phys. Rev., 91, 513.
123)   Merz, WJ. (1954), Phys. Rev., 95, 690-98.
124)   LANDOLT-BORNSTEIN "Numerical Data & Functional Relationship in Science & technology" Series : (Hellwege, KH & Hellwege, AM Edtd.) Springer-Verlag, Berlin).

 (a) Mitsui, T. *et al.*, (1981,1982), "Ferroelectrics & Related Substances" (Oxides) III /16 , 1981, 683 pages. (Non-oxides) 1982, 992 pages

(b)   III/28, Supplement to III/16 (Oxides) 1990 (Non-oxides) 1990

(c)   III/3   "Ferroelectric & Antferroelectric Substances" ,1969, 584 pages

(d)   III/9   Supplement to III/3 1974, 496 pages

(e)   Sub-volume a 1989, 430 pages

(f)   Sub-volume b 1994, 149 pages

                                          Other useful volumes

(g)   III/1   "Elastic, Piezoelectric, Piezooptic & Electro-optic constants of Crystals" (1966) 100pages

(h)   III/2   supplement and Extension to III/1 232 pages (1969)

(i)   III/II   (1979) 854 pages

(j)   III/18   (1984) 589 pages

(k)   III/29.   {(a.) (1992) 743 pages, (b.) (1993) 543 pages}

(l)   III/30   (1996) 497 pages.

125)   Morrell, G., S. Devanarayanan & RS. Katiyar, (1991), J. Raman Spectroscopy, 22, 529-34, 1991.
126)   Naussbaum, A.(1967), "Electronic and Magnetic Behaviour of Materials" (Prentice-Hall)1967.
127)   Newham, RE., (1975), "Structure - Property Relations" (Springer Verlag, New York, 1975).

128)   Novakovic, L.(1975), "The Pseudo-spin Method in Magnetism and Ferroelectricity" Vol. 77 (Pergamon, Oxford) 1975.

129)   Nye, JF, (1957), "Physical Properties of Crystals" (Clarendon, Oxford, 1957).

130)   O'Donoghue, Michael,(1982), "Beginner's Guide to Minerals" (Butterworth, Newnes Technical Books, London) 1982.

131)   Peercy, PS. (1975), Phys. Rev., B 12, 2725.

132)   Perelomova, NP and Tagieva, MM, (1983) "Problems in Crystal Physics with Solutions" (Mir Publ, Moscow).

133)   Raman, CV & TMK. Nedumgadi (1940), Nature, 145, 147.

134)   Robinson, MC. & AC. Hollis Hallett (1966), Canad. J. Phys., 44, 2211-30.

135)   Samara, GA. (1966), Phys. Rev., 151, 378-86.

136)   Samara, GA. (1971), Phys. Rev.Lett., 27, 103.

137)   Sanjurjo, JA., Lopez-Cruz & G. Burns (1983a) Solid State Commun., 48, 221

138)   Sanjurjo, JA., Lopez-Cruz & G. Burns (1983b) Phys. Rev. , B28, (12) 7260..

139)   Sawyer, CB & CH. Tower (1930), Phys. Rev., 35, 269.

140)   Scaife, BKP & Scaife, WGS., Bennett, RG., Calderwood, JH, (1971), "Complex Permittivity Theory and Measurement" (English Univ. Press, London) 1971.

141)   Schmidt, V. Hugo, (1969), "Ferroelectricty experiment for Advanced laboratory", Am. J. Phys., 37, 351.

142)   Schubring, NW., Nolta, and RA. Dork, (1964),"FE hysteresis tracer featuring compensation and sample grounding" Rev. Sci. Instrum., 35, 1517-21.

143)   Scott, JF., (2013) ,"Prospects for Ferroelectrics", PSRN Materials Science, Vol 2013, Article II, 24 pages.

144)   Sheeba, P.K., (1992),"Ferroelectric Phase Transitions at High Pressures", M.Phil. Dissertation, University of Kerala, Thiruvananthapuram, Nov 1992.

145)   Shirane, G., F. Jona and R. Pepinsky. (1955), Proc. IRE, 43, 1738-93.

146)   Shilds, J.P., (1966),"Basic Piezoelectricity" (Howard W. Sam & Co. Indianapolis) 1966.

147)   Shilds, J.P., (1958), 'Progress in Dielectrics' (Heywood, London) 1958.

148)   Shilds, J.P., (1970), "Proceedings of the European Meeting on Ferroelectricity, Saarbrucken, 1969", Wissenschafftliche Verlag MBH, Stuttgart) 1970.

149)   Sikka, SK., Hema Sankaran, Surinder M. Sharma, V. Vijayakumar, BK. Godwal & R. Chidambaram, (1989), Indian J. Pure Appl. Phys., 27, 472.

150)   Sirotin, Yu I and Shaskolskaya, MP (1982) "Fundamentals of Crystal Physics" (Mir Publ, Moscow).

151)   Shuvalov, LA., (1963, Acta Crystall., 16 (Part 13).

152)   Shuvalov, LA (19630, Kristallografiya, 8 (4) 617-24.

153)   Shuvalov, LA. (1964) Sov. Phys. Cryst., 8 (4) 495 -500.

154)   Steward, EG. (1952), J. Sci. Instr., 29, 214.

155)   Stewart, JW. (1967), "The World of High Pressure"(D Van Nostrand, NY).

156)   Takeuchi, K., D. Damjanovic, T. R. Gururaja, S. L. Jang, and L. E. Cross, (1986), Proc. 1986 IEEE Symp. Appl. Ferroelectrics, 402 (1986).

157)   Tiwary, HV, S. Devanarayanan, & PS Narayanan,.(1969), Ind. J. Pure Appl. Phys., 7 (9) 655 (1969).

158)   Yagnik, CM, JP Canner, R Gerson & WJ James (1969), J Appl Phys, 40(12), 4713-15.

159)   Valasek, J., (1921), Phys. Rev., 17, 475.

160)   Volger, J. (1952), Philips Res. Reprints, 7, 21.

161)   Wada, Mitsuo, A. Sawada & Y. Ishibashi (1981), J. Phys. Soc. Japan, 50 (6), 811

162)   Weller, EF.,(1967), "Ferroelectricity" (Elsevier Pub. Co., Amsterdam, 1967).

163)   Wood, Elizabeth A., (1963), "Crystal Orientation Manual", (Columbia University Press, NY,) 1963.

164)   Wooster, WA, (1957), "Experimental Crystal Physics" (Oxford Univ. Press, London).

165)   Wooster, WA & Breton, A.(1970), "Experimental Crystal Physics", (Clarendon, Oxford, 1970).

166)   Wul, B. and I.M. Goldman, (1945), Dok. Akad. Nauk, SSSR, 46, 139.

167)   Xu, Y.(1991), "Ferroelectric Materials & their Applications" (Elsevier Sci., Amsterdam, 1991).

168)   Yagnik, CM, JP Canner, R. Gerson and WJ. James (1969), J. Appl. Phys., 40 (12) 4713.

169)   Zheludev, Ivan Stepanovich, (1971), "Physics of Crystalline Dielectrics ", Vol I, "Crystallography and Spontaneous Polarization", pp 1 – 326, Vol. II, "Electrical Properties", pp 337 – 620) (Plenum Press, NY) 1971.

170)   Zheludev, I.S, (1973), "Seignettoelectricity", (Atomizdat, MIR, Moscow) 1973.

171)   Zwikker, C. (1954), "Physical Properties of Solid State Materials", (Pergamon, London).

172)   Proceedings of the 1997 Williamsburg Workshop on Ferroelectrics, Williamsburg, Virginia 2-5 February 1997.

Part I: Perovskite type ferroelectrics. Single crystals of potassium niobate. Solid solutions of potassium tantalate niobate. Lead, zinc and magnesium niobate ferroelectrics with a smeared phase transition.

<u>Part II</u>: Tetragonal ferroelectrics of potassium tungsten bronze-type structure. Single crystals of barium strontium niobate. Barium sodium niobate single crystals. Other crystals with tetragonal potassium tungsten bronze-type structure. Nonlinear optical crystals with lamellar structure.

<u>Part III</u>: Some physical aspects of oxygen octahedral ferroelectrics. Photorefractive properties of oxygen octahedral ferroelectrics. Relation between ferroelectric and nonlinear optical properties of oxygen-octahedral ferroelectrics. References. Index

173)   XU Guisheng 1,2 , LUO Haosu 1 , WANG Pingchu 1 , QI Zhenyi 1 & YIN Zhiwen (2000)

Relaxor ferroelectric single crystal *Chinese Science Bulletin* Vol. 45 No. 15 August 2000

&&&&&&&&&&

About the Author

Prof. S. DEVANARAYANAN, Ph.D. (IISc); D.Sc. (USA), Dip (Uppsala)

Dr. S. Devanarayanan was educated at the University College, Thiruvananthapuram, Indian Institute of Science, Bangalore, and Institute of Physics, Uppsala, Sweden. He had a brilliant academic career throughout. He was the Professor and Head of the Department of Physics, University of Kerala during 1993 – 2000; and has 37 years of teaching / research experience in Physics and materials science.. He has to his credit over 80 published research papers in standard scientific periodicals and 42 presentations in National and International Science events. A Monograph entitled THERMAL EXPANSION OF CRYSTALS (1979) and book "QUANTUM MECHANICS: Principles & Applications" (2005), "QUANTUM CHEMISTRY"(2013), "PHYSICS IN A NUTSHELL: Companion for Success in Competitive Tests" (2016), and "A TEXT BOOK ON NUCLEAR PHYSICS" (2016) were authored by him. He has served as a Professor in Physics at the University of Puerto Rico, USA, during 1989 –91. He was awarded the SIDA Fellowship and worked at The Institute of Physics, Uppsala, Sweden, during 1970 – 71. A Life Member of various academic bodies like the Indian Physics Association, American Physical Society, and Fellow of the Indian Cryogenic Council, his biography has found place a number of times in the publications of Marquis' Who's Who (USA), International Biographical Centre (UK), American Biographical Institute, Refacimento International, *etc.* The Govt. of Kerala appointed him as a member of the Commission of Enquiry on the working of the University of Kerala, in 2000.

As an experimental physicist / materials scientist his research specializations include Phase Transitions in crystals, Mössbauer Effect, Crystal growth, Vibrational spectroscopy and Atmospheric physics. Devanarayanan's early studies were on phase transitions in ferroelectric crystals at the Indian Institute of Science, where he had used Fizeau's optical interferometer and cryogenics. The work on single crystal sodium trihydro selenite is the outstanding among them. He continued the studies on magnetic transitions and magnetic structute in Fe_2P, down to liquid helium temperatures, at the Institute of Physics, Uppsala, using Mossbauer spectroscopy. This work and the extensive studies on ordering on the role of concentration of silicon in Fe-Si alloys form important contribution. Yet another investigation was on a series of phase transitions at cryogenic temperatures in crystalline lithium cesium sulphate using Raman Spectroscopy at University of Puerto Rico. He had encouraged his students to start work on in the field of biological crystals, vibrational spectroscopy and q-deformed oscillators in Solid State. He was responsible for analyses of rocket-sonde temperature data over a period of solar cycle in the middle atmosphere over Thumba, and at mid- and high- latitudes.

A number of students received Ph.D. and M. Phil. degrees under his research supervision. He had the special honour of being invited by the Royal Swedish Academy to submit proposals for the award of the Nobel Prize in Physics for 1995. Devanarayanan believes in Sir C.V. Raman's (NL) advice that one can become a good scientist only when one takes up research along with teaching at a University level. He has made academic visits in Sweden, Finland, Leningrad (USSR), The Netherlands, Germany, France, Australia, Czechoslovakia, Hungary, Austria, England, and USA. For more details, Website: <researchgate.net>.